Garden Earth

From Hunter and Gatherer to Global Capitalism and Thereafter

Garden Earth

From Hunter and Gatherer to Global Capitalism and Thereafter

Gunnar Rundgren

Regeneration, Uppsala, Sweden
www.gardenearth.info

First English edition, rev 3.1 (24 May 2013)
Translated and adapted from Trädgården Jorden,
published 2010 by Gidlunds, Hedemora Sweden

Cover design: Kolbjörn Örjavik
Front cover illustration: John Nyagah
Back cover author photo: Nicola Phillips

ISBN: 9163710102
ISBN-13: 978-9163710100
Garden Earth (Regeneration), Uppsala, Sweden
www.gardenearth.info

Table of Contents

Part IV What Lies Ahead of Us?   **309**

Foreword

Believe those who are seeking the truth. Doubt those who find it.

(André Gide)

In 1977, at the age of 19, I moved to the countryside with my life companion and a few good friends. Our objective was to live simple, in harmony with nature, and to use as environmentally sound and safe technology as possible. We bought Torfolk Farm[1] in a sparsely populated, forested part of Sweden. We started to restore the derelict buildings and to learn the many skills needed to run a self-sufficient organic farm, including welding, churning butter and tanning of skins. We lived very simple; produced milk, butter, cheese, meat, potatoes, vegetables and a lot more for our own consumption. For a few years, we also grew grain, but the quality of flour was never up to the mark so we stopped growing grain. Among many useful items, the forest gave us wood for heating and for construction. You learn a lot from such a life—there is a big difference between mastering skills needed for survival and having an ordinary, atomized job and living in an apartment.

For one reason or another—perhaps because we had children and didn't want to force them to live as we had chosen, or because we had to pay taxes (also on our own consumption), or simply because we wanted a more comfortable life—we thought we needed more money to be able to buy things. Thus arose a need to develop one or a few of our many trades into a commercial venture—to specialize and com-

---

[1]   The farm is still active and supports a small-scale food processing company (www.torfolk.se).

mercialize. You can't make money from a diversified production such as ours, unless you want to run an open-air museum. To specialize you need machinery, and to pay for the machines they need to run as much as possible. We decided to grow organic vegetables. The more commercial we became, the more we had to drop unprofitable activities, including the production of things for self-consumption. Then, we needed more money to buy those things. We mechanized more and expanded, and suddenly our own labour wasn't enough, so we had to employ people. That changed our view on work; earlier there was really no difference between work and leisure. This classification of work and leisure is meaningless in a self-sufficient economy, as also the distinction between private and belonging to the company. Our way of management also had to change; previously, we had no hierarchy, but the employees found it hard to cope with our collective leadership style.

You could say that we simulated more than 100 years of societal change in just 10 years. We got insights into the pros and cons of the two ways of organizing production, and the drivers that rule us, in ways one is often not conscious of. I became aware of the forceful nature of the capitalist thought pattern or paradigm and how it changed the way we worked and related to each other, with consequences going much further than we first realized. This is perhaps the real invisible hand. A gnawing thought is that even if every single choice was rational and reasonable, the end result needn't necessarily be desirable.

I have been engaged in environmental and social issues since I was a child. The pollution of my summer paradise, Lake Mälaren, triggered the first awakening. In my early teens, I became involved in the environmental issues of my hometown, Uppsala. Our association had at its prime almost one tenth of the city's population as members, and I can note now, almost 40 years later, that our plan for a bicycle path network in the city has almost been realized. I was 15 years old when the first global environmental conference was held in Stockholm in 1972.

Over the last 15 years I have worked extensively in developing countries, in particular in Eastern Africa. Largely, I have been engaged in market-oriented efforts to improve livelihoods of small farmers by linking them to organic markets, a window front example of how to use capitalism to support sustainable ecological, economical and social development, where the farmers have substantially increased their income by exports to high income countries. Many of my assumptions were tested and modified in these developing countries. While people (including myself) in high-income countries, use too many resources, a large part of humanity lives in conditions that lack basic amenities such as electricity, and where the provision of these could do wonders; for example, the use of electric light and the use of other technology for cooking rather than an open fire could mean improved health, education and quality of life. At the same time, promises of 'development' ring hollow in many countries; are there really any circumstances for them to climb even to a middle-income level? In what field will they be competitive? Who will supply them with cheap raw materials and labour as we have been supplied?

During 2001–2005, I was President of the International Federation of Organic Agriculture Movements, an international umbrella organization for organic farming, with 800 member organizations in more than 100 countries. It sets standards for organic farming, accredits certification bodies, makes a code of conduct for companies and implements various development projects, over and above the normal advocacy work of NGOs. This gave me an insight into the global civil society, its strengths and weaknesses, a phenomenon that some believe is the seed for a new kind of society. Although I am not sure of that, I admit it is fascinating to experience how human beings, in a purely voluntary way, outside the frame of the nation-states, can organize cooperation and regulation of complicated matters among people from all walks of life.

When starting to write this book, I went back to read things that I read in the 1970s, when much of my worldview was shaped. It was actually frightening to realize how much of what was written then is still relevant today, frightening because it shows how much was already known at that time, and how little has actually been done since. The first energy crisis in 1973 initiated a broad debate on 'sustainability' issues, even if the word was not in use at that time.

Development of solar, wind and other alternative technologies began, along with the development of the organic farming movement. This continued for decades but, to a very large extent, fell off the public agenda to resurface at the time of some particular new scandal. However, criticism of the system that developed in the 1970s was not given much space. Even the Green parties that won seats in the parliaments of Europe soon realized that 'political realities' didn't allow for much radical change, and catchphrases such as 'zero growth' were forgotten. When climate change surfaced as a main issue, when oil prices broke the 100-dollar mark and when food prices surged, for a short while I had the feeling that things might change. But then came the financial crisis of 2008 — in my view, just another example of the same extractive and exploitative system — caused by banks lending money to people to buy houses they couldn't afford, which were expensive because the banks had lent money to people for buying them . . . and, of course, environment was forgotten.

Some statements made in the 1970s were certainly far too 'alarmist' in nature. Many of them, for example, those made by the American ecologist Paul Ehrlich, the Club of Rome and the Swedish food-scientist Georg Borgström, predicted difficulties in feeding the population by the turn of the century. People were starving for sure, but that was not rooted in environmental constraints or lack of land but in inequality. The miscalculations were mainly caused by the lack of faith in human ingenuity and the lack of understanding of the force of the market economy to satisfy demands. On the other hand, if these alarmist statements had not been made, many necessary changes, such as the regulation of chemicals and pesticides, limiting emissions to air and water, etc., would not have happened.

The Swedish scientist Svante Arrhenius observed climate change and the link to carbon dioxide in the atmosphere already in 1896. At that time, it was more of a scientific observation than an identification of a potential environmental disaster. In the 1950s, meteorologists rediscovered the issue, and in the 1960s, environment debaters took it up, with surprisingly accurate descriptions. In 1964, the American environmental and social activist Murray Bookchin wrote:

> It can be argued on very sound theoretical grounds that this growing blanket of carbon dioxide, by intercepting heat radiated from the earth, will lead to rising atmospheric temperatures, a more violent circulation of air, more destructive storm patterns, and eventually a melting of the polar ice caps (possibly in two or three centuries), rising sea levels, and the inundation of vast land areas. (1971: 22)

Was this an alarmist statement? Would it not have been better if it were taken seriously at that time instead of 30 or 40 years later? These years would have made a great difference in our ability to cope with the problem.

The other side of the debate, the technology and market optimists are also often wrong. Pan Am, once a glorious American airline, since long bankrupt and gone, took advance bookings for round trips to the moon shortly after the first moon-landing. Today, techno-optimists say that nuclear fusion will be commercial around 2050; in the 1960s, they said it would be commercial at the end of the century. Large-scale breakthrough of solar power is always some 15–20 years ahead. Chemical fertilizers and pesticides were supposed to eliminate hunger and malnutrition 50 years ago; today, genetically modified organisms shall do it because chemical fertilizers failed. Constantly, new rabbits, or promises of new rabbits, are pulled out of hats. Some rabbits are alive, but many are mere fantasies. And even when problems are solved, they often don't address root causes; instead, we end up reforming sick systems.

Car traffic is a case in point. Already some 40 years ago, there was a lot of criticism against car traffic. Society responded, not by questioning the system (very few citizens would have agreed to that) but by addressing each problem with technical solutions such as bike paths, led-free petrol, catalytic converters, noise walls, highways and tunnels. For each step, more was invested in the car system instead of developing alternatives. The costs mounted all the time. If all these costs and the end result had been presented to citizens some 50 years ago, along with the costs of alternative solutions, humankind probably would have had another situation today. Now we are stuck with our version of the statues of the Easter Islands, and have lost 40 or 50 years of development by not starting with alternatives in time.

The term sustainable development has been predominant for many years. It means that development should be socially, economically and environmentally durable and stable, and that these issues are strongly interconnected. Still, almost all issues are discussed in isolation. I am convinced that we cannot discuss global warming separated from poverty, or the exploitation of natural resources separated from the economic system, or economy and environment isolated from how we relate to each other, that is, from the society we have developed.

Many thoughts circulate repeatedly, often with new words but essentially with the same content. I don't know whether I, with this book, have been able to formulate any single thought that is in itself entirely new.[2] I consider my biggest accomplishment to be that I have put together thoughts and facts to make a more comprehensive, and at least in my view consistent, picture—a picture that not so many have seen and that some have seen but have not backed up with facts to build a credible case. This also coincides with my profession of the last 20 years, that of a consultant. I used to say, half-jokingly and self-ironically, that consultants steal ideas and knowledge from others and then repackage them for sale.

---

[2]   At the beginning of each chapter, the quotes without a reference are mine, or at least I believe so. Admittedly, I could have picked them up from someone inadvertently.

Let me introduce some of the thinkers and their works that have shaped my thought and therefore this book. For the description of the long-term developments that has lead us to where we are and a view of the future, rooted in history, I want to mention Swedish professor of agrarian history Janken Myrdal's *Framtiden — om femtio år* (the Future — in fifty years). My view of the steps determining the transformation of an agrarian society to a capitalist industrial market economy has been largely influenced by the economist Karl Polanyi's *The Great Transformation* (1944) and by the sociologist Giovani Arrighi's *The Long Twentieth Century* (2010). The perspectives and opinions of Adam Smith, Karl Marx and Joseph Schumpeter on this strange creature, capitalism, have played a big role. I quote them, as well as others, in this book, sometimes because I agree with them and sometimes just because they express a certain perspective in an eloquent or perhaps provocative way. Other thinkers who have influenced me in my views are Amartya Sen, Jonathon Porritt, Gregory Bateson, Ivan Illich, Peter Kropotkin, Rolf Edberg and Murray Bookchin. I am also wholly indebted to all the researchers who find out things about our world and publish papers that I devour ravenously. Myself, I was too lazy to become a researcher!

The Swedish version of this book, *Trädgården Jorden* (Garden Earth), was largely written in the first part of 2009 during a bicycle trip I made from Sweden to Turkey and back. Kari Örjavik, Maria Gardfjell and Bo Malmberg commented on drafts of the book. Sophia Twarog and Inger Källander helped me to sharpen my thoughts and reasoning by listening and asking questions when I related the content to them. Krister Gidlund, Anders Svedin and Sara Gidlund at my Swedish publisher, Gidlunds, helped to shape the book as well as correct my many errors and typos. *Trädgården Jorden* was published in April 2010.

This English version was largely written in the last half of 2010, and first months of 2011, in places such as the Youlgrave (England), Resö and Uppsala (Sweden), Arusha (Tanzania) and Suva (Fiji). It benefited from the input and friendly criticism of Nicola Phillips, and was reviewed by Kelvin Richards, Ulrich Hoffmann, Diane Bowen, Grace Gershuny, Daniel Hedin and John Stansfield. The manuscript in English was edited and improved by Ayesha Chari, under the guidance of Vidhu Panicker, Senior Project Manager, Chillibreeze Solutions Pvt Ltd, India. I have translated all quotes from non-English-language sources, unless mentioned otherwise.

I always intended to write an English version, and although it is based on the Swedish version, it is not at all a literal translation. I have reworked the text, included some new parts, tried to sharpen the red thread in the book and left out some excessive data or thoughts that lead to a cul-de-sac. The feedback to and reviews of the Swedish version, the many questions and debates I have had since the publication, as well as my own thoughts have led to quite a few changes. I also managed to read some works I had only heard about earlier (e.g. Bateson, Arrighi and Hornborg), which influenced me and this book. The last part in this English edition is more mature than that in the Swedish one, and the critique of the capitalist market economy is more clearly formulated.

Uppsala, December 2012

Introduction

The golden rule as formulated in the Bible is: 'So in everything, do to others what you would have them do to you, for this sums up the Law and the Prophets' (Matthew 7:12). It is found in various forms in most religions and philosophies; for example, in Hinduism: 'One should never do that to another which one regards as injurious to one's own self. This, in brief, is the rule of dharma' (Mahabharata, section CXIII). Can our (the high-income countries) lifestyle be applied by 7 or even 10 billion people and future generations without harming the ecosystem and our own humanity in an irreversible way? It is one of the starting points of this book. One could call it the global golden rule. If the answer is 'no', and I'm afraid it is so, we can conclude that our lifestyle is not sustainable and we should face the consequences.

The human being is a great colonizer. She has spread all over the planet and, almost everywhere, she has managed to find an ecological niche. In many of these niches, her place is well defined, her numbers governed by the limitations of the natural environment, and often she has had no opportunity to radically alter the terms thereof. Therefore, traditional communities who survived on gathering, hunting and fishing remained relatively stable. A new phase began with agriculture. Human beings changed the ecosystems to a much wider extent than ever; it also led to changes in social organization with the introduction of rulers and subordinates (slaves, serfs or tenants). Agricultural societies were in many ways more sensitive to disturbances, and the impact of natural phenomena could be dramatic. Soil degradation was a recurring reason for the collapse or gradual decline of societies.

The next major change occurred when human beings began to harness fossil fuels. This freed them from depending on their own metabolism and on biomass that they took from others in nature; huge resources could now be brought to work. Ecosystems were reshaped in an unprecedented way. This coincided with industrialization and the introduction of a capitalist market economy. Industrialization increased human control of nature, in principle, to all corners of the globe. We go further, climb and fly higher, dive and drill deeper, all in order to find more resources to exploit. This led to a marked expansion of the areas in direct service of man, particularly forests and oceans. Industrialization also provided new force to the farming man. This was followed by a sharp increase in the input of energy, so that agriculture, from being a net producer of energy, became a net consumer of energy.

Modern society has freed us from old chains and limitations and has created enormous wealth. Labour productivity is ever increasing; we can produce more in an hour than our ancestors could produce in a week or even a month; we can also destroy more at the same break-neck pace. Old prejudices and oppression have been broken down, and freedom and human rights have risen sharply. The idea of equal worth and rights of all people is—at least in theory—universally accepted. Several components in modern society have contributed to this development. Predominant of these are political democracy, capitalism, the great use of external energy, mainly in the form of fossil fuels, and of other natural resources, and industrialization with a far-reaching division of labour.

Now, we have reached a point where the system exhibits reduced marginal utility, or in some cases diminishing returns. What previously were positive forces—the restless search for new countries and new areas to colonize, the liberation of man from tradition, affirmation of the individual over community, a far-reaching division of labour, that everything can be bought and sold on a market—now are destructive ones. More capitalism, more oil and coal, more chimneys were, perhaps, good slogans during the 1800s; they were probably good in many parts of the world even during the 1900s; but today, fewer and fewer areas need these slogans, and in some parts of the world, they simply haven't worked. Since 1950, oil production has increased tenfold and the overall economic activity in the world has increased sixfold. Meanwhile, carbon emissions have quadrupled, world population has grown two and a half times, and the economic gap between the poorest 20% and the richest 20% has more than doubled. The global socioeconomic and ecological system is thus rapidly changing and the pace of change is increasing constantly. This means that we live in an unstable system (Helmfrid and Haden 2006).

I intend to illustrate this further and make a clearer link between the ecological situation, our well-being, our society, our use of energy and technology and the economic system—capitalism. I summarize the most salient points below:

1. Our relationship with nature is characterized by over-exploitation of raw materials and natural resources. Certainly, history shows that we have a remarkable capacity to overcome the limitations of our natural environment. Nevertheless, our use of both mineral resources and living resources, with their origin in photosynthesis, is now at a level that simply cannot be sustained. This can be expressed in different ways such as that we are soon reaching 'peak oil'; that our ecological footprint is already one and half Earth; that we are already using more than half of the entire biosphere; and that wilderness is now only marginal compared to human-dominated ecosystems.

2. We know that the release of greenhouse gases, largely the effect of extraction of fossil fuels and degradation of natural resources, will lead to marked changes in climate, resulting in great human suffering and material costs. In addition, we have caused severe reduction of biological diversity, the very web of life on earth. We have manipulated other life processes, such as the nitrogen cycle, to such a scale and in such a way that it will most certainly lead to severe disruptions.

3. We have unleashed 100,000 chemicals, but we have no idea how they affect us. We take medicines, eat food additives and indirectly consume pesticides we spray on crops. Drinking water and the air we breathe are full of man-made chemicals. We have very little knowledge about the long-term impact on us of the cocktail of chemicals we spread, and we know even less about how it affects our living environment.

4. The production and productivity revolution, which we explain with entrepreneurship, the superiority of capitalism and an individual's strive for personal gain can equally be interpreted as the result of one thing: access to external sources of energy, primarily fossil fuel. Use of energy is also the cause of many problems, most obviously the greenhouse effect. At least equally serious is that energy is the engine in almost any other resource depletion and degradation of ecosystem services. But the cheap energy sources are drying up and, more importantly, their efficiency is dwindling.

5. With the commercialization of farming and the introduction of chemical fertilizers, we no longer have to reproduce our production ability and capacity within the system itself. To replenish natural capital, fertility, labour and genetic resources are bought over the counter, thus separating production and reproduction. This system

is commercially efficient, but very inefficient in its use of energy; it is threatening the long-term fertility and capacity of the soil. Despite the emphasis on production, a billion people go to sleep hungry every day.

6. The system does not have sufficient self-correcting feedback loops to keep in check income inequality; the only thing that can keep it running is a constant economic growth so that those at the bottom, after all, will be a little better off every year. This is not primarily an economic problem but a social and moral problem. The difference between reality and rhetoric (in which all are said to be equal and to have the same opportunities) is simply too great.

7. The capitalist model of development was most successful in countries that industrialized first, followed by countries that had different comparative advantages in certain development stages. But large parts of humanity are outside of the development, including both the individuals who are left behind in industrialized countries and those whole countries that have no comparative advantages to exploit. Just like not all communities could make the transition from gathering and hunting to farming, many communities today cannot make the transition to a capitalist society because they lack the right conditions.

8. The values, attitudes and mechanisms that form the foundation of capitalism are obstacles to the harmonious development of society. Competition is promoted at the expense of cooperation, self-interest at the expense of solidarity and commitment to others, private property at the expense of commons and exploitation at the expense of stewardship or nursing.

9. The combined effects of those values, industrial technology, capital accumulation and market competition drive endless growth and consumption. It is certainly within human nature to be a little dissatisfied, to want more, to explore new opportunities. Without those properties, we would probably never become human beings in the first place. It is in society's nature to contain these desires so that they continue to benefit us all. With the capitalist takeover of society, these restrictions don't work any more, and we are all trapped in a treadmill of ever-increasing consumption. This consumption, driven by several of the factors mentioned above, has made us healthier, wiser, more beautiful and perhaps happier to begin with, but it has long crossed a line after which 'more things' do not mean more well-being.

10. That things are not a lot worse than they are is a result of the deep and strong properties of humanity and of our larger organism, our society. We do our best to mitigate the ills produced by our economic system. When old social institutions break down, we build new ones; when old values and culture lose their meaning, we develop new ones. Ultimately, we will also build a new world. We have now reached a stage where we need to divert more attention to building a new world than to fixing the old one.

By and large we can look upon capitalism, both the economic system and the values on which it builds and advocates, as a system and an ideology of colonization. It is about colonization in the classical sense, that is, conquest of other cultures and countries, but it is as much about the colonization of larger and larger parts of nature. Our exploitation of nature is now nearly complete; we are reducing the capital, the stocks, of raw materials that the bio-geosphere built up over millions of years, so that we become poorer each year. It is also about colonization of bigger parts of our lives and of the life of our society. More and more functions that were managed earlier by us, by our families, by civil society or by the public sector have now been taken over by markets. Contrary to claims, the so-called invisible hand can't manage the economic system, let alone the ecological and social systems that are even more important for our survival. Capitalism is dependent on a thriving society and on the exploitation of natural resources for its existence. At the same time, capitalism undermines both these fundaments of its own existence. This is the real crisis of capitalism.

Now we are here. We have reached the end of the road. There is nothing left to colonize. Therefore, we must replace the mechanisms and the logic that drove us here with new ways of thinking that are better suited to where we are now. Many of the required changes have already begun, albeit often on a small scale. A change in regime or shift in our paradigms usually occurs after small gradual changes, even those changes that sometimes manifested by sudden events such as Rome's surrender to the invading Vandals in AD 455, or the French revolution in 1789, or the fall of the Berlin wall 200 years later. Some 50 countries have now stable, or even declining, populations; many of these also exhibit very moderate economic growth that is not so materially resource-intensive. Organic farming, passive heating, electric bicycles and solar energy represent some technologies that are better adapted to a resource-efficient society. The wild has found new ways to coexist with human beings and vice versa. Interest in local communities, direct democracy and participatory organizational forms is growing. On the international scene, we see much more dynamism and development of civil society platforms for cooperation than in the frameworks of nation-states. Finally, a change of values has started; more and more people want to jump off the treadmill and focus on what gives real well-being. The change in values is reinforced by the change in population age structure.

As a contrast to the capitalist ideology, I offer *Garden Earth*. It expresses that we consider the globe as a garden that we take care of. We simply have to go back to the 'garden of Eden'. But we can no longer do so as the innocent creatures we were when we were driven out. We can no longer be there as equals with the giraffe, the carrots and the lamb. Whether or not we like it, the future of the earth, of Eden, is entirely dependent on us, and not just our own fate but also that of other species is in our hands. While the rich world would do well to tighten the belt a notch or two, many people still live in poverty without the most basic amenities. It is most important that Ethiopia's rural poor are able to control their land; that they are able to get electric lights, clean water and effective sanitation; that the Chinese peasant is able to say what he wants and live where he wants without risking punishment; that the Afghan girl needn't have to wear the burqa and can work where she wants and marry whoever she wants; and that the Brazilian landless gets access to the country's vast resources.

This book is organized as follows.

Part 1 provides a brief description of human development. One should not discuss the current situation or the future without a perspective or background. We are still by and large a product of our history and, even if we want to think of ourselves as modern, it is remarkable how stuck we are in the conventions of our thought and lifestyle. My description of history acknowledges that economic incentives have played a small role in the development of humankind and that markets and money were marginal phenomena until the mediaeval city-states, whereas ecological conditions, forms of energy, technical progress and social conditions were crucial, both for society at large and for the individual's behaviour and values. With the introduction of agriculture followed notions of private property, elitism, oppression, the state, religion and several other institutions that characterized the world for a long time. Later on, capitalism came with the liberation of human beings from the shackles of the past; however, they were subjected to a new, simpler, but no less cruel, logic, 'produce or die'. With capitalism, not only is the distribution of goods built around a market, but the market will eventually, in many small and some big steps, take over the entire society. Instead of the market serving society and humankind as a whole, society serves the market.

The importance of fossil fuels for the last 200 years of development can hardly be exaggerated. It is unlikely that anything even close to the development we have seen could have taken place as a result of capitalism or industrialization *without* fossil fuels, just as earlier civilizations could not have developed without agriculture. The easiest way to understand the importance is by translating the energy levels we have at our disposal to human labour units. Using that approach, we have 200 billion slaves working for us—and the good thing is that we do not have to house and feed them all.

Part 2 describes many of the environmental and provisioning challenges that humanity faces. These challenges can be broadly grouped into over-exploitation of resources, poisoning ourselves and nature, pressures on entire ecosystems and the services they provide, and disturbances of major biophysical processes such as the carbon and nitrogen cycles. Fossil fuels and our population exacerbate these challenges. Agriculture is dealt with in detail because of its central role in our survival; it takes up much of the planet's surface and how we run it is a management system for much of the planet. Services delivered by the nature that existed before the fields must be replaced. Agriculture is also very sensitive to the disruption of major ecological cycles (with the climate as a very clear example). It produces our food and must also, increasingly, produce bio-energy and ecosystem services. Food and agriculture play a big role in shaping society, despite the dwindling proportion of the gross domestic product (GDP) of that 'sector'. The aim is not to be complete; numerous challenges are not described; for example, depletion of the ozone layer, deforestation and over-fishing (which is just dealt with in passing).

I also introduce the view of nature that guides me, a vision that is opposed to the capitalist view of nature as a source of raw materials and as a waste dump, but also is opposed to ideas of man being an animal like any other and of nature regulating itself in a 'good' way.

Part 3 is devoted to the social organism we have created—society. I introduce society as part of the human organism, and upbringing and culture as part of evolution. Society and culture lift us above the short-term-ness and near-sightedness of genes and self-interest. If we see society as a 'meta-human', as an extension of the individual, the eternal opposition between heredity and environment can disappear, and our society is then a part of the people and culture; values and norms and other environmental factors are like society's genetic code that is passed from generation to generation.

We rush through both work and leisure. Are we happier? Nothing actually suggests that. As with natural resources, human well-being is not accounted for. It is a bit of a residual item, something you can discuss when all economic discussions are finished. Why is that? How did that happen? Even the most materialist of us love and want to be loved, enjoy a good book or music, see beauty in falling leaves, and so on. In fact, these things are much more important than the purely material. To live in a society, in a social context, is crucial for all people, and investment in social capital may be among the most effective in creating the foundations of happiness and well-being. Capitalist economy and values, however, tend to break down the social fabric.

Large parts of humanity, about 900 million, live in extreme poverty (less than a dollar a day), and 2 billion live on less than two dollars a day. Eight hundred and fifty million are malnourished. Average life expectancy is below 50 years in 32 countries. Broadly speaking, poverty is located in what we call developing countries, almost all of them colonies of European countries a century ago. Despite all the rhetoric, the capitalist system has not managed to eradicate poverty. The differences in wealth do not appear to be diminishing at all, but if anything increasing, which is immoral and unsustainable in the long term.

The sections on economics and how society works can be provocative in an age where capitalism's final victory has been declared, and all possible alternatives are discredited. Thirty years of environmental policies show that, despite all the rhetoric about sustainable development, capitalism as a social system does not have the ability to manage the fact that natural resources are limited. Fifty years of discussions, aid and development measures also show that the global economy offers no easy path to success—and perhaps no way at all—for those countries that are lagging behind. I criticize the inflated self-image of capitalism, but that does not mean I agree with the anti-capitalist demagogy. Above all, I do not buy conspiracy theories, with obese men in top hats (perhaps gradually superseded by slender, botoxed and top-trimmed women in costume) knowingly destroying nature and abusing workers. Of course, this happens sometimes, but it is a result of the logic of the market and capitalism and not of evil people.

Part 4 is a summary of conclusions and policy options. By now, it should be clear that I doubt we can solve the challenges we have within the capitalist paradigm. I demonstrate a number of trends that point in the direction of change. Particularly important are the demographic changes and the changes in the values attributed to them. I point to the importance of society for human welfare. It is in the organization of society that we shall find the key to how we organize ourselves in the future. Of the spheres of the market, government, civil society and the private, capitalism has strengthened the market and the private and changed the role of the state. The future is not to let the state take over the production and distribution, but to further develop various forms of local stewardship, and to let communities and civil society play a larger role.

The prevailing political discourse is mostly about tax rates, marginal adjustments to existing mechanisms, and so on. But the great challenges are to change the self-reinforcing processes, the positive feedback loops that exist, and even more to challenge the entire system of thought that is the basis of our view of society. I distrust utopias, but I give at least some ideas about how a radically different system might look. That is a work in progress, which hopefully can grow stronger through the reactions I get from readers of this book.

PART I
THE GREAT NARRATIVE

It is not without difficulty that we have succeeded in making ourselves so unhappy.

(Jean Jacques Rousseau,
On the Origin and Foundations of
Inequality among Men, 1754)

Hunter and gatherer societies, agrarian societies and industrial societies succeed each other. We think we are modern, but in reality we are to a large extent a product of our origin. That holds true for our genes as well as for the institutions of society and for the ways in which we think. At the end of the Ice Age, most people lived in societies of hunters and gatherers. As late as the beginning of the sixteenth century, less than 20% of land surface was subdivided into nation-states with a central authority, and almost all societies were based on farming. Today, 500 years later, the whole land mass is under the control of states, most of which are based on industrial technologies.

The first great civilizations emerged in areas from Egypt to Mesopotamia and from the Indus Valley to China. Apart from the Greek and Roman empires, Europe was a backwater, a periphery of the mighty Arabian, Mongol and Turkish empires. Colonization of the

Americas, and later of large parts of Asia, the Industrial Revolution and capitalism made a few nations of Western Europe leading powers, later to be overtaken by their descendants in North America. Which factors drove this development? By studying the transformations and the societies that succeeded and failed, we understand better the forces and mechanisms that pushed the different cultures ahead and those that made them crumble. Therefore, it is fitting to start our discussion with a broad overview of the development of human society.

We look into the functions and mechanism, the institutions and drivers, and the close connection between the ecological conditions and the societies they nurture. Some observations are orientated on the enormous impact of energy—one can actually discuss and understand the three major forms of societies from an energy and ecological perspective. We also look into how the twins market economy and capitalism emerged as a system to govern industrial society, what drove this, and the extent to which a 'market society' is something new.

– 1 –
Setting the Stage

(The Instruction of Ankhsheshonq)[3]

We tend to believe that most features of today's society are hard-wired into humankind (whether in the genes or in the culture is another discussion that we will come back to later) because previous societies were similar. Nothing could be more wrong. The chapters in this part aim to give us the background to understand where we are today and where we might be heading, to understand the commonalities of earlier cultures and the differences, and to understand which parts of modern society have existed all along and which parts are new. We have very often been wrong in identifying the steps of development. Let's look at a popular case study of Easter Island:

> The morning of 5 April 1722, the Dutch captain Jacob Roggenveen and his crew sighted a small isolated island in the South Pacific. The history of Easter Island, or Rapa Nui, as it was called and its giant statues spread rapidly to Europe. Until today, this example, more than any other, has symbolized how over-exploitation by humans leads to total collapse. The Polynesians inhabited the island in the sixth century, built sculptures and cut down forests. The population grew rapidly and reached 12,000 to 15,000 in the fifteenth century. When the last tree was cut down, they could no longer build more boats and they lost their main source of proteins from the sea. Conflict and internal strife contributed to a downward spiral towards a complete collapse in the middle of the seventeenth century when just a few dozen inhabitants remained. (Elmqvist 2007)

---

[3]   Ankhsheshonq was a sage of ancient Egypt. This quote has been taken from Lichtheim (1980: 176).

In 2006, new explanations that invalidated these steps were published by Terry Hunt, an anthropologist from the University of Hawaii. Hunt noted that a stowaway in the canoes, the rat *Rattus exulans*, most likely was the cause of the extinction of the giant palm *Jubaea disperta*, the predominant plant on Easter Island, in the eighteenth century. This rat multiplied to numbers between 1 and 3 million on the little island. Almost all seeds retrieved of the *Jubaea* palm were gnawed. Further, the rat was a predator of the bird species that pollinated the palms, a lorikeet. Hunt didn't believe that the human population ever exceeded the 3000 mark—the number at Roggenveen's visit. The following decline was caused by disease brought by Europeans, slave trade and emigration (Elmqvist 2007).

The history of Easter Island has helped to form our perception of human societies and that over-exploitation of natural resources and social collapse are inevitable results of population growth. It also shows the inability of human societies to understand the impact of what individuals are doing. 'What did the Easter Islander who cut down the last palm tree say while he was doing it?' asks the American professor of geography and physiology Jarred Diamond in his book *Collapse: How Societies Choose to Fail or Succeed* (2005). History has possibly a slightly different take from that of Diamond's—that invasive species can wreak havoc in nature and in man's environment. Nevertheless, history tells a story about human failure. It was human beings who brought the rat to the island; they lived there for 1000 years without understanding how the ecological situation was gradually changing to a state where there would be no space left for them. It was also human beings who brought disease onto other human beings and who put their brothers and sisters into slavery. Let us keep in mind the diverging stories; despite our exploration of the moon, of the North Pole and the depth of the oceans, our knowledge and understanding of human development are still surprisingly shallow and erroneous.

Hunters and gatherers

Many seem to believe that hunter and gatherer[4] societies were characterized by misery and starvation, but few observations support this

---

[4]   I use this term here because it is the standard term. I rather prefer the term 'capturing societies', which would also include fishing. In most cases, the food gathered was likely much more important than the food hunted. Still, the hunt perhaps drove and shaped the development of societies in a different way.

and plenty others show the opposite (Livi-Bacci 1992). Today, most of these people are pushed back into the hardest and most difficult environments, though even in these environments they do pretty well. The San people (the so-called bushmen) of Africa have a calorie and protein intake that surpasses recommendations and it is improbable that they will suffer from mass starvation caused by famine similar to those that harmed agrarian societies over centuries (e.g. the Irish potato famine in the 1840s, the Bengal famine of 1943 that perhaps killed 3 million, or the 1984–1985 Ethiopian famine). And the hunters, fishermen or gatherers in ecosystems more fertile than the Kalahari Desert must have had a richer livelihood.

Hunters and gatherers not only ate well but they also worked little, much less than the later farmers and labourers. Their societies appear to have been egalitarian, at least compared to societies that succeeded them; there were no classes, economic specialization or slavery. There was no private property. A certain division of labour was based on gender or age, not on the fact that some could appropriate the fruits of others' labour (there are exceptions to this). The absence of state and hierarchies is another characteristic of the hunter and gatherer societies. 'Leaders' existed, but they did not have the powers of a formal chief nor the possibility to use violence or force on members of the band, the privilege of the state (Clastres 1974; Farb 1968; Diamond 1997). 'There are no leaders other than those who have earned the respect of their peers by being models of good conduct, and who can only advise and not dictate,' says biologist and anthropologist David Sloan Wilson (2003: 21).

Some claim that the standing of women can be traced to the ecological conditions of societies: women had a weak position in societies with high meat consumption (i.e. where hunting was important) and, conversely, they had a stronger position in societies that were predominantly vegetarian. Similar thoughts about the importance of ecology are formulated with regard to the sharing of food. In most hunter and gatherer cultures, the band shared food among its members. Sharing seemed to be more prevalent in cultures with food shortages and many sources of food, and less common in those where food came from a few abundant sources. Still today, a universal phenomenon, somewhat paradoxically, is that people in poor societies and in cultures developed under hardship are more generous than those living in affluent societies.

Many of these societies were nomadic, which meant that women could not give birth to many children because it is difficult to carry more than one child at a time. Breast-feeding prolonged the period between births, and they most certainly used herbs and other natural remedies to influence fertility and induce abortions (Austen 1987). This meant that populations were fairly stable. A nomadic life didn't support accumulation of wealth until the nomads became cattle keepers, thereby building up assets in the form of herds and carrying more things with them, on the back of the animals.

There is no consensus about the extent to which these societies were peaceful. Most claim that war was rare, and at least three arguments favour such an assumption. Let us imagine a band of some 25–30 people of which perhaps 5 are warriors. It is quite apparent that such a band cannot engage in wars with deadly results very often; if they did, the whole band should be extinct in a few years. Today, one can observe the Dani people of New Guinea fight in a ritual way from morning to dawn, with breaks for siesta and when it is raining. Non-combatants stand at the sides and cheer, and actual casualties are low. It is thus more a ceremonial fight than a lethal war (Goldstein 2001). Death caused by a weapon was also uncommon, as deduced from the skeletons of hunter and gatherer societies. Finally, fights over territory are rarely lethal in other species, simply because the costs and risks are too high compared with the possible gain.

Having said that, there are indications that war existed in the period just preceding the agrarian civilization. Between 12,000 and 8000 BC, there was a revolution in weapons technology; the bow, the sling, the dagger and the mace were developed, and cave paintings show their use against men as well as animals. The settlement of Jericho might have started in an oasis as a basis for hunting. For protection, walls were built. This chain of development has been proposed by Joshua S. Goldstein in *War and Gender: How Gender Shapes the War System and Vice Versa* (2001). Ljungkvist (1987) sees the development of settlements and wars to be the result of an aggressive culture caused by some external factor such as ecological degradation, an earthquake or climate change. When one culture shows an aggressive behaviour, surrounding cultures have to adapt themselves and become militaristic. This creates a situation where people will yield powers to a headman or a king; protection and security are so important that freedom is sacrificed for them, and, in a battlefield, an authoritarian hierarchy mostly works better than democracy. In this way, two main curses of human society emerge: an outer threat in the form of war and aggressive neighbours and an inner threat in the shape of an authoritarian leadership, mutually supporting each other.

In some ways, the hunter and gatherer societies seem to have been a secure and rather pleasant way of life, but most of us would certainly not swap our lifestyles voluntarily with such a way of life. In any case, that discussion is meaningless, humankind simply can't live like hunters and gatherers unless we were to reduce population to a fraction, perhaps a hundredth, of what it is today.

Transformation to an agrarian society

Agriculture developed some 12,000 years ago through processes where plants and animals were domesticated gradually. The first animal to be domesticated was the dog, or rather the wolf that became the dog. Sheep and goats were tamed some 11,000 years ago, pigs 9000 years ago and cows 8000 years ago. Turkeys were domesticated in what is Mexico today and hens in India. Wheat, rye, beans and peas were grown in West Asia; buckwheat, soya, onion and eggplant in East Asia; Sorghum, cotton, okra and banana in Africa; bananas, taro, coconuts, squash, rice and sweet potatoes in South East Asia; and chilli, potato, tomato, peanuts and pineapple in the Americas (Tansey and Worsley 1995).

Archaeological findings show that in many places the first farmers were shorter and less well fed than the hunters and gatherers they pushed aside. If they had realized the consequences of food production, they might have never done it (Livi-Bacci 1992; Diamond 1997). No strong reasons exist to believe that human beings were forced to farm to feed an increasing population. It was rather that improved methods of hunting and fishing or life in particularly rich border zones, such as an oasis or coastal flats, enabled people to settle in certain places.[5] Once settled, farming developed both as an opportunity and as a necessity. An interesting, and perhaps odd, case is represented by the Norte Chico settlements (just north of today's Lima in Peru) where city development is more than 5000 years old. They were ostensibly engaged in farming, not of food, but of cotton, needed primarily for their extensive fishing (Mann 2005).

Once settled, population increased as a result of better care of sick people, fewer violent encounters with enemies or dangerous animals and more children per woman. An emerging power, a chief, might also

---

[5]   There could also be settlement around a strategic resource such as a flint or obsidian, in which case specialists would settle and trade, for instance, tools for food.

see benefits in an increasing population, for instance, to strengthen its settlement compared to that of others—more arms wielding an axe was probably how strength was defined. Rousseau (1964) showed in the 'Discourse on the Origin and Foundations of Inequality among Men' (or the *Second Discourse*) that society must have emerged before farming because many functions and institutions need to exist to make farming possible. We can also envision that a more subordinate role of the woman developed in this context; instead of being an equal partner in life, she became a child-breeder.

Those who think that farming emerged because it was a more efficient method of feeding the population should think about how it was for the first so-called farmers. Hunters and gatherers knew the basics of reproduction and they saved breeding stock of the animals they hunted (Karsten and Knarrström 2002). They also assisted the growth of certain crops using near-cultivation measures; the Paiute in Mexico watered wild plants (Ljungqvist 1987) and the people in Sweden coppiced and pruned hazel, to increase nut yield but perhaps, more importantly, also to get good construction materials (Karsten and Knarrström 2002). Still, we can't compare the methods of agriculture of the first human beings with those of the farming societies that emerged after thousands of years of trial and error. That would be like comparing floating on a log with sailing in a boat. Farming wasn't a single ready-made technology waiting to be picked up or discovered; it was a complex of technologies, biology and human knowledge and, ultimately, a society.

Crops still were mainly wild, a far cry from the cultivated varieties grown today, and the farmers had no tools for cultivation. It is clear that these first farmers actually could not feed themselves from farming alone to begin with; if they had tried, they would have starved to death. On the contrary, farming was most certainly developed by settled people who still got most of their food through the old ways of hunting and gathering. They farmed certain crops that were uncommon and difficult to find, such as medicinal herbs, or crops that they fancied a lot[6], more like gardening. Gardening is probably older than farming, and some cultures that have still not converted to

---

[6]   Indulgence is underrated as human motivation. For some reason, need, struggle and hardship as the basis of development appeal more to our analysis than pleasure, love and adventurousness. We say 'necessity is the mother of invention' even if the world every day proves the opposite. It is in affluent societies that innovation and change are quicker whereas in poor societies with limited resources and limited space for pleasures there is little innovation.

farming cultures keep gardens (Radkau 2008). When settled populations grew, farming changed from being an opportunity to a necessity. This is also paralleled in the transition from an agrarian society to an industrial one.

Two early forms of farming were swiddening (slash and burn) and farming flood plains. Emergence of both is easily understood from the conditions under which they developed and the realities of farming. Big rivers flowing through dry plains gave rise to unique possibilities for farming. Because it was dry, there was little forest on the plains, and because there were plains around the rivers, they easily flooded. When they flooded, the land was covered by fertile mud and this meant that two of the most labour-intensive parts of farming were done by nature—soil preparation and application of fertilizers. What was left was just to sow and to weed, somewhat simplified. Later on, this led to the development of irrigation farming. Still, after a few thousand years, the same flood plains continue to provide billions with food if they do it right; of course, in some cases, the same flood plains have become salinated or have turned into swamps.

In a forest fire, ground-covering plants are killed and most trees and shrubs are killed or at least set back considerably. The only land preparation needed after swiddening is using a stick to make a hole where seeds are dropped. The few weeds that grew there before are not serious competitors to the chosen crop as they are not adapted to an open landscape. There are no, or very few, seeds of the aggressive weeds that are everlasting companions of and a curse to farming; the pressure of pests is also low. Many nutrients are released, partly from ashes from burnt wood and foliage, and partly through a slow decomposition of dying roots in the soil. Swiddening thus helps clear the land, prepare it for sowing, regulate weeds, pests and disease, and supply nutrients. But all these benefits decrease over time. The ease in land preparation lasts only the first year and becomes increasingly hard as trees re-grow and weeds spread out. Regulation of weeds and decreased availability of nutrients limit the returns; therefore, new land has to be cleared by burning and the land is left to regenerate. However, as long as conditions allow, slash and burn farming is likely to continue because it requires little work.[7]

---

[7]   Under similar conditions in Cameroon, work effort per hectare

Despite agrarian societies not providing their members, or at least not the majority of them, with a better standard of living, they had competitive advantages over neighbouring tribes living on hunting and gathering. These included a bigger population, defence constructions, accumulation of wealth and a societal organization. Hunter and gatherer societies were small because ecological adaptation didn't support high population densities. Agrarian societies could therefore squeeze them out wherever the conditions for farming were good.

Today, hunters and gatherers only remain in ecological margins, where farming has not been successful. Fights between agrarian societies and hunter and gatherer societies, as well as nomadic pastoralists,[8] have been a never-ending story, with prominent examples being the conflicts between farmers in the American prairies and the first Indians and then the ranchers, those between Swedish settlers and the reindeer-keeping Sami and those between the expanding Bantu people in Africa and the Batwa (pygmies). Even the story of Cain and Abel is about this same conflict. And in each case the hunters and gatherers and nomads lost in the long run, even if the nomadic herders, such as the Mongols, ruled for a period.

Gradually farming technology developed, and, equally importantly, so did technologies for storage and preparation of food. The latter is often overlooked, perhaps because later observers saw it as part of consumption as opposed to production—a completely artificial construct, or perhaps because food preparation, by and large, has been a female occupation and therefore has been undervalued, like most female occupations, by those (men) writing history. Iron played a role not only in war. Its importance in farming can hardly be exaggerated; with iron mountings on ploughs, spades and hoes, the efficiency of these tools increased manifold. Improved tools increased productivity in soil preparation, weeding and harvesting—the three most labour-demanding elements in most agricultural systems.

The development of farming, and in particular the development of irrigated farming systems, with increasing population densities is intrinsically tied to a more complex societal organization. A new institution, the city, emerged in Mesopotamia some 6000–8000 years

---

jumped from 700 hours for swiddening to 3300 hours for permanent cultivation (Livi-Bacci 1992). Even today, one can observe that populations, once they get land in abundance, might revert to shifting cultivation, because they can feed themselves with less work (Radkau 2008).

[8]   Pastoralism is the lifestyle of herding, nomadically or semi-nomadically, domesticated or partially domesticated animals.

ago. Some of the characteristics of this institution are the differentiation in the population (not all residents grow their own food, leading to specialists), payment of taxes to a deity or king and a state religion; those not producing their own food are supported by the king and there is trade and import of raw materials. The institution saw of the rise in landlords, who didn't farm themselves but did so with a farm-bailiff, through tenants, or with slaves or serfs. All this was possible with the surpluses generated by farming through irrigation and ploughing. We can now see the strong connections between class differences, the division of labour, the state, the accumulation of wealth and new technology.

In Egypt, the pharaoh imposed taxes in the form of goods delivered to the state. In the Inca Empire, taxes were mainly in the form of forced labour: smallholders got just sufficient land to grow food for the family and all extra labour mastered would be directly in the service of the Inca. All land and livestock were basically the property of the Inca. Although a certain degree of local barter was allowed, the state regulated the distribution of every important product. This meant that no accumulation of wealth could occur outside the state; individuals lacked both incentives and real possibilities to develop their productivity (Mazoyer and Roudart 2006). Many privileges developed, the most disquieting (albeit disputed) of which was the systematic use of human sacrifices for nutritious purposes by the Aztecs. All through history the rich have had access to more and to higher quality food than the commons, a self-reinforcing process as they and the children of the rich thereby become stronger, healthier, smarter and more attractive. Other rules or taboos have prescribed certain diets for men and others for women.[9]

Distribution of products

Within the hunter and gatherer societies, there was no market. This was most likely also the case in smaller farming communities. Between bands and between regions, there was mostly some kind of exchange and trade. These exchanges were hardly organized as a 'market' but more often as gifts or sometimes by barter. Gifts are a very old form of creating relations, and presenting a gift is still how you would approach a tribe for the first time. The reciprocal exchange of gifts is a

---

[9]   I have personally noted that women in the north of Uganda are prevented from eating chicken, which is reserved for men.

socially important act. (If you are in doubt, look into the private lives of families.) Gifts were also the foundation of status. Instead of extracting wealth from others, traditional leaders were supposed to give as much as possible to their peers or to organize big parties. A gift creates mutual dependency.

The great kingdoms and empires were strictly planned bureaucracies; societies were more or less 'socialist' in the sense that everything was produced for society and the rights of the individual were minuscule. Production was recorded (therefore, the emergence of letters and numbers), and what wasn't needed locally was moved to central storage facilities. Other types of society developed other types of stewardship, distribution and division, but markets didn't play any major role in any of them for the distribution of daily necessities and in only a few of them for the organization of society itself (Polanyi 1944; Austen 1987). At the most extreme, the Inca had neither money nor market (Mann 2006).

Notwithstanding, international trade is certainly nothing new. One of the oldest occupations is that of a trader, and already some 3000 years ago there were specialized trading nations, such as the Phoenicia (with its centre in today's Lebanon) that traded in timber and metals. Land routes, such as the Amber and Silk roads, and sea routes, such as those in the Indian Ocean and the Mediterranean Sea, were used. The Silk Road, established during the Roman Empire, provided a continuous connection from Europe to China, even if it was unlikely that many traders travelled the whole stretch by themselves, for no other reason than that the trip took several years. Early trade routes were established for trading fish and shells from West African shores, pastoral products from the desert and metal goods (Austen 1987).

However, long before this distant trade, the origin of trade can be traced to exchanges between local societies. This exchange was not mainly an economic activity but an ecologically motivated one. Trade made it possible for human beings to establish themselves even where some basic resource was absent. One tribe had access to a resource that the other was missing. In some rare cases this situation might have led to war, but more often it led to trade. The same situation prevails today; most oil, fish and coffee are peacefully traded and not obtained through war. In some places flint or obsidian was abundant, and in other places hunters had no access to those stones for making spears and arrows. In other areas there was no salt, which was important for the preservation of meat and curing of skins. The best flint in Europe is found in Sweden and Denmark, and that was the basis for extensive trade already 5000 years ago (Schilling 2008).

This ecological origin of trade is also the reason for why loss of trade links can have devastating effects on some societies. The role of trade in ecological adaptation has, in some cases, meant that communities have been able to specialize in forms of production that are very well adapted to their ecological context. Through trade with the plains, the peasants of the Alps could shift entirely to pasturing livestock and their Mediterranean colleagues to viticulture without the need to plough fragile mountain slopes. In this way, trade in agricultural produce, even staple food such as grain, can help sustainability (Radkau 2008).

When the great civilizations emerged, trade expanded to luxury items and symbolic products (gold, jade, shells) for the elite and strategic products for the military, including high-value products that could cover costs of long-distance transportation. Some staple products were subject to long-distance trade, for example, grain from Egypt and Sicily to Rome and Greece. At the time of the Peloponnesian War (431–404 BC), between one-third and three-fourths of the food of Greece was imported (Montgomery 2007). Development of shipping allowed bigger and bigger quantities to be traded from longer and longer distances. The Romans spent a lot of energy in safeguarding their food supply. They were to some extent dependent on private merchants and made some of the first attempts to regulate food prices. They failed, as did all later attempts to regulate food prices for longer periods. In AD 301, the Romans fixed maximum prices for 1000 different products, with the death penalty for violations. Disaster came rapidly. Already in AD 304, Rome had to buy wheat from Egypt at prices that were ten times costlier than those maximum prices (Mazoyer and Roudart 2006).

From around AD 1000, Eurasia was increasingly integrated into one international network. Direct sea routes existed between China and India; the monsoon sailing ships in the Indian Ocean connected India, East Africa and the Arab world; and the Arab civilization stretched through North Africa to Spain. Northern Europe was connected through Viking trade both via the Atlantic Ocean and the Mediterranean Sea and via the great Russian rivers and lake systems. Finally, the vast empire of the Mongols provided safe access to the Silk Road. Parallel to this trade integration, agriculture practices and innovations spread along with other things: the geographic extent of Islam's core areas from Morocco to Indonesia also reflects this breakthrough in trade (Myrdal 2008).

Emergence of the Industrial Revolution

After the collapse of the Roman Empire, Europe was a peripheral area with regard to population, culture, economy and technology. Up to the seventeenth century, India[10] and China had much bigger populations and economies than Europe. From the middle of the nineteenth century, Europe has dominated, followed by the United States. Between the years 1000 and 1913, the size of the economy has grown 6-fold for India, 9-fold for China and 90-fold for Europe (Maddison 2001; see Table 1.1). How can this be explained?

Table 1.1 Gross production (in billion international dollars).

	1500	1600	1700	1820	1913	1998
India	60	74	9	114	204	1,703
China	62	96	83	229	241	3,873
Western Europe	44	66	83	164	906	6,961
United States	0.8	0.6	0.6	13	518	7,395
World	247	329	371	694	2,705	33,726

Source: Maddison (2001).

There are many attempts to explain why Europe, and later the United States, got the dominating role it had. The discussion is a minefield where all explanations about societal organization, mentality and culture are subject to accusations of Eurocentrism or possibly racism. It is 'safer' to look for physical or geographic reason or simply for chance. But let's accept the challenge and look into this matter. The purpose is not to make a final statement about this but to place further discussion in a perspective. Studying the transformation of an agrarian society to an industrial, capitalist society tells us a lot about the main features of our society. The European rise was clearly linked to its development into an industrial capitalist society.

At the collapse of the Roman Empire, Europe slumped. This was followed by a slow recovery that was again set back by the plague in

---

[10]   India should here be understood as a geographic demarcation of the South Asian subcontinent and not as a state or empire. Only occasionally in the course of history was this great India a unified empire, such as the Mauryan Empire of 321–185 BC.

the fourteenth century. It took almost 200 years until the population recovered from the plague (see Table 1.2). The population of China was estimated to be 100 million by the year 1000, 2.5 times as high as that of Europe, but at the start of the Industrial Revolution, it was 180 million, which was just slightly higher than that of Europe. A population explosion in Europe occurred during 1700–1900, as shown in Table 1.2. We should also keep in mind that some 50 million Europeans emigrated and multiplied rapidly in their new countries.

Table 1.2 Estimate of the population of Europe from the year 400 BCE to 1900.

Year	Population (in millions)	Year	Population (in millions)
400 BCE	23	1300	73
0	37	1400	45
200	67	1500	69
700	27	1600	89
1000	42	1700	115
1100	48	1800	188
1200	61	1900	401

Source: Bath (1976).

With technologies that existed 1000 years ago, Europe simply could not compete with India and China. The basis for wealth at that time was still, almost exclusively, agriculture, and the conditions for farming in the fertile river valleys of China and India were superior to those in Europe. The Mediterranean area was dry and the Greeks and Romans, to some extent, had destroyed the soil with bad management. Further north, the climate was cold and vast forests covered the ground; many good soils were badly drained and needed drainage and forceful ploughing to be productive. Also, almost all crops and livestock were domesticated in warmer and drier climates: it was a slow process to adapt varieties and breeds to the damp conditions of Western and Northern Europe. For instance, sheep are from dry hot highlands and were now bred on moist cool lowlands. Even today, this causes problems with internal parasites in sheep.

Technical development outside farming was not a major driver of change 500 years ago, except in the two important areas of war and shipping. Many of the higher cultures and vast empires showed little interest in technical development. The Arabs developed science to new levels. Through their trading network they could observe developments in most part of Eurasia; still they seemed to be unable or uninterested in developing more productive technology or even economic development in general. The same goes for the Mongol Empire — on the surface, the biggest empire ever seen. Despite its enormous power, the Mongol Empire never became a centre for economic and technological development.

Looking at technological development, without doubt China was more advanced than Europe early on. Before 1450, the Chinese invented the lock gate, cast iron, the compass, the harness, book printing and gunpowder. Europe was no leader in almost any technology at that time; it got most of its knowledge from the Arabs, who in turn made a bridge spatially to China and India, but also temporally to the Greeks and the Romans. The Chinese fleet, on the other hand, sailed all the way to the east coast of Africa 50 years before the big voyages of the European ships.

During the fifteenth century, the Ottomans shut off the Europeans from the trade routes to Asia, which initiated a febrile development of shipping. This coincidentally led to the 'discovery' of America, which gave the European economy a very strong injection of silver, which was most sought after by the Chinese. Later on, the New World played a big role in supplying raw materials. It must be more than a coincidence that all the most powerful nations in this period were shipping nations: Venice, Genoa, Portugal, Spain, France, Great Britain and the Netherlands. The early development of capitalist society was more about trade than about industrial production.

Diamond explains the success of Europe compared to that of China. China was an old homogeneous empire that was unified in 221 BC; it had one script and a common language, culture and religion. This was its strength, but it was also its weakness when contrasted with the restless innovations and dynamism of Europe. At the onset of the Industrial Revolution, Europe was divided into 500 small city-states, countships or fiefdoms. They were not politically unified, but there was some economic integration. Cultural and scientific impulses spread rather freely (e.g. the Lutheran reformation), and innovations spread rapidly. The story of Christopher Columbus from Genoa, who first tried to persuade the King of Portugal to finance his expedition and, having failed, then turned to the French and the Spanish monarchs, and finally succeeded when Isabella of Spain agreed, is also a history of how ideas and innovations flowed between the European states. In Europe, the states were competing with each other and there were many parties to go to if you had a good idea to share. In China, everything was assessed by whether or not it was beneficial for the emperor (Diamond 1997).

In addition, the emperor of China had huge population resources to exploit, and therefore had little incentive for technical innovations. The ruler of a European city-state had a very small population to tax if he wanted to live like the emperor of China, and therefore was more interested in innovations and exploration. David S. Landes (1998) emphasizes the importance of free markets and legally protected property rights. The Chinese government meddled in private enterprises, took over some and prohibited some good businesses and manipulated prices. Landes also points out that the innovation of glasses, in Pisa in the thirteenth century, was very important. With glasses, craftsmen and artisans could now work with fine mechanical things, such as watch-making, and with other technical instruments for 20 years longer than they could without them. The water mill, a very important new form of energy, spread in China as in Europe in the early Middle Ages. But in China its development halted because of the competition for water with irrigation. In Northern and Western Europe, the need for irrigation was less and landowners, who controlled the streams, could in any case decide to use the water for the mills rather than letting the peasants use it for farming (Radkau 2008).

The importance of markets

A factor that, in my opinion, is not emphasized enough is the access to markets. If all other countries were poor, European expansion would never have happened or been possible. We are used to seeing the colonization of already poor people as a strong feature of European expansion, combined with the exploitation of cheap labour and natural resources. But it was, on contrary, the incredible wealth and manufactured products in other parts of the world, in particular in China, which was the attraction. As we have seen, most of the world's wealth and production was in Asia and not in Europe. The Europeans were interested both to buy those products and, if possible, to control trade with them. Already in 1594, the Peruvian vice-king wrote to Madrid complaining about Chinese goods being much better and cheaper than Spanish ones. Approximately three-fourths of the silver the Spanish took from the New World ended up in China, mostly via English or Dutch merchants. The Dutch didn't only bring spices to Europe, they used them for trade in the Indian Ocean. India alone consumed twice as much as Europe (Arrighi 2010). Trade was certainly at the centre of European expansion and it is no coincidence that early colonization was driven by semi-commercial corporations such as the Dutch and the British East India Companies. This was also most likely the reason why Africa was not colonized at the first stage, despite it being closer than both the Americas and certainly India; there was simply not enough wealth in Africa, or products of interest to buy or buyers of European products. Africa was left with trading its people.

The purchasing power of each country was not adequate incentive for industrial development, in the same way as exports is driving China's development today. Through 'efficient' production of cloth, tools and pearl, wealth could rapidly move from India to England. India was the predominant actor in the world market for textiles in the eve of the eighteenth century, but Great Britain outdid it, first in the world market and ultimately also in the Indian market. In the early eighteenth century, the British government had banned the import of Indian cotton cloth, but in the early nineteenth century trade reversed and rapidly grew, from 7 million metres in 1820 to 90 million metres in 1840. This led to a remarkable deindustrialization in India (Marks 2002; Clingingsmith and Williamson 2005) and was the basis for the English industrialization. In 1813, there were 3000 power looms in England and a staggering 200,000 handloom weavers; around 1860, there were 400,000 power looms and no handlooms left. As Arrighi (2010) stated: 'It is hard to imagine how this great leap forward in the mechanization of the British textile industry could have occurred at a time of stagnant domestic and foreign demand for its output except for the conquest of the Indian market and the consequent destruction of the Indian textile industry.'

Trade in the Indian Ocean, the centre of the world, before the arrival of the Europeans was largely peaceful, but competing European countries occupied major trading ports and limited trade by others. The combined military and commercial aggression was simply a very efficient strategy. The fusion of state, capital and trade was a European speciality (Arrighi 2010).

Why did England, or rather Great Britain, become the leader? In the sixteenth century, Italy[11] was the largest economy of Europe, followed by France, Spain and Germany. In 1870, the economy of Great Britain was clearly the largest of them all, 35 times bigger than that in 1500 (Maddison 2001). What had happened?

Spain, starting as a leader and with the discovery of the Americas, seems to have been a strong candidate. In 1492, the Arabs were driven out of the Iberian Peninsula and Columbus discovered America.

---

[11]   Here, Italy refers to the area that *today* constitutes the nation of Italy; at the time referred here, no Italy existed.

Cordoba was, thanks to the Arabs, the knowledge centre of Europe. The wealth of Spain did indeed grow rapidly as a result of the conquest, but most silver was used to fund Dutch arms dealers or Genovese or English financiers, who in turn used the money for profitable trade in Asia, while Spain was locked out from trade in Asia, at least until they took Manila and could trade directly from America to Asia. In addition, the Spanish upper class looked down on entrepreneurship and trade (Danielsson 1977; Marks 2002).

The Spanish and the Portuguese brought feudalism to their colonies and had mainly a territorial approach, whereas the Dutch and the English brought capitalism to their colonies (Weber 1986). The Dutch were ahead of them all when it came to drawing benefit from the expansion of trade, and thus Amsterdam became the trade centre of the world, but Holland didn't pursue a very expansionist territorial policy. Arrighi (2010) states that the Dutch were 'more capitalist' than the English, and therefore more successful in their original expansion, which was both to the East Indies and to the Baltic Sea trade. Once the other states in Europe, and in particular England, adopted the Dutch capitalist approach combined with a territorial expansion they took over the business. When the triangular trade on the Atlantic (see below) became important, English ports could take over from Amsterdam as the *entrepôt* (trading post) of the world (Arrighi 2010).

England, just like China, had reached ecological limits; the land could not provide sufficient food, firewood or clothing. But there was coal in shallow layers just close to London, which initially was extracted for replacing firewood. Between 1580 and 1680, the annual supply of pit coal to London increased from 20,000 to 360,000 tons[12] (Radkau 2008). When pits grew deeper, there was a need to pump out water from them, and because coal was abundant in the coal mines(!) the steam machine was first developed. The manufacture of iron and steel was facilitated by the use of coal instead of charcoal, and, together, coal, iron and the steam engine were the basic building blocks for the Industrial Revolution and supported each other (Weber 1986; Marks 2002). Of course, questions arise about cause and effect: was it coal that actually drove the development or did other factors increase the demand for coal? That access to abundant energy sources is no guarantee for strong economic development is demonstrated by Saudi Arabia, Venezuela and others.

---

[12]   Causing smog already at that time.

The colonies of Great Britain also played an important role and the triangular trade between England, Africa and North America was an important feature. Copper, liquor, guns, cloth and beads were bartered for slaves in Africa. The slaves were sold in North America and the Caribbean and then ships filled with cotton, tobacco and sugar turned to England. Apart from the direct profits of trade, cotton was needed for the expanding British textile industry (see above) and sugar supplied the workers with cheap rum and sweet tea to keep them happy and working.

In general, the commercial climate in England was 'good for business'. English farmers were commercially orientated compared to their continental colleagues. They could also keep more of their surplus and thus invest in land and technology. The ruling class of England did not have the same powers as on the Continent to extract wealth from surplus labour of farmers. On the other hand, they had larger landholdings, even before the enclosures (see below), which meant that the prevailing model of farming was developed by tenant farmers who had to pay their rents in cash, and even in many cases compete for the leases. The American political scientist Ellen Meiksins Wood sees a distinct agrarian origin of English capitalism. The industrial system was, according to her, a result of the agricultural developments: 'Without a productive agriculture sector, which could sustain a large non-agricultural workforce, the world's first industrial capitalism would have been unlikely to emerge' (Meiksins Wood 1998).

Great Britain was also earlier a 'nation' with a common identity and loyalty. The emerging capitalism was paradoxically, which will be further discussed below, also strongly entangled with the nationalist project and benefited from a strong state. A comparative political liberty was also beneficial for trade and entrepreneurs. Wealth in England appears to be more evenly distributed than in many other countries, which meant that there was a substantial middle class of bigger farmers, traders, manufacturers and craftsmen. Not only were they industrious, but they also constituted an emerging 'mass' market compared to the countries dominated just by nobles and serfs. In the fifteenth century, most local customs had been abolished, which also meant that there was one national market.

The so-called enclosures, the transformation of land, over several hundred years, from common management and ownership to privately owned property is considered an important factor for the development of English capitalism. An increased demand of wool meant that it was more interesting to get income from large-scale sheep farming than from extracting rent from the land; so tenants and others were driven off the land and the sheep took over, something that still can be seen in the landscape of the British Isles. This process led both to the emergence of a new industry (of wool manufactures) and also to the emergence of a 'proletariat', a more or less new class of people compared to earlier societies where most people had their life and roles determined at birth. The wool industry in turn was the overture to the real Industrial Revolution that came with the cotton industry. Cotton was a raw material that could be produced in all the colonies and it was a fibre that lent itself to mechanization better than wool (Polanyi 1944). The mental impact of the enclosures was substantial when the land became private and the sharing among neighbours was reduced, and thus every household became an island. This change in mentality, the decline of the sense of community, might well be as important as the economic effect.

As we discussed earlier, it was not because of desperation or need that human beings went from hunting and gathering to farming; on the contrary, the communities that started farming were successful communities living in areas with rich natural endowments that allowed them to settle. In a similar way, it was not primarily the lack of something that brought about the Industrial Revolution and capitalism. Dutch and English farmers were among the most advanced in the world; they were also comparatively free, which stimulated agricultural development that became the foundation for developments to come. At that time, China was clearly reaching a number of ecological limits and had severe problems with soil erosion, whereas the ecology of England was fairly stable despite large-scale deforestation. In England, with easy access to pit coal the lack of forest was less dramatic a factor, and because of the humid climate the lack of forest didn't lead to soil erosion as it did in southern Europe or China. From its position of a dominating trader and global power, England could export its environmental demands, and associated degradation, to other countries, through big imports paid by substantially smaller, but higher priced exports.

Why some fail

As interesting as it is to look at successful societies, so it is to look at those failing. But we need to qualify this a bit, as all societies fail sooner or later. The only realm that has remained intact for a very long time is China, but even China has had many shapes and forms, heights and declines, through millennia. It was ruled by the Mongols for a while and it shifted nature dramatically through the revolutions in the twentieth century. Now it has converted itself to a state capitalist society. This ability to adapt is what has made it survive. If there is one reason for why societies fail it is their inability to adapt—to adapt to new technologies, to external or internal changes. And China proves this: though it never totally collapsed, it certainly lost out on development a few times in history, mainly because of its inability to adapt.

In *Collapse* (2005), Diamond examines societies that failed and collapsed. He sees environmental degradation, aggressive neighbours, climate change, loss of trade links and the inappropriate response to these by the cultures themselves as determining factors of failure. Of environmental factors causing collapse, Diamond mentions deforestation, erosion or other soil destruction (salinization, water-logging), water regulation, over-exploitation of fish, shellfish or game, invasive species, and population growth. The inhabitants on Pitcairn Island failed because of deforestation by their major trading partner. The Anasazi in North America failed because of deforestation and drought. Environmental degradation, loss of trade links, climate change and hostile neighbours brought down the Viking settlement in Greenland. We already mentioned how most early trade was ecologically motivated and not driven by profit motives; this was the reason for loss of trade links having such devastating effects. It meant that some essential product would be missing, a product such as salt, a metal or certain foods important for nutrition. Thus the loss of trade links is a kind of ecological disaster. This is perhaps easiest for us to understand if we imagine that the supplies of oil, coal and gas were cut off.

The Maya could not adapt to climate change and hostile neighbours. They also caused substantial deforestation and erosion (these two are often twins in development), combined with population growth. 'When the Maya culture came closer to collapse, wars increased and so did the construction of new temples, despite both just exacerbating the problem. They responded to their problems by turning to the gods for meaning and understanding. They didn't understand that they just enforced the collapse,' writes Myrdal (2008). Andres Ciudad, an anthropologist, ventures that the ultimate collapse came when the citizens of Maya realized that their kings were no gods (*MercoPress* 2006). Others point to the fact that the increasing complexity in civilizations as they develop is an important factor for collapse. For instance, the American historian Joseph Tainter (2009) says,

'Societies adopt increasing complexity to solve problems, becoming at the same time more costly. In the normal course of economic evolution, this process at some point will produce diminishing returns.'

Environmental degradation has followed man since his first attempts to farm, well even before farming. Hunters too caused substantial degradation, for example, by the extermination of large mammals in North America or of many bird species in the Pacific. Soil erosion is one of the most prevalent and harmful kinds of environmental degradation. It was of concern for rulers of ancient Greece as early as the sixth century BC. The lawmaker Solon proposed a ban on cultivating hillsides in order to prevent erosion. The ruler Peisistratus rewarded peasants for planting olive trees instead of cutting down forests and grazing livestock. Two hundred years later, Plato wrote of land devastation in Attica:

> . . . in the primitive state of the country, its mountains were high hills covered with soil, and the plains, as they are termed by us, of Phelleus were full of rich earth, and there was abundance of wood in the mountains. Of this last the traces still remain, for although some of the mountains now only afford sustenance to bees, not so very long ago there were still to be seen roofs of timber cut from trees growing there, which were of a size sufficient to cover the largest houses; and there were many other high trees, cultivated by man and bearing abundance of food for cattle. Moreover, the land reaped the benefit of the annual rainfall, not as now losing the water which flows off the bare earth into the sea, but, having an abundant supply in all places, and receiving it into herself and treasuring it up in the close clay soil . . . (360 BCE)

If not dramatic collapse, at least the slow decay of most civilizations can be traced back to soil erosion and wasteful agricultural practices. It is a sobering insight that very few agricultural systems have proven to be really sustainable, and for many of those it is thanks to external factors that they are sustainable, not to the ingenuity of human beings. The fertility of the Nile valley and other flood plains is mainly a result of erosion upstream bringing new soil and new nutrients downstream. David Montgomery, an American soil scientist, states in *Dirt: The Erosion of Civilizations* (2007) that most historic cultures lasted between 500 and 1000 years. Civilizations start off with fertile soils created by natural processes. Fertile soils in hills with good rainfall are cultivated as population increases, but soil erosion makes fertility fall leading to the cultivation of marginal lands where fertility plummets and, finally, the whole civilization collapses and the area is depopulated until nature replenishes the soils again. Montgomery (2007) shows that

Schwarzwald (Black Forest, today dominated by forests as its name indicates) in Germany has gone through three such cycles.

The environmental factors of failure mentioned by Diamond (2005) are still critical. New environmental factors that could lead to collapse are climate change caused by green house gases; imbalances in the nitrogen cycle; the spread of toxic compounds in nature; energy shortages; over-exploitation of the planet's photosynthetic capacity by human beings, leaving little space and energy to the rest of nature; and loss or shortage of a critical raw material (e.g. oil, copper). We will discuss all of these in Part 2.

– 2 –
Development of Modern Society

Whoever has will be given more, and he will have an abundance. Whoever does not have, even what he has will be taken from him.

(Matthew 13:12)

In this chapter, we discuss the most important components and processes of modern society. By 'modern society' I mean the mix of industrialization, market economy, capitalism, nation-state, global trade, democracy and material and economic growth that characterized most parts of the world throughout the twentieth century. It originated from an increase in trade that was the foundation for a new class of tradesmen and financiers in the city-states of Northern Italy and of the Flanders (we noted earlier also the possible agrarian root of English capitalism). Craftsmen too were involved in rather large-scale pre-industrial manufacturing. Both these groups were separate from feudal society and from the landed nobles. They hired labourers for work. 'Free' markets emerged where producers sold their goods in competition. This was the social context, but it was with the introduction of new technology and, in particular, with the use of a new energy source (coal) that these groups were able to unleash the combined powers of industrialism and capitalism, and, seemingly, free themselves, and slowly the whole of society, from constraints imposed by the biological production of the land.

With capitalism, most aspects of life could be managed as markets. The ideology of capitalism and market economy—that most things are best organized by a market, with governments just providing a framework—penetrated society fairly successfully. However, many aspects of life and society were not easily cast into the capitalist framework, and in the end compromises had to be made by the capitalists, the state, the old nobility and 'the people'— all safeguarding their own interests.

Confusion among the terms market, market economy and capitalism is prevalent, both by the Left and the Right. As Keith Hart (2000) put it, the fine distinction between them is that capitalism is 'making money with money' and the market is 'buying and selling with money'. Others would say that the market is about exchange, and exchange doesn't assume profit, whereas capitalism is about ownership, property, profit and control.

Capitalism: from city-states to empire[13]

At the end of the thirteenth and the beginning of the fourteenth century, a trade network from China to England and from Scandinavia to Africa had been established and had expanded substantially. In Europe, the main beneficiaries of this were the Italian city-states, in particular Florence, Milan, Genoa and Venice. Florence and Milan were engaged in manufacturing and trade in the northwest; Genoa was engaged in silk trade in Central Asia; and Venice was engaged in spice trade in South Asia. They were initially not competitors but complementary to each other, especially as long as times were good. The silk trade of Genoa was estimated to generate three times the revenue of France. At some point in the early fourteenth century trade tapered off, triggering a period of fierce competition between these city-states. And these states were mercantile states where traders and manufacturers had a lot of influence, so competition was not only commercial but extended to a series of wars. The wars strengthened the position of the merchants even more as the state had to borrow money from them to conduct the wars (which were on behalf of the merchants in the first place).

In Genoa it led to a private takeover of the finances of the cities by the Casa di San Giorgio, whereas in Florence the most powerful family, de Medici, took over the state as a whole. Now the wars themselves become a commercial undertaking. 'What capitalist groups could no longer invest profitably in trade, they now invested in the hostile takeover of the markets or the territories of competitors both as an end in itself and as a means to appropriate the assets and the future revenues of the state within which they operated' (Arrighi 2010: 92–93). Parallel to this development, and in particular after the Peace of

---

[13] This builds to a large extent on Giovanni Arrighi's *The Long Twentieth Century* (2010).

Lodi in 1454, the Italian city-states generated high finance, financing in particular war-making governments. It is in Florence and later in Genoa that the toolbox of managing money and transactions developed; it is there that cheques and bills of exchange were used: most transactions were made through bank transfer and the art of managing client accounts to make the capital work in one's favour was taken to new heights. It was the Genovese who bankrolled much of Spanish expansion and, then later, converted the silver from America into gold coins to pay the armies.

Meanwhile, to the north, Dutch traders had taken control over the trade of grains from the Baltic Sea, an important and profitable trade. The profits far exceeded what could be reinvested in the trade, and so Dutch capital sought new uses. Some of the money went into farming and was one of the reasons for the modernization and increased productivity of Dutch farming. A second use was to back a local territorial organization, the House of Orange, seeking protection and state-making, thus converging capitalism and territorialism. The House of Orange engaged in a long battle with Spain. The Dutch managed to steer a lot of trade to Amsterdam where they could continue to make profit on the ever-growing stream of goods coming in and going out. But Amsterdam became the centre not only of trade but also, perhaps more importantly, of money and the capital market of the European economy. One means to accomplish this was to establish the first stock exchange in permanent session.

A final successful initiative was the formation of the Dutch East India Company. It was not only a business enterprise but also a war-making and state-making organization acting on behalf of the Dutch government. This is what we mentioned earlier, the unique fusion between the state and the capitalists that became a determining factor in the conquest and domination of Asia. Well, to cut a long story short, after a while England, adding industrial might and colonial empire to the fusion, took over Dutch leadership and expanded and exported capitalism all over the globe.

Until the Industrial Revolution, production had been outside the primary reach and interest of the capitalists. Some capitalists did engage in farming or other production, but this wasn't as significant as trade and high finance. With industrialization, enormous scope for profit opened up in production itself. Karl Marx identified wage labour as a critical component for the emergence of capitalism; it made possible the accumulation of capital via the profits made from the 'surplus value'. The industry owner, the capitalist, could exploit the worker and take the surplus value in the production process (Marx 1867). Max Weber agreed and stated that capitalism would only appear when there are property-less people who are driven by hunger to offer their labour (Weber 1986).

Everything for sale

In pre-capitalist societies, markets played a much smaller role than we often realize. There were markets, but there was no market economy. In agrarian civilizations, only small parts of the harvest, if any, were sold in the market. Most cultures despised money and trade and those involved in it were politically marginalized, even more so if they were from another ethnic origin. The story of the Jews is well know, but the same held true for the Greeks under the Ottomans, where the Ottomans considered trade an unworthy occupation and left it to the infidel Greeks, something that most certainly paved the way for the resurrection of Greece.

Karl Polanyi (1944) makes a plausible case for how mechanization was a driving force for market development and liberalization. Before mechanization, farm households in the villages manufactured a lot. Traders organized production by supplying raw materials and buying the products. Trade was mostly mutual. Raw materials (say yarn) were sold to the farms and ready-made products (say shirts) were bought. Thus, if the farmers produced they earned money, if not they earned nothing. For the tradesman this was not so important, as hundreds of farm households were doing the same thing and there was no reason for the trader to get involved in trying to control how and when they produced. His (it was mostly a man) profit didn't rest in the efficiency of the work but in the margin between the raw materials bought and sold and the ready-made products bought and sold.

With mechanization this changed. Machines were expensive and there was no chance that the farms could have purchased them. In addition, it was essential that the machines ran as much as possible, so that investment costs could be spread out over several products. In farm-based manufacturing, production was concentrated in the low or off-season, when farmers had little else to do. In high seasons many other tasks took priority over the production for the trader. Therefore, it was necessary to organize the operation of machines. This led to the development of factories, and gradually also to the labour market. Salaried labourers also constituted a new market for food, clothing and housing, in particular. Industrialists had an interest in freeing the markets, perhaps not so much for what they sold, but more in order to easily access raw materials for the factories and also to indirectly keep the costs of labour low — it was in their interest to keep food prices as low as possible to be able to keep salaries as low as possible. This led to the notion that capitalism is about 'free markets'.

Polanyi (1944) describes how pre-capitalist markets were 'embedded' and separated from each other; there was not one market but many markets separate from each other. There were local markets for the exchange of food and goods between city and land. But we shouldn't believe that most food was distributed in that way; on the contrary, most (surplus) food was given to landlords as rent, to the church as tithe or to the crown or was used for big events.[14] The markets were also not particularly free; there were many regulations for prices and for who could trade and where they could trade. There was also a very weak connection between local and international markets; traders who engaged in international trade were largely shut out from local markets. Grain, the most important commodity, didn't become a fully tradable commodity until the end of the nineteenth century in Asia (Clingingsmith and Williamsson 2005).

Agrarian civilizations had people working for a salary, but that was mainly in the government, church and other institutions where they were appointed to their jobs; thus no free labourers offered labour in the modern sense. Free labourers did very little work in agriculture, and most craftsmen and their apprentices were paid by the piece. At the end of the Middle Ages, perhaps up to 25% of the work force was salaried in the most advanced economies, but in many places this proportion went down when servitude expanded again (Lucassen 2005).

A secure supply of labourers could only be guaranteed by slaves, indentured labourers[15] or wage labourers. The use of slaves was becoming politically impossible, and a major disadvantage was that with slaves came the responsibility of their whole family, of their children, of education, marriages and deaths. In addition, slaves were also a responsibility when there was no work; in times of lower demand of

---

[14]   I did a job in Samoa in 2009, and realized that in Samoa most of the meat was used for gifts, parties and other forms of non-market exchange.

[15]   This version of forced labour was particularly common in the English colonies: '. . . not less than half, and perhaps considerably more, of all the white immigrants to the colonies were indentured servants, redemptioners, or convicts' (Hofstadter 1973: 34).

production, it wasn't easy to just sell them. All in all, salaried workers better fitted the needs of industrialists. In the nineteenth century, for the first time a majority of the work force was brought into wage labour,[16] for factories in trade and in the growing administration and services of the nation-state (Lucassen 2005). The early division of labour was mainly between professions and not so much within professions. With industrialization, the division of labour *within* a profession was essential. The capitalists and industrialists needed to control and supervise labour and they wanted to be less dependent on skilled labour comprising mostly well-paid and independent-minded craftsmen. The division of a process into small simple tasks allowed them to employ unskilled people and made it easier to supervise the process, thus getting people to work instead of pretending to work. It also made it harder for workers or outsiders to discern the profits generated.

The emerging labour market was not very free. On the one hand, it was still entangled in rules from the guilds and from feudal society, and, on the other hand, new rules were developed to ensure a supply of labourers. Polanyi (1944) states that one could not speak about a real labour market in England until 1834. Still today, labour markets are heavily regulated in most places. Salaried labour should be motivated by a salary, but, in reality, not too much choice exists for most of workers. Almost all societies have various mechanisms for taking care of orphans, the elderly or sick. This holds true for the hunter and gatherer society to the capitalist society. Many social institutions were deliberately taken away or lost in the societal transformation that took place with the emergence of industrial capitalism. The threat of starvation motivated people to seek employment, not the promise of a better world. That is the reason why Adam Smith, David Ricardo and Karl Marx calculated the cost of labour as merely the cost of survival of the labourer and perhaps of his family and not more. Slowly social security was reintroduced with the modern welfare state. It was perhaps seen as something new but was merely a return to the normal — we take care of our brothers and sisters.

The very strict laws against vagrants and the unemployed underlined that seeking employment was not voluntary. From the

---

[16]   We will see later that, in fact, even today a majority of the global work force is not engaged in salaried labour.

Middle Ages up to 1920, in Sweden, the law mandated work. Penalties were lashing, cutting off the ears and, finally, banishing (Zetterberg 1977). In England, the measures were even more draconian: in 1530, under the rule of Henry VIII, a vagrant was tied to a wagon and lashed, the second time half an ear was cut off and the third time he was ordered to be executed as an enemy of society (Marx 1867). The abolition of these kinds of laws had perhaps as much to do with a need to encourage mobility in the work force as with a need for compassion and social progress. In *Oliver Twist*, Charles Dickens mixes realism and satire to describe the effects of industrialization on nineteenth-century England and to criticize the conditions of the poor and the harsh Poor Laws. Oliver is trapped in a world where his only options seem to be the workhouse, a gang of thieves, a prison or an early grave.

Other ways to force people into salaried labour was the introduction of taxes that had to be paid in cash, a system that turned farmers into labourers. As the Kenyan environmentalist and winner of the Nobel Peace Prize Wangari Maathai describes:

> When the British decided to collect revenue and finance local development, they did not want to be paid in goats. They wanted cash. They also wanted to create a labor force, but they did not want to force people to work. So they introduced an income tax for men in most parts of the country that could be paid only in the form of money. This created a cash-based rather than a livestock-based economy. Of course, the colonial government and the British settlers were the only ones with money in their hands. So the local people, especially men, were indirectly forced to work on settlers' farms or in offices so they could earn money to pay taxes. (2006: 13–14)

Land formed the basis for military, judiciary, administrative, religious and political systems, and was also the economic unit of importance in an agricultural society. The right to land was regulated by laws and customary rights in feudal society. In many parts of the world, land is not sold in the market, and customary or state land is still very common. Converting land from public or communal stewardship to private ownership is just the first, but essential, step toward making land a tradable commodity, and part of the transformation of society into a market society. Speaking of the primitive nature of indigenous peoples in the nineteenth century, US Commissioner of Indian Affairs, T. Hartley Crawford stated, 'Unless some system is marked out by which there shall be a separate allotment of land to each individual [. . .] you will look in vain for any general casting-off of savagism. Common property and civilization cannot co-exist' (in Kinney 1975: 109).

Because land was a fundamental building block in society, one first had to separate economy from society in order to allow a land market to develop. Indirectly, the transition of land to a market was one of the driving forces behind the integration of agriculture into the market economy. Farmers needed cash income for rents and land leases, which forced them to produce for the market, and once that was done there was no going back.

With the innovation of money—a kind of virtual property and a representation of real values—property took on other attributes. Vast opportunities emerged for the accumulation and distribution of wealth in both time and place, to the next generation, allowing a person to move to a new place without physically moving belongings and still remaining as prosperous in the new environment. In early societies, you had something because you made it or you needed it, and the group you belonged to (the band, the family or clan) safeguarded your rights to it—or took it away from you if you misused it or if someone else needed it more. With money, property was individualized in a new way. You could own more than you needed and you could carry a lot more. So even if money by itself didn't mean capitalism, the innovation of money certainly paved the ground for accumulation of capital. Money and private property both reinforced the need for a state. Trade needs to be safe, so either the tradesmen arm themselves, such as the Vikings or the British East India Company, or the state protects their interests. And the bigger and more powerful the state, the more protection it can offer.[17]

The money of the state first represented a real exchange value, for example, the metal content of the mint. The state issued bills of debt that had value as long as the public had confidence in them. But if the government issued too many, the value of each bill shrank. In many cases, the bills were a hybrid, like during the era of the gold standard, which worked as an international currency. It was established by Great Britain in 1816 and gradually expanded to most major economies by the end of the twentieth century. Polanyi (1944) sees the gold standard as one of the building blocks of this period's globalization, but he also

---

[17]   Some would claim that with the ascent of transnational corporations this is no longer the case. Certainly the nation-states are somewhat weakened by them, but in the end the emergence of the World Trade Organization (WTO) exactly proves this point. The transnational cooperation needs the state, and in this case a coalition of states, to set the rules in order to protect their investments from the state, from competitors or from 'the people'. The job of the WTO is to ascertain, globally, property rights, rules for investments, etc.— the same things that the nation-states do for companies.

claims that the system was to blame for the Great Depression of the 1930s. The gold standard represented a way to put money outside the control of the governments, as a commodity that can be sold and traded, just like any other commodity. Thus the monetary market was created, which could be exploited by the capital owners, the capitalists.

The French historian Fernand Braudel argues that this is the real playground of capitalism. He speaks of the three economic layers: material life, market economy and capitalist economy. Capitalism has little to do with daily life and work. Small shops or factories operate in a market economy but their activities are not 'capitalist' as such even if they are affected by capitalism. It is in the third layer that 'the great predators roam and the law of the jungle operates. This — today as in the past, before and after the industrial revolution — is the real home of *capitalism*' (Braudel 1982: 229–230, emphasis added; quoted in Arrighi 2010: 10). Both Arrighi and Weber also point to the huge importance of high finance in the development of capitalism, in particular the bankrolling of wars and expansion of states (exploration, colonization): '. . . the wealth and power of the Dutch capitalist oligarchy rested more on its control over world financial networks than on commercial networks' (Arrighi 2010: 45). Arrighi sees financial expansion as the initial and concluding moments of the capitalist cycles. He points to four such cycles in the past linked to the rise and fall of the dominating capitalist powers: the Italian city-states, the Dutch, the English and now, finally, the Americans.

In the early stage of capitalist development, one saw very few links between the individual as a producer or a labourer and as a consumer or a buyer of goods. Labourers just needed a bit of food and clothing to work, but they were not 'the market'. In *An Inquiry into the Nature and Causes of the Wealth of Nations*, Smith says, 'A man must always live by his work, and his wages must at least be sufficient to maintain him. They must even upon most occasions be *somewhat more*; otherwise it would be impossible for him to bring up a family, and the race of such workmen could not last beyond the first generation' (1776: 31; emphasis mine). Marx and Ricardo agreed with this. But this view of the labourer proved to be erroneous for several reasons. One was the organization of labour demanding higher salaries; another was that the state and the capitalists realized that the population was not only a source of labour but was also an important market. In order to constitute a market it needed purchasing power. This held true not only for the labourers but also for the farmers and farm workers. A broader consumer basis meant more demand for agricultural tools, paint, wood or coal furnaces, lamps, porcelain, clothes and, later on, even sewing machines and bicycles. West European countries like England and the Scandinavian countries benefited from this compared to Eastern European countries and Russia where a small elite took all the proceeds from the export business and bought imported luxury items (Gadd 2000). The tremendous increase in productivity allowed salaries to increase rapidly and softened the conflict between workers and industrialists, because it was possible to combine increased exploitation and increased standard of living.

The capitalist ideology

With capitalism, markets came to the fore and gradually society adapted itself to them. Thus we had not only a market economy but also a market society. Not only goods but also land and people (labour) became commodities to be sold and bought—for profit. Before this revolutionary change, property and money were certainly important, but primarily as tools to gain respect and position and not the other way round. No society stated economic growth as the goal. Confronted with other paradigms of a more egalitarian or communitarian nature, the emerging capitalists often had to resort to compulsion to change people's behaviour and slowly also their values.

For instance, in many traditional societies, there was no or little private property and traditional forms of redistribution ensured that no individual amassed too much wealth. Authorities in the United States and Canada banned the practise of potlatch, a festival ceremony practised by indigenous peoples of the Pacific Northwest Coast. The essence of the potlatch feast is the redistribution of wealth. Potlatch was made illegal in Canada in 1884 through an amendment to the Indian Act and in the United States in the late nineteenth century, largely at the urge of missionaries and government agents who considered it wasteful, unproductive and contrary to 'civilized' values: '[It] is not possible that Indians can acquire property or can be industrious with any good result, while under the influence of this mania' (Patel 2009: 104).

The modern capitalist ideology is based on the perception that all human beings act in self-interest (which is quite plausible even if it is less plausible that it would be the *only* reason to act) and that that self-interest, in the first place, is economic (which is not at all plausible). With this ideology, capitalism, over and above its superior economic performance that only a fool could doubt, could also be made morally justifiable and superior to any other ideology. The whole package of capitalist ideas contained, among others, the following:

Goods should be traded freely.

Labour should be regulated by contract between labourers and those buying labour and people should freely move to where employment can be found.

Land should be a marketable commodity.

Money, currencies and financial instruments should also be tradable commodities.

The role of the government is to protect private property and to act as an arbiter in contractual disputes, and it should, by threat of violence, uphold the right to trade.

Apart from these ideas, which mainly concerned the economic sphere, capitalism also paved the way for other ideas, mainly politically liberal ideas, which fitted well without necessarily being part of the capitalist ideology. Among those, there was the view that each person is born equal and that it is through one's own qualifications and work that one should get standing and wealth in society, not by birthright or tradition, and that women had the same value as men. Earlier society had very little social mobility and rigid gender roles, all seen as 'given by God'. In this way, capitalism also paved the way for democracy, even if nothing in capitalism presupposed democracy.

These views were also supported by scientific advancement, in particular by Darwin's theory of evolution. The idea of a market that, without design or control, still regulates itself and allocates resources where they are best needed, through individuals working out of self-interest, has a lot of similarities with the natural selection process that drives evolution. In that way, capitalism could be seen as at least given by Nature, if not given by God. How strong this narrative is can be seen by how hard it is for us, after just a few generations, to think about labour outside the category of wage labour, and how natural it is for us to buy and sell land. As the journalist–activist Raj Patel says, '. . . the transformation not only changed society, it also changed us, by changing the way we see the world and our place in it' (2009: 18).

The nation-state and capitalism

The notion of a nation-state is based on one people, one territory and one state, which represents the people and which is responsible for maintaining unity and defending the territory and the people. For many this may mainly have an emotional, cultural and political meaning, but one should not forget the role of the nation-state in the economy. Nation-states in Europe seemed to form in two waves, one was in the Middle Ages, which coincides with the establishment of the Dutch republic in 1579, the crowning of Gustav Vasa in Sweden in 1523 and the establishment of Portugal in the thirteenth century, sometimes claimed to be the first European nation-state. After this initial wave, many countries were torn in conflicts between the kings who tried to strengthen their power and the feudal structures, for example, the English Civil War (1641–1651).

If we limit our discussion to Europe, nation-state formed through another wave in the nineteenth century (for instance Germany and Italy). The nation-state created unified economic and administrative conditions in a larger territory: local customs were abolished; governance was through direct rule and not through county nobles or other feudal masters; a governmental representation was established in provinces; conscription armies were established; citizens were counted and registered; and compulsory school was introduced. Direct contact between the state and its citizens made it more important to have a common language and to ensure that people could read and write. A growing number of people were directly employed by the state (Hobsbawm 1991).

There were diverging interests between the landed classes and the new industrial capitalists. This was clearly expressed with regard to the issue of free trade, where the capitalists supported free imports of food and raw materials for their factory, whereas the landlords were interested in high raw material prices and were thus against imports. The factories and the rapidly growing proletariat, however, caused a social unrest that the capitalists couldn't handle. They broke off from the old society, but, in essence, they were still dependent on the state to defend their property and to quench rebellion and civil unrest. As discussed above, the state needed to take measures to supply the factories with 'willing' workers. The capitalists were not the state-bearing class and, therefore, had to turn to the nobility, the army, the church and the crown for stability and protection. Nationalism and religion, deeply entrenched in the agricultural society, formed the glue to keep society together (Polanyi 1944; Hart 2003).

In the 1860s, most leading nations of the West went through upheavals that strengthened and institutionalized capitalism. It was the emergence of what could be called state-supported capitalism. The state, originally a remnant of the agrarian civilization, donned a new role and, today, is as much a part of the capitalist system as the capitalists themselves. In some countries, there were coinciding political interests between the workers and the capitalists, for example, to abolish privileges for the nobles. The development of a class consciousness among the workers, with a clear edge against the capitalists, was stronger in continental Europe than in England and the (former) English colonies. In England, there was less tension between the nobles and the capitalists; in fact, some nobles got involved in industrial production. This meant that the capitalists didn't need the workers as allies against the nobles. Social and economic mobility was more developed in England than in the continent, which meant that successful workers could become tradesmen or industrialists.

This was even more the case in the United States where there were no nobles or feudal structures to get rid of. The constant influx of new workers starting at the bottom of the ladder made it hard for a class consciousness to develop as it did in Germany or France. But also in those countries, workers never took the revolutionary steps that Marx predicted but were content with the bureaucratic welfare state, with its armies of doctors, teachers, social workers, lawyers and economists. The communist revolutions were not driven by the working class but by farmers and the intelligentsia (like in China and Russia). It was only by conquest that any industrial country was converted to a communist one (I here think about Eastern Germany, Poland, the Baltic states, Hungary and Czechoslovakia).

Development of the market society

Historically, political power, be it a king, a chief, a clan leader, the senate of Athens or village communities, took and gave rights and privileges and there was nothing distinctively 'political' as opposed to 'economic'. The state also had no separate existence as a corporate entity apart from the community of citizens (Meiksins Wood 1995). Early liberalism, defined as the assertion of inviolable *liberties against the state*, was a result of the state becoming an independent entity, separate from the community. Without that separation, those liberties made no sense, both because they were not needed and because there was no one to enforce them. Political liberalism originated from the assertion of the rights of feudal lords versus monarchs, which is what the famous Magna Carta[18] is about. A parallel struggle by the village communities was shown by the many peasant uprisings in late

---

[18]   An English charter, originally issued in 1215.

mediaeval times. Later, the bourgeoisie took up many of the liberal ideas and many also became part of social movements.

Already the Greeks had a democratic ideal, but both women and slaves were excluded. The medieval city-states revived this old tradition, but, largely, kings and feudal lords managed to stifle most democratic developments. Because of the revolutions in England and France, people connect capitalism with democracy. But this connection is not very straightforward. The capitalists certainly demanded freedom of speech, voting rights and parliamentary democracy — for themselves that is. They normally resisted freedom and voting rights for workers, tenants, beggars and women. The French revolution in 1789 introduced voting rights for men, but this was soon abolished. It was reintroduced again in 1848, but not until 1944 did women get the right to vote in France. In many countries, such as Great Britain, the United States and Sweden, the right to vote was limited to those with property, which meant that workers and women stood outside the system. New Zealand was first to implement universal suffrage in 1893. Also with regard to other democratic rights, such as freedom of speech, freedom to meet, etc., capitalists were vigilant about such rights for themselves but were hesitant or outright hostile toward such rights for workers.

Through parliamentary democracy, a third institution is brought into play, a platform for democracy. One of the building blocks emerging in the new society, however, was that private property and associated powers, rights and liberties were brought *outside* of the political realm. And as the modern capitalist economy was based on private property and the right to accumulate capital, and dispose one's wealth according to one's wishes, this meant that, for the first time in history, economy was separate from politics. In this way, a very important part of society was not only brought outside the whims of a still semi-despotic ruler, the monarch, but also brought outside the control of democracy. So while democracy advanced in the political arena, the scope of the political arena was severely reduced. In this way, parliamentary democracy posed little threat to the new capitalist class. The high respect for the right to property that is enshrined in western, and particularly Anglo-Saxon, legislation protects individuals against the state far better than it protects public or common goods against private encroachment.

The main achievement of capitalism in the democratic arena is that it weakened feudal institutions and, through wage labour, increased the self-confidence of women. This worked in two ways. In agrarian societies, peasants were organized in village communities that had substantial clout, mainly in local matters and in relation to landlords, but they could, and often did, also confront the state. These communities were, however, broken or made redundant by capitalist development. The privatization of land and commons took away parts of their reason for existing; the introduction of labour markets and factories created new relationships and made their whole system redundant, and gradually village communities became almost anachronisms.

In pre-capitalist societies, markets played a limited role for distribution of products and also for the organization of work. Most natural resources, including land, were commonly owned or owned by the state and were not traded in the markets. We have noted how capitalism developed, by its own irrepressible energy, partly in conflict with the old feudal society, but also partly in alliance with the state for protection of property, for supply of labour and for expropriation of common goods.

Ultimately, the capitalists and old powers and institutions found a compromise, a compromise we are still living in 100–200 years later. We have also noted that this transformation could only take place by compulsion and with the application of substantial coercion; the resistance of craftsmen and peasants to this change was rather strong. To change common land into a commodity and convert human activity into salaried labour was not a free choice for most people. An important aspect of this narrative is how little the birth of capitalism has to do with the entrepreneurs or small tradesmen that are the poster children of capitalism. Instead, we see high finance and the state acting together. Capitalism can certainly exist without entrepreneurs—but it can't exist without the state. And it could have never become the dominating system unless it allied itself with the state.

In the process of developing this market society, old 'subsistence economies' have lost out. It is important to realize that the weakness of these economies is not ecological or that they didn't provide sufficient wealth to their members; the weakness is that they didn't generate the potential for power and war craft, that they were not competitive. Therefore, subsistence economies, local, regional or even national, easily fall under external control of military might or of taxes and dues.

How Technology and Energy Shape Our World

This is a sad hoax, for industrial man no longer eats potatoes made from solar energy; now he eats potatoes partly made of oil.

(Howard T. Odum,
Environment, Power, and Society,
1971)

Bronze is an alloy of copper and tin, both of which are abundant but often not located in the same place. Long trade chains were developed for these raw materials and for trade in bronze products, for example, between southern Scandinavia and the eastern Mediterranean. This influenced the societies involved in these chains. Bronze was expensive and soft, which meant that its use for daily chores and for farming was limited. Only elite troops could be equipped with bronze weapons because of the high costs. With the discovery of iron everything changed. Iron is abundant and can be found in most places. It can be used for agricultural improvement and thus for increasing the productivity of farming, the main economic activity; historically, this meant more surplus for the upper classes. Iron can be shaped into cheap swords, which formed the basis for the first mass armies and great empires—first came the Persians, followed by the Romans and then the Chinese. This was also the first period of formulation of great ideas and religions[19] (Myrdal 2008)—a coincidence?

We are all familiar with how technology shapes the world. The introduction of the steam engine, based on the use of coal that was easily available in England, paved the way for the British Empire, for railroads and steamships, that is, for the first wave of mass communi-

---

[19]   In this period lived Siddhartha Gautama (Buddha), Isaiah, Jeremiah, Ezekiel, Daniel, Confucius, Lao Tzu, Zarathustra and the Greek philosophers Aristotle, Socrates and Plato.

cation. We can't see a certain technology or a tool as separate from the society wherein it operates or could operate. Largely, a technology, a tool or a machine, plays a role in a societal and technological context. This context can be seen as a technological complex, where all parts are needed and where the parts mutually reinforce each other, and in the end form society.

Technology, ideology, economy and society patterns interact with each other. The ones that prevail reinforce each other; others will wilt and be thrown on the kitchen midden. Of course, there are technologies that are 'neutral' and tools that can be used for many different purposes, but the prevailing technological complex or systems are not at all neutral. That is why a hunting society is different from a society based on gathering wild plants, and both of them are very different from an agrarian society. Not only the political system is different, but also religion, science and gender patterns change, as shown earlier. This shouldn't lure us into a total determinism where a certain technology alone defines society—the influence is both ways. In the same way as a certain technology drives certain societal changes, innovations in society can also drive technological changes, and both interact with economic organization and human values. The link between technology and power relations too should not be underestimated.

Which technology will be used?

The Greeks knew the power of steam and built the first steam engine as a plaything; they even knew the principle behind electricity. Water toilets were developed in Egypt some 5000 years ago and were used by the Romans, but it took thousands of years before they came into common use. The wheel was known in Latin America, but was not used productively before Columbus. The first faxes were developed in the nineteenth century but it took another 100 years for them to become common.[20] Fuel cells, which are seen as technology for the future, have been around for more than 150 years, but never reached the stage of a technological breakthrough. The first steam engines in Sweden stood idle, collecting dust, mainly because they were not profitable and partly because there were no competent operators (Hård and Jamison 2005), a situation similar to that of a lot of technologies that are, mistakenly, in-

---

[20]   To soon be forgotten again. Young readers will probably have to look up 'fax' in Wikipedia to check what it is.

troduced in developing countries today. All these examples show clearly that the existence or the knowledge of a technology doesn't mean it will be used.

Many reasons exist for why a technology is or is not used. The word technology comes from Greek *techne*, meaning 'art', and we know that a lot of technology was (and still is) used for pleasure, for the demonstration of power or for religion. Realms with a large population and a small elite whose wealth is based on extortion of the masses most likely developed very few productive innovations, the majority being developed to amuse and entertain the rulers. Therefore, technological development was very slow in the big agrarian empires.

War and conflict have been forceful drivers of innovation, comprising examples of the importance of government-supported innovation. The Egyptians knew of the wheel, yet their slaves used sledges to pull boulders for the pyramids. But the use of the wheel rapidly spread in its military application in the form of the chariot. One can also compare the spread of the saddle and the spur with the spread of the harness. The saddle and the spur were of critical military importance and gave a marked advantage to the armies using them, so they spread rapidly. The harness was mainly used for draught animals in farming and spread across the world very slowly (Lönnroth 1977). Modern examples of innovations with a military origin are pasteurization, the Internet and antibiotics. Communications is another area in which innovation was (and still is) very strong. Innovations in shipbuilding or navigation played a major role in determining which people would rule. The supple Viking ships could reach into narrow fjords and rivers and were key to the Viking expansion; later the cogs of the Hanseatic League took over, owing to greater capacity (Lönnroth 1977). Thus, societies that developed quicker or safer means of communication or those of higher capacity had a marked advantage over others.

We have already discussed how the Chinese emperors were sceptical to technological development as it threatened stability. The Japanese are another example. Japan first came in contact with guns in the sixteenth century. In the early seventeenth century, it produced more and better guns than any other country in the world. However, the samurai, the military nobility, felt threatened by this foreign invention and gradually managed to curb production and licensing. Ultimately production ceased altogether, and resumed only when the first American warships appeared in Japanese waters in 1853 (Diamond 1997). In France, the scythe was not allowed to replace the sickle for a long time because there were communal rights to graze cattle on stubble, the value of which would be lost if a scythe were used (Boserup 2005).

Technology is not neutral. It serves the interest of the social group that develops it. With its symbolic power it supports the legitimizing ideology of society and the worldview it represents. Think of the steam engine or the space ship not to speak of the atomic bomb. Schiermeier *et al.* write that nuclear energy has 'benefited from decades of expensive research, development and purchases subsidized by governments; without that boost it is hard to imagine that nuclear power would currently be in use' (2008: 18). Technology has also enabled development of the modern city and the relative independence of its hinterland. The early cities were, with a few exceptions, built on the relationship with the surrounding agrarian landscape. A lot of production occurred in the farms and the economy of the city and its hinterland was interwoven. With the Industrial Revolution, the transport revolution and capitalism, cities could free themselves from ties to their hinterland for raw materials, labour and markets. Thus, workers in Manchester spun cotton from the United States, picked by African slaves, and the fabrics were sold in India.

Technology also creates or cements power relationships and patterns of exploitation in various ways. Mostly, technological developments lead to suppliers of raw materials getting a rough deal compared to those who use the products of a certain technology. This is the basis for accumulation of capital, for profit, in industrial processes where 'raw materials + work + cost of production' has a lower value than the product resulting from these inputs. The same holds true in international trade or in the international division of labour. Alf Hornborg (2009) goes as far as stating that technology in a fundamental way is about such price relationships and a tool for exploitation; there is no technological rationality separate from issues of distribution. We will come back to this discussion in Part 3.

Contrary to many beliefs, during most of human development, science didn't play a big role in innovation and social change. Very few of the early groundbreaking technologies were products of science (before the scientific method was developed, what constituted a scientist was perhaps not that clear either). With the Renaissance, science began to play a bigger role in influencing and legitimizing society; it greatly influenced ideas and the way human beings related to each other and to nature. Its force to remould society was more linked to the fact that it drove us to question many 'truths'. Technical and industrial development was driven more by craftsmen and entrepreneurs than by scientists. Rather, the development of technology enabled natural science to develop, especially the making of better instruments for measurement. Industrialization of science was a real breakthrough, the division of labour allowing for the study of many separate topics one by one and for the application of methodological experiments. In some production, particularly in farming, direct technological development and trial and error seem to still have more importance than the systematic application of science.

From fire to oil

To look at the history of humanity through an energy lens is as interesting as looking at it through an economic, political or socio-biological lens. Because of the close relationship between energy and ecology, the energy lens also partly becomes an ecological perspective. We can describe the whole history of humankind as the consecutive conquest of new forms of energy sources and technologies that made the usage of them efficient and effective. The arrowhead and the stone dagger made it possible for us to hunt large game and to use the energy of our arms better. Fire let us get energy out of wood, which made metallurgy and pottery possible as well as settlement in colder areas. Farming meant taming photosynthesis (a form of solar energy) to meet human needs. Animals were domesticated not only for food but also for transport and power in agriculture. The use of wind for sailing dramatically extended our reach; new areas could be settled in and trade allowed settlement in places that could not be settled in before. For each new form of energy, a technological process can control and direct the energy form in a better way. We increase the heat of a fire with bellows; we develop harnesses, wagons, saddles, reigns, spurs and bridles for horses, each one of them essential to 'harness' the energy. Technology can, with this perspective, largely be seen as a process to 'direct' energy.

In the infancy of industrialization, human beings learned how to convert water and wind power to mechanical power, and thereafter to use fossil fuels. For each of these steps, human beings have used bigger proportions of the planet's energy sources. And for each of these steps, the population and consumption levels have jumped. Also, our societal structures can be interpreted from the perspective of energy. The agrarian revolution led to a society where some, a few rulers, could appropriate the energy of the masses for their own purposes, through slavery, serfdom or forced labour. With the availability of more and cheaper external energy sources, the interest in forced labour substantially reduced and, today, it prevails mainly in sectors where energy and mechanization doesn't play much of a role, such as in service occupations including prostitution. Another way of looking at the same development is that 'the landed gentry who owned large solar-collecting estates were replaced at the top of the financial and social ladder by the new industrialists who directed new production systems using the more concentrated energies of coal and then oil' (Hall and Klitgaard 2006: 13).

We can contrast the energy content in human beings with the external energy sources conquered, to give us an idea of the importance of the development of external energy sources. Let us do a back-of-the-envelope calculation.[21] A human being needs around 2500 calories per day (2.9 kWh[22]). We use around 80% of that just to stay awake, think, eat, sleep, breathe, etc. So a bit less than 600 Wh is perhaps left for work. Hard-working people could perhaps use 800 Wh, but they will also eat more. On a yearly basis, the average person consumes around a million calories, which corresponds to 1.16 MWh. The average American uses 7.71 tons of oil equivalents (toes)[23] of energy (the average British 3.8, the Swede 5.65 and the Senegalese 0.25), which corresponds to 90 MWh, that is, the energy in the annual food consumption for 77 persons, or if we calculate on the basis of the 20% that we actually use for work it would correspond to roughly 380 persons. On the other hand, energy losses also occur for the conversion of external energy to useful work. If we assume that this is 40%, we arrive at a figure of 150 persons.

So, depending on how we count, each American has somewhere between 75 and 400 'energy slaves' working for him or her, and the richer ones have thousands. The global average energy use is of course lower, and represents figures between 18 and 90, which translated to the total population would mean somewhere between 120 and 600 billion energy slaves. From this perspective, it is certainly no surprise that human beings have taken over the planet! Another way of looking at it, from an economic perspective, is that a barrel of oil represents the energy of 25,000 hours of human toil, that is, 14 persons working

---

[21]   Calculating energy is tedious because so many different units are used and sometimes in-energy and out-energy in process are confused or it is not clear what is referred to. So there are many possible errors in the calculation presented here, and the main reason for presenting the calculation instead of just giving a figure is to be transparent (perhaps a brain better than mine can present more solid figures in the future).

[22]   1 kWh=3.18 MJ=860 kcal.

[23]   A *toe* is a common unit for energy and expresses the amount of energy released when a ton of oil is burnt. 1 toe=42 GJ=11 MWh=10 Gcal.

round the year with normal Western labour standards. The cost for pumping the oil is not more than a few dollars per barrel, and even with an oil price of many hundred dollars per barrel, it is very cheap compared to human slave labour. The beauty of the oil slaves is that they don't have to be feed at all. The slaves of the past, even under huge oppression and extortion, used a lot of the food themselves just to survive. From this perspective, our current wealth can be easily understood and demystified. Even hundreds of years ago, long before industrial society and capitalism, the person who had hundreds of others working solely for him or her could lead a comfortable life.

Some 100,000 years ago, human beings started to use fire. It was the first defining moment for humanity to use external energy for itself. Fire gave heat, but it also meant that human beings could eat things that they couldn't eat before. Even for the things they could eat, the conversion to energy in the body substantially improved. Heating food made it softer and easier to digest: even if the food didn't contain more calories through cooking, less energy was used for digestion and more of the energy in the food was absorbed. Some claim that cooking was a prerequisite for developing the human brain, which uses 20–25% of all energy in food (*The Economist* 2009a). Fire can then be considered an external assistance to our metabolism. A cubic metre of firewood represents about 1 MWh, that is, more or less a human being's annual energy consumption in the form of food. To make a campfire to cook a deer you need about 50 litres of wood (i.e. 50 kWh). This corresponds to 15 people's daily energy consumption in the form of food.

The introduction of fire meant that we took substantial energy sources under our command. Even today, 100,000 years later, firewood or charcoal (made from firewood) for cooking is a very substantial part of the energy source of the poor. As mentioned earlier, fire also played a very important role in the development of tools, from its use in breaking rocks to melting ores and burning pottery. In farming, fire was used for clearing land and for getting rid of waste and pests. Pastures were maintained and rejuvenated through regular burning. This is still a common, and necessary, practice.

The introduction of draught animals some 6000 years ago was another form of conquering for human needs. Still today, some 400 million draught animals are used for the cultivation of 50% of all agricultural land. Of these, some 300 million are cattle, including buffaloes, and 80 million are horses, donkeys and mules (IFRTD 2009). The power of the animals varies from some 200 W for a donkey up to 650 W for a camel.

They also endure in varying degrees: 4 hours for a donkey and up to 10 hours for a horse (cattle certainly endure, but they spend so much time to eat and ruminate on the food that they can't work such long hours). This means that their daily energy output is somewhere between 1 and 6 kWh (FAO 2000) compared with the 0.5–1 kWh output of human beings. Draught animals thus substantially extended the power (energy) that human beings could use. However, natural and biological conditions limit the use of these animals and their efficiency. In most agricultural situations, they can only be used (or are only needed) for a rather short period of the year. Most importantly, they can expand the land brought into cultivation by increasing the amount of land cultivated, especially in the countries with very seasonal farming (in temperate or dry climates). For trade, the use of these animals is advantageous: one human can easily lead five animals, each carrying a heavier load than a single human.

Draught animals have to eat, so their use is limited when land is scarce. Large numbers are kept in areas where there is plenty of pasture that grows year round. In temperate climates, where the grass doesn't grow in the winter, substantial areas of farmland has to be set aside for growing food for the animals. In slash-and-burn cultivation systems there is little space for draught animals, as the roots and rubble in the fields make it difficult to work with tools that are pulled. In addition, the fire does a lot of the work in the first place, so the need for animal power is much less. In Africa, south of the Sahara, only in modern times were draught animals brought into use. A study from Mali and Senegal shows that draught animals represent 90% of the mechanical energy in farms. The use of these animals was introduced in Senegal, Guinea and Mali at the end of the 1920s, and the main effect was that the proportion of cultivated land increased from 30% to 40% in Senegal and from 40% to 70% in Mali (Fall and Faye 1999).

Sail ships can be found on 5500-year-old Egyptian vases, indicating already the use of wind energy. Wind is a good example of how it is not enough to merely know about energy sources to be able to use it. Anybody walking through a storm can feel its power; it is only when we know how to master and direct it that it becomes useful. In its simplest form, wind energy was used (and is still used) for winnowing, separating the grain from the chaff by throwing up the grain in the air. The wind took the lighter chaff, twigs and other impurities with it. It took a long time for water and wind power to became important in production processes. Wind power was used mainly for milling and pumping and water wheels ran pumps, mills and, later, sawmills and trip hammers.

The most important use of these energy forms was in communication systems. The streaming water in rivers was used for transportation of various goods, but could only be used in the natural direction of the rivers. Consequently, the invention of sailing was revolutionary. It allowed human expansion and increased trade tremendously; it also gave a distinct advantage to coastal areas for economic development, a trend that holds even today.

Wood has many uses. It has been the totally predominant fuel. It has been used for shipbuilding and housing and in all kinds of crafts and production from pottery to mining. The energy intensity in wood can increase by making charcoal, which simplifies transportation and storage and provides for more efficient combustion and higher temperatures; the latter was very important for the development of metallurgy. Most civilizations simply would not have existed without forests, and some of them even collapsed because of collapsing forests. The deforestation of the Mediterranean is one such example.

In Europe, forest resources approached a collapse in the nineteenth century. Wood was used for more and more purposes and trade expanded rapidly. Even a sparsely populated and forest-rich country such as Sweden saw widespread deforestation in its mining areas in particular. In more densely populated areas with iron works such as in Luxembourg the price of wood increased by a factor of 24 from 1611 to 1790, and in Hunsrück in Germany it increased almost 100 times in less than 150 years. This was also the background to the early forest conservation efforts in Germany (Gerbehaye 2002). As oil price hikes threaten to destabilize the world economy today, scarcity of wood leading to the increasing price of wood threatened society 200 years ago. This changed when profitable uses of fossil fuel, in particular mineral coal, were introduced (SvD 2008a).

A similar situation was apparent in Japan and China. Japan was largely deforested and the government took harsh measures to protect existing forests and to plant new forests. In some regions of Japan, saws were banned. One of the reasons for the decline of China can very well be a combination of soil erosion and deforestation (which often go hand in hand and feed off each other). In 1949, Japan had restored forests to cover 68% of the land, whereas China had restored forests to cover 8% of the land (Radkau 2008).

The use of fossil fuels has increased 800 times since 1750 and 12 times during the twentieth century (Hall *et al.* 2003). Fossil fuels represent some 80% of all energy supply today. There is no principal difference between coal, natural gas and oil, even if the practical usage differs a bit. They are all created by photosynthesis and geological processes. Coal comprises some 70% of the total stocks of fossil fuels and 26% of the global energy supply. It is mined and used in all continents. We have already discussed the importance of coal in the early industries. Up to the middle of the twentieth century, mineral coal was the predominant fossil fuel along with gas, and, to some extent, a liquid fuel was produced from coal.

Petroleum (oil) has been known for thousands of years and was used as a fuel from the mid-1800s. The comparative advantage of oil over coal is that it is more energy dense (50% more energy per weight units) and as it is a liquid it is also easily regulated and fed into an engine. The energy content per kilogram of oil is four times higher than that of biomass (Helmfrid and Haden 2006). Its primary use is in the transport sector where petroleum-derived gas, kerosene and diesel, runs all kinds of vehicles, including planes and ships. Today, 95% of transport energy comes from oil and oil represents 34% of the energy supply.

Natural gas is cleaner than coal and oil; at combustion, carbon dioxide, steam and nitrous gases are the main effluents. For this reason, natural gas is the most rapidly increasing energy source and, today, represents some 20% of the energy use (IEA 2008). However, the usage is limited as natural gas is bulky and it has to be compressed for transportation and most uses.

– 4 –
What Drives What?

The ancient Egyptians built pyramids because they believed in an afterlife. We build skyscrapers because we believe that space in downtown cities is enormously valuable.
(Donella Meadows,
Leverage Points: Places to Intervene
in a System, 1999)

There is an eternal debate about what is driving development. A specialist in a certain area has a tendency to exaggerate his or her own field of expertise as being the determining factor. Clearly, certain technologies are predominant in certain kinds of societies, and certain technologies and certain forms of energy belong together. Think of the human body—as long as it is the sole source of power only certain forms of technology can exist; limits are set by our strength and our environment. Introduce draught animals and suddenly a plough can be used; threshing tools and mills can be further developed. When the wind is tamed for sailing it represents a society dependent on trade and fishing. When the wind is harnessed in a windmill, it represents an agricultural society with surplus production (which enables and motivates investments in such a device). Harnessing energy and the use of this energy in various technical applications is one of the most characteristic and defining human activities. Technology is not neutral and one kind of technology requires or fits into a certain society. And not only that, certain technology can only be developed if another technology is in place. Therefore, different technologies and societies hang together in technological complexes and it is not possible to change, or say no to, solely one component.

Some believe that technology, and its associated organization, shapes how the rest of society is organized. Others believe that power relations shape technology and the organization of labour. Looking at the transformation of society from hunting and gathering to agriculture, I tend to believe that it was driven more by emerging powerful groups than by technology, but once agriculture was established it contributed to those power relations. If we look at the emergence of the modern capitalist world, the mechanized factory simply required a labour market, a market for land and a market for goods. But, again, power relationships controlled the development of those factories as a new way of capital accumulation.

We should not forget that competition from a more vibrant and aggressive society constitutes a kind of natural selection among societies. Agriculture first took off where conditions were good. Once established it spread not only to all places where farming was possible, but unfortunately also to areas where farming was not possible or sustainable in the long term. This occurred as a result of successful farming communities and aggressive colonization by settler farmers. Even if a number of factors led to the establishment of the first farming civilizations, later conversion to farming may have taken other routes.

Change may occur either as a result of such aggressive expansion or through the example a successful society. This is very clear when we look at the spread of modern capitalism. We can spend years trying to agree on the factors that determined the rise of capitalism in England: Was it the eviction of peasants from the common lands or the bourgeois political revolution or the steam engine and the coal pits or the colonies or simply the entrepreneurial spirit of the British society? Even if we, unlikely, were to come to an agreement, it would be true only for England. In Thailand, Peru and China, capitalism developed in different ways and through other routes. Some countries 'chose' a path because of external force (like most of the colonies), others because of foreign investments (such as the industrialization of Sweden), and yet others because of the power — and threat — of the example (as in the case of Japan). Capitalism was introduced in some countries where hardly any of the conditions that were at hand in England existed, and in others where only a few of them existed.

This is also, most likely, a major reason for why capitalism and industrialization failed in so many places. Even today, not all communities have made the first revolution, the first step, from a hunter and gatherer society to an agrarian one. It is not out of ignorance that these communities continue with their old lifestyles. The main reason is that farming has simply not been successful under the prevailing conditions. Similarly, it might be the case that capitalism and industrialism have already conquered all places where they work well, and the so-called developing countries simply have nothing to gain by trying to jump onto that bandwagon. We discuss this further in Part 3.

My discussion might have neglected the importance of ideas. Without doubt new ideas can by themselves drive change, although only ideas that are launched at the right time will be viable. However, at that point, they act as forceful instruments to guide and focus human energy and ambition. Most of the time people live, eat, love, laugh and cry, but these activities don't have a particular direction when it comes to society. It is here that great ideas come into play: it can be a religion, a political vision, a scientific attitude or something else that gives meaning and direction. And at such moments ideas and ideologies can be major drivers of development. Of course, this can also work the other way round. When people realize that the prevailing ideology is unattractive and lacks relevance it can spur them to action, as discussed earlier in the case of the Maya and the collapse of the Soviet Union. And most likely this is now the situation for the capitalist paradigm. Its sales pitch is increasingly undermined and its strong points are no longer strong. The appeal of more of the same wanes with increasing material wealth.

I believe that things are interconnected and influence each other rather than one factor determining all the others. To decide that one factor is the determining one is only interesting if we argue that we have to change that one factor to change the total situation. But just because one factor perhaps was the determining factor in one stage of development, it doesn't follow that the same factor will be the determining factor in another stage. Ultimately, this also relates to the question of which are the underlying drivers of human activity. Is it the egoistic wish of genes to multiply themselves as stated by socio-biologists or is it our economic self-interest as claimed by the economicists?[24] This discussion will be taken up in Part 3. In the end, only those civilizations where technology, ideas, societal structures and classes, ecology and energy 'cooperate' and reinforce each other will be successful, regardless of whether one factor contributed to the original impulse. This is then the secret of any future successful society taking over from ours, which we will discuss in Part 4.

---

[24]   The term 'economicist' is used here for those who explain all, or at least most, of human action through economic discussions and drivers. This is in contrast to economists who, at least in theory, limit the application of economics to the field of economics. Economicists seem to believe that their studies in economics give them a unique competence to state what is good politics, what is natural human behaviour and what are good social structures — yes, in general to have a supposedly scientific opinion about all in the world.

PART II
ECOLOGICAL CHALLENGES—
A PERFECT STORM?

Conscious man, as a changer of his environment, is now fully able to wreck himself and that environment—with the very best of conscious intentions.

(Gregory Bateson,
Steps to an Ecology of Mind, 2000)

The growing human population, market economy, capitalism, industrial technology and fossil fuel together form a force of such dignity that we can now speak about *Anthropocene*, a development stage of the planet, where human influence has become a determining power for the whole planet, for the biosphere, for the atmosphere and even for the geosphere. In this part, we will discuss what can be called 'natural resource' and environmental problems. It is not an attempt to cover all issues of relevance. Many pressing environmental issues, for example, ozone layer depletion, deforestation, destruction of corral reefs and general pollution, etc., are not dealt with or mentioned in passing. The purpose is to show how big and far-reaching environmental issues are, and the issues covered do just that. Ecosystems, energy and farming are given large space as they are of critical importance.

Many, yes most, environmental issues are multifaceted and can't be properly understood in isolation. For instance, because of the bleaching effect of warmer water, elevated nutrient levels from pollution, over-fishing, sediment deposition as a result of inland deforestation, acidification and other pressures, tropical coral reefs worldwide increasingly become algae-dominated with catastrophic loss of biodiversity and ecosystem functioning, thereby threatening the livelihoods and food security of hundreds of millions of people (CBD 2010).

Despite industrialization, farming and other uses of land are at the core of almost all important debates, including those about global development. Many people say, or think, that industrialism has made human beings less dependent on nature. That is an illusion. On the contrary, compared with any previous society, current society is dependent on much bigger parts of nature. Hunters and gatherers were certainly dependent on nature, but they foraged on the surplus of limited parts of the ecosystem. They didn't use all the mineral and fossil resources that society uses today; they were not dependent on a lot of physical infrastructure as society is today, and they used less of the ecosystem services. Even compared to the agricultural civilization that preceded current society, human beings today are as or more dependent on the supply of products and services from nature. True, society today produces a lot more per square metre, but the population too has increased so that the entire surplus is needed. There is no more food surplus today than before, rather the opposite. One difference perhaps is that, today, food can be shipped from surplus areas to areas of deficiency with the help of cheap fossil fuel and modern technology, meaning that local food shortages should be less devastating. But society doesn't do that. One billion people are short of food despite all the progress.

The illusion that current society is less dependent on nature is caused by human beings living further away from nature. But this only distorts our perspective. That we get electricity via cables and petrol from tubes at petrol stations doesn't mean that we are less dependent on nature for energy than the hunter sitting around the campfire or the farmer putting another log on his hearth.

Industrialism was introduced to profit from the labour embedded in manufacturing, and later moved into agriculture, fisheries and other natural resource-based industries. A contradiction exists, however, in that while the population goes through the roof and natural resources are getting scarcer and ecosystem services are stressed, we still act as if manpower is the most limited resource and needs to be saved at the cost of others. The flaw of this perspective starts to show first in the sectors that directly use natural resources. Production in the future will be more limited by the availability of fish rather than boats or nets, of trees rather than chain saws and of topsoil rather than ploughs or genetically modified organisms.

That things have 'worked out well' so far should not be a big comfort. Even if industrialism has been around for 250 years, and some believe (I don't) we live in a post-industrial society, this explosion of resource use is fairly recent. In 1961, human beings used a little more than half the earth's bio-capacity; in 2006, human beings used 44% more than was available. The estimated annual environmental costs from global human activity were equated to 11% of global GDP in 2008 (UNEP-FI 2010). And at the same time human beings use so-called environmental services provided by nature valued to two times the volume of the economy. The real value of the third part of what nature provides, minerals and fossil fuel, has not been estimated at all.[25] An overall question is of course how do we calculate this and what do we want to accomplish by doing it.

---

[25]   Ecologist Jeff Dukes calculates that there are 98 tons of biological materials embedded in a gallon of oil. Expressed in another way, the *daily* use of fossil fuel in the world corresponds to the total biomass growth in a year (Siegel 2003).

– 5 –
Ecosystems

If we have any doubts about how to value a 500-year-old tree, we need only ask how much it would cost to make a new one from scratch. Or a new river? Or a new atmosphere?
(Paul Hawken,
Natural Capital, 2005)

All species on earth and the environment they live in are connected in complicated cycles and interactions, and they produce services essential for human life, for survival as a species even. These so-called ecosystem services are often taken for granted; they don't show in the balance sheets or in national budgets. Nevertheless, human beings are totally dependent on oxygen, plants and potable water, and all of these are provided by ecosystems. Ecosystems not only provide what is needed, they also take care of what human beings don't need—waste, from the carbon dioxide exhaled to the sewage from the cities. For most ecosystem services there are no substitutes; it is very difficult to artificially mimic them, and often very costly too, and we will mostly fail.

This was pointedly demonstrated in 1991–1992 when the scientists managing the Biosphere 2 experiment in Arizona discovered that they could not maintain oxygen levels for the eight persons who lived in the 200-million-dollar project to mimic nature (Hawken *et al.* 1999). The earth does this for free for 7 billion persons every day. Mounting evidence suggests that gradual shifts in central conditions such as biodiversity, soil quality and nutrient cycles can lead to sudden dramatic changes when critical threshold values—so-called tipping points—are reached (Rockström *et al.* 2009). Human use of nature has two relatively distinct parts—ecosystems and mineral resources. Ecosystems and their services are, at least in theory, renewable; whole minerals are not, at least not in the time horizon of the human civilization.

What nature does for free

The Millennium Ecosystem Assessment (MEA) was initiated by the Secretary General of the United Nations in 2000. The report of 2005 was a result of the cooperation between 1300 scientists in 95 countries and is the largest study ever of the links between ecosystems and human welfare. The study was orientated to determine the capacity of the ecosystems to supply human beings with essential life-supporting goods and services. MEA was the cooperative effort of governments, NGOs, UN organizations, leading research institutions and the business community.[26] Over half of global employment is directly based on resources of the ecosystem.

The board of MEA summarizes, with politically balanced words,[27] that everybody on the planet is dependent on nature and ecosystems for a healthy and secure life. Human beings have made unprecedented changes in ecosystems to meet the demands of food, water, fibre and energy. These changes have improved life for billions, but simultaneously weakened the ability of nature to supply other important services such as pure air and water, protection against disease and access to medicines. Some of the main conclusions of the MEA report are:

> Many fisheries are vulnerable, have collapsed or are on the brink of collapse.

> Two billion people are very vulnerable to loss of critical ecosystem services such as access to water and threats by climate change and pollution by nutrients.

> The impact we have had on ecosystems can lead to sudden and potentially irreversible changes with dramatic effects (such as dead seas and climate change).

The MEA notes that there are winners and losers when ecosystems are changed and that, hardly surprisingly, women, the poor and indigenous people are among the losers. Many changes in ecosystems have involved privatization of resources that were earlier common, and those living off them earlier have often lost their livelihoods. This development continues even today, for example, as seen with the conversion of mangrove swamps to shrimp farms or tourism, the ploughing of pastures of nomads, the flooding of grazing lands and forests caused by large dams, etc. The use of these ecosystems by the

---

[26]   For more information, see www.millenniumassessment.org

[27]   Like all UN-led processes, the texts are cautious and balanced so as not to annoy its members.

poor is mostly absent from economic statistics, but the few times this is measured, one sees substantial contributions of these ecosystems to the lives of the poor (MEA 2005).

A study from 17 countries shows that 22% of the income of rural people is from sources not included in national statistics, such as wild plants. Those sources were proportionally much more important for the poor than for the rich (MEA 2005). In Ghana, a fifth of all food comes from the wild; in northern Thailand, this is half. In western Kenya, 100 different plants are collected; in India, between 150 and 200 species are collected for food, animal feed, medicine and fuel (Pretty 2007).

The MEA classifies ecosystem services in four groups: provisioning (production of food, water, fibre, etc.), regulating (food regulation, water purification, etc.), cultural (recreation, aesthetic experiences, etc.) and supporting (nutrient cycles, soil formation, etc.) services. The production of plants, livestock and wood has increased, which has also increased wealth and provided income to billions, but this comes at the price of degradation of many other services, some of them critical for the continued supply of those products. Of 24 ecosystem services surveyed, 15 are threatened globally and 5 are threatened in some places. The supply of food from agriculture, a service that is often the centre of debate, is not considered to be under direct threat (MEA 2005).

In the middle of the 1990s, a study estimated the value of ecosystem services to be some US$ 33 trillion, double the size of the world GDP (Hall and Klitgaard 2006). The MEA doesn't try to assign values to ecosystems and ecosystem services, but it shows with numerous examples that if we did calculate such values, many things considered to be profitable today would not be profitable and vice versa. This is because profit is calculated by the entrepreneurs running a company. As long as they can use ecosystem services for free and as long as the degradation of the services is not paid for, the latter will not appear in any economic calculations. Some examples of estimated values of ecosystem services are:

> In Belize, mangrove and coral reefs give rise to fisheries and tourism and provide coastal protection for a value between US$ 395 million and 559 million annually, according to studies of the World Resource Institute (WRI) and the World Wildlife Fund (WWF). This can be contrasted to the whole GDP of Belize of US$ 1.3 billion (WRI 2008).

> The value of biodiversity in New Zealand was estimated in 1997 to be

US$ 152 billion in contrast to the GDP of US$ 56 billion. The valuation didn't include theoretically potential services but only those that are known (OECD 2001).

'Business-as-usual' deforestation and land-use changes cause *annual* losses of natural capital valued at between €1.3 trillion and €3.1 trillion, a sum exceeding the total financial losses of Wall Street and London during 2008, their worst year ever (IUCN 2010).

Deeper and deeper

Because human beings are a land-living species, the exploitation of land has gone further than the exploitation of oceans, but oceans are now rapidly exploited. Only 8% of the biological production of the oceans is used; notwithstanding, numerous fisheries have collapsed. On the temperate continental shelves, approximately 35% of production is used. The fisheries target a few species, normally at the top of the food chain (cod, tuna, salmon). Global predatory fish biomass today is only about 10% of pre-industrial levels. At first glance, one might believe that it is quite safe to take the fish at the top of the chain, but this can also lead to major ecological changes. In addition, some fisheries have vast by-catches of other species, sometimes up to a third of the catch. Fishermen thus have to go deeper and deeper to get the fish they want (see Figure 5.1) and the fishing methods themselves, in particular bottom trawling, destroy fish habitat. In the UK, landings per unit of fishing power have reduced by 94% (17-fold) over the past 118 years. This implies an extraordinary decline in the availability of bottom-living fish and a profound reorganization of seabed ecosystems since nineteenth-century industrialization of fishing (Vitousek *et al.* 1997; Thurstan *et al.* 2010; *The Economist* 2009c).

The combination of increasing population, subsidies for oversized fishing fleets, too generous a fishing quota, profit-seeking and new technologies has led to massive overfishing. Most indications point to us having already passed 'peak fish'; that is, global catches of fish are bound to decline (Zeller *et al.* 2009). Between 1950 and 2005, commercial fish and shellfish landings in the United States increased by almost 90%, which could indicate that there should be no worries. But since 1990, Alaskan waters have accounted for the bulk of commercial landings in the United States, and Alaska is the only region where landings have increased since 1978.

Figure 5.1 The average depth of fishing (1950–2001).

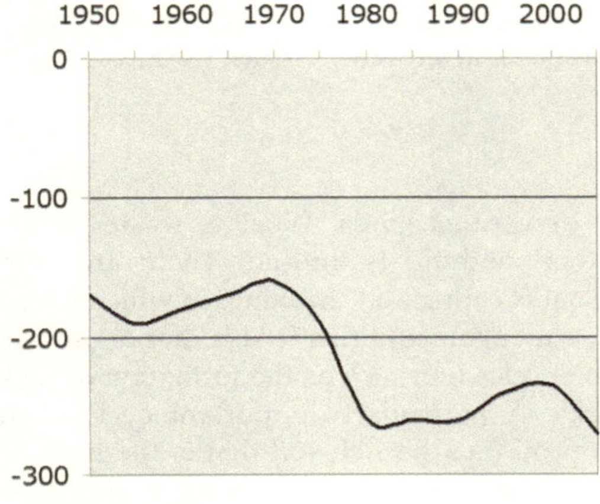

Source: Millennium Ecosystem Assessment (2005).

Between 1978 and 2005, landings decreased in the West Coast and Hawaii, the Gulf of Mexico, and the North, Mid- and South Atlantic (H. John Heinz III Center 2008). Within the European Union, 88% of fish stock is subject to over-fishing and 93% of cod is caught before it is sexually mature, and will therefore never reproduce (Thurstan *et al.* 2010; *The Economist* 2009b). The world trade in fish has increased more than 10-fold in 30 years from 1976 to 2007, from US$ 8 billion to over 90 billion (Ababouch 2009). Ironically, fisheries are also subsidized to a large extent. The United Nations Environment Programme (UNEP) estimated the total value of global fisheries in 2005 to be US$ 17 billion while subsidies amounted to US$ 27 billion (UNEP 2011).

The industry is ready to commit suicide at any time
'The world has passed "peak fish" and fishermen's nets will be hauling in ever diminishing loads unless there's political action to stem the global tide of over fishing,' says Dr. Daniel Pauly, Director of the Fisheries Centre at the University of British Columbia in Vancouver, Canada. 'We don't need more science. This is a message that's different from [that of] many of my colleagues. Of course, we need to learn more about fish. But research is often publicly funded on the grounds that this is an alternative to other political action. We know enough to act to prevent the continued decimation of global fisheries.' Dr. Pauly is adamant that pulling back from a global fisheries collapse—one on par with the collapse of various regional fisheries, such as the Atlantic cod fishery off Canada's Newfoundland coast—requires recognizing what

> he describes as a deep divide between the fishing industry and those who eat fish. He argues that fisheries companies' actions show that they're primarily interested in maximizing short-term profit, with little or no regard for the long-term sustainability of fish stocks. 'The industry is ready to commit suicide at any time,' he says. (NSERC 2006)

The important cycles

Nature is based on cycles of various kinds. What is waste for one organism is food for another; nothing is unused. There are quick cycles, such as when an animal is eating and drinking, in which a lot of what was taken in is soon let out again and this feeds other organisms. There are very slow and long cycles too, such as the formation of rocks and the weathering of rocks. A particularly important cycle is the formation and erosion of topsoil (i.e. the rich soil that is the basis for farming). Both processes are natural; erosion can be slow or very rapid (as in landslides) whereas soil formation is mostly very slow, except in cases of dramatic flooding; the landslide that represents rapid erosion can lead to soil formation downstream. Other important cycles are the carbon, water, nitrogen and phosphorus cycles. We ought to realize that cycles of minerals and other things that might have been considered unproblematic suddenly can prove to be of immense importance. In some cases, we try to avoid recycling and, instead, work with a linear process; for example, mining uranium, using it in nuclear power plants and, finally, locking in the waste products in such a way that they are prevented from entering natural cycles. It is exactly this linear way of working—foreign to nature—that is at the core of the nuclear power problem. If materials are so hazardous as well as impossible to recycle, we should simply refrain from using them.

Several carbon cycles are interconnected. The one most discussed lately is the turnover of carbon dioxide in the atmosphere. Carbon dioxide is released through respiration and from the decomposition or burning of organic materials. Through combustion of fossil fuels, the carbon cycle has radically changed, resulting in rapidly increasing concentrations of carbon dioxide in the atmosphere. Land use changes too have caused great changes in the balance. This is discussed more in the chapter about climate change (Chapter 7). There is also a carbon cycle in the soil, which is discussed more in the chapter on the problems of agriculture (Chapter 12). The quantity of carbon in soils greatly exceeds that in living plants.

In the article 'Planetary boundaries: Exploring the safe operating space for humanity, Johan Rockström *et al.* (2009) identify the nitrogen cycle as one of three areas (together with climate regulation and biological diversity) where human beings have surpassed a threshold for stable development. Nitrogen is the main building block of proteins, amino acids, and nucleic acids like RNA and DNA. It is the most common compound in air, which is composed of some 78% nitrogen. Most of the nitrogen, however, is biologically inert. It can be converted into active forms, such as nitrate and ammonia, through thunderstorms, fire and biological processes such as symbiotic nitrogen fixation (by Rhizobium bacteria in the roots of leguminous plants) or fixation by blue-green algae.

The quantity of biologically active nitrogen released annually into the biosphere has increased ninefold in 100 years; most of the increase has occurred in the last 50 years (see Figure 5.2). Increased use of chemical fertilizers is the dominating reason. The increase is also projected to continue from 165 million tons in 2000 to 270 million tons in 2050 (MEA 2005).

Figure 5.2 Active nitrogen in the biosphere (1900–2000).

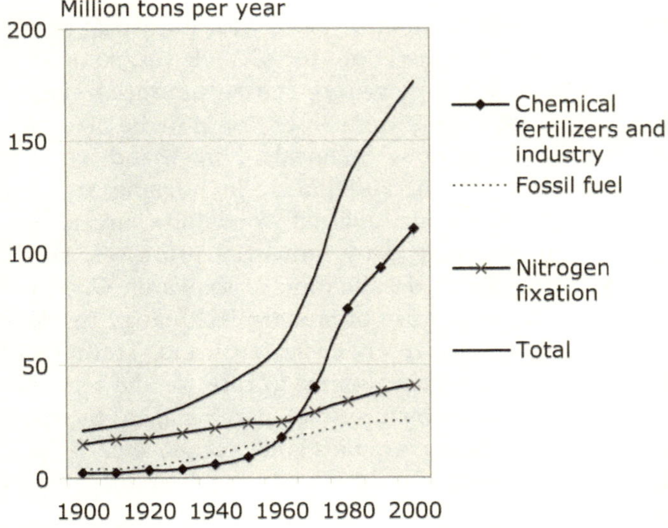

Source: Millennium Ecosystem Assessment (2005).

According to the MEA, the levels of nitrogen and phosphorus are two of the most important factors for changes in ecosystems. The effluent of nitrogen to the sea has increased by 80% between 1860 and 1990. This run-off leads to eutrophication with tremendous effects on the composition of species, and, in particular, it stimulates the bloom of algae and the associated dead zones, such as part of the Mexican gulf and the Baltic Sea (MEA 2005). The extent of dead zones in the oceans has doubled every 10 years since the 1960s. About 400 coastal areas are now periodically or constantly oxygen-depleted as a result of fertilizer run-off, sewage discharge and combustion of fossil fuels (UNEP 2010b). The use of chemical fertilizers has led to higher and higher levels of nitrates in drinking waters. Nitrogen also plays a role in the formation of tropospheric ozone, which leads to damage of crops and plants (MEA 2005).

In summary, there are good and frightening reasons to follow closely the development of the nitrogen cycle. We should not be surprised if effects and costs associated with disturbed nitrogen cycles are found to be as dramatic as those of the carbon cycle. What is certain is that there will be an impact. Considering how farmers and farm-lands have become 'addicted' to the use of chemical fertilizers, it could be thrilling to reduce nitrogen effluents.

But can't factories do all that?

Is it not possible to swap ecosystem services with industrial services and natural resources with synthetic products? Well yes, some of the services and products we get from nature can be produced syntheti-cally. Many products earlier made from wood can now be made from oil (e.g. plastic, tars, resins, various chemicals, fibres), and wood has been replaced by metals, cement and plastic in housing and ship-building. Oil and mineral coal have replaced wood and charcoal. Some ecosystem services can be replaced by industrial processes, such as purification of water and air or desalination of seawater. One of the most prominent examples of the use of modern technology to replace an ecosystem service is the use of chemical fertilizers. Traditionally, farmers needed to reproduce their system, to care for the fertility of soils within the limits of their own system, for instance, by having good crop rotations, integrating animals and plants, etc. With the advent of chemical fertilizers, fertility could be bought 'over the counter' instead.

There are discussions of physical ways of capturing carbon from the atmosphere and getting it out from the carbon cycle (see more in Chapter 7, 'Global Warming'). These artificial processes, however, are mostly more expensive than the services that nature provides. They are also, often, a lot more sensitive to disturbances and need constant surveillance and calibration. Mostly, they are also dependent of the use of non-sustainable resources such as oil. Most ecosystems are multi-functional and multi-dimensional, simultaneously providing many ecosystem services. The artificial replacements are almost uniquely one-dimensional.

Increasingly, systems are developed for payment of maintenance or new production of ecosystem services. Already in 1996, Costa Rica introduced a system by which landowners could get compensation for carbon sequestration, biodiversity, water regulation and aesthetic values. In 2001, the payments in this programme had reached US$ 30 million for a total of 280,000 hectares. Farmers in the European Union and the United States are paid for all kinds of environmental services to protect landscapes and water sources, maintain or recreate biodiversity.[28] The United States pays US$ 1.4 billion annually for 13 million hectares in a conservation reserve programme (FAO 2007).

The two possibly best-known markets for environmental services are the markets for the reduction of emissions of greenhouse gases: one is government controlled and is regulated by the Kyoto Protocol; the other is a voluntary market for companies, organizations or individuals who want to reduce their environmental footprint or make themselves 'carbon neutral'. The Clean Development Mechanism under the Kyoto Protocol, the EU system for emission rights, and the voluntary 'carbon market'[29] were worth some US$ 143 billion 2009 (World Bank 2010b). In Brazil's first so-called REDD (Reducing Emissions from Deforestation and forest Degradation) project, each family in the Juma Sustainable Development Reserve in the Amazon receives US$ 28 per month if the forest remains uncut (UNEP 2010b). Farmers in the Scoltel Té project in Chiapas (Mexico) sell carbon sequestration, in the soil and in vegetation, for US$ 3.30 per ton. Of this 60% goes directly to the farmers, representing an increase in their income from between US$ 300 and 1800, big sums for households where the average income is about US$ 1000 (World Bank 2007).

New York gets most of its water from an area stretching 200 kilometres north and west of the city. The authorities concluded that it was more efficient to manage the catchment area for good water quality

---

[28]   The link between the payments and the supplied service is rather weak, in particular in the European Union. The levels of compensation are not calculated in any scientific or economic way, but are rather just the results of horse-trading between farmer organizations and the EU member states.

[29]   The real 'carbon market' is of course the coal, gas and oil market and not the carbon dioxide–greenhouse gas emission reduction market.

than to invest in a purification plant. Such a plant would cost between US$ 6 and 8 billion. Instead, the implementation of a catchment management programme cost US$ 1.5 billion. This includes measures such as purchase of some critical lands, education and support to farmers for reducing use of chemical fertilizers and synthetic pesticides (FAO 2007). Insurance and shipping companies are financing the reforestation of the Panama Canal Zone to restore freshwater flows and avoid increased costs caused by canal closures (TEEB 2010).

Taking the idea of getting paid for *not* using a resource to new levels, the President of Ecuador, Rafael Correa, said in 2009 that his government is prepared *not to extract* nearly a billion barrels of oil from Yasuní National Park, a part of the Amazon rainforest of extraordinary but fragile ecological and cultural richness. To do so, however, Ecuador will need to be compensated by the international community for its loss of oil revenue — to the tune of US$ 350 million per annum for the next 10 years. A report stated:

> A conservative estimate is that exploiting the oil would bring in $5.7 billion in present value terms, or 10% of Ecuador's GDP. [. . .] The World Bank has estimated the abatement[30] cost for carbon dioxide at $14 to $20 per ton [. . .]. The cost to the world to abate these emissions will be between $1.7bn and $2.4bn for the extraction and burning, and $909m for deforestation, for a total between $2.6bn and $3.7bn. (*The Guardian* 2009b)

As with other new programmes, especially when they involve money, there are potential problems and side effects, some of which are not seen initially. For example, in India, poor landless people have been harmed by community forest management plans because the plans limited their access to resources to which they before had unrestricted access. Other examples can be found of communal land being appropriated by private owners (lately referred to as 'land-grabbing') who suddenly see a commercial interest in these environmental services. In Guyana, a private equity firm has bought the rights to 20% of the value of environmental services from a 370,000-hectare rainforest reserve

---

[30]   'Abatement' and 'mitigation' are used interchangeably to describe the processes by which greenhouse gas emissions can be reduced or compensated for by some other mechanism, such as carbon sequestration or carbon sinks. 'Adaptation' refers to the process of adapting to climate change, that is, not changing it or doing anything about it, but learning how to live with it.

anticipating that its carbon storage, water storage, biodiversity maintenance and rainfall regulation services will become a valuable asset in the future (IUCN 2010).

Should they have a price?

We have seen the importance of ecosystems and the services they supply. Mostly, we are not aware of the services that are provided by nature for human benefit. The composite value of these services is very high; notwithstanding, many of them are of such importance for human survival that their value simply can't be calculated, such as the supply of oxygen. We can't even imagine how to organize the same services if the current ecosystems were to collapse. Ecosystem services can be made more efficient, as human beings influence them all the time. Agriculture is of course the most prominent example where human beings over millennia have increased the value (for themselves, that is), the yield, from a certain area. But the example also highlights the dangers of looking at only one ecosystem service, in this case the supply of food. It is exactly this narrow perspective that has led to degeneration of the agricultural landscape, to erosion and to pollution at the same time as it has given rise to the possibility of the population growing 10-fold.

Trade in ecosystem services of various kinds is a booming market.[31] There are carbon offsets, payment for ecosystem services, and wetland banking (discussed again in Chapter 26) and species banking.[32] For instance, by putting your money in the Florida Panther Conservation Bank you can get the right to encroach on the habitat of the panthers in some other place. 'Through a combination of comprehensive large-scale planning and a coordinated mitigation strategy, the Florida Panther Conservation Bank reduces conflicts and costs between development and conservation aims, thereby increasing the ecological effectiveness of our conservation strategy' (Florida Panther Conservation Bank 2011). The only 'rational' way capitalism is able to manage ecosystem services is to convert them into commodities than can be traded. This comes with a large chunk of problems. First, it is the valuation itself; as we discussed, it is very difficult to appraise

---

[31]   At least it is portrayed as such by proponents of a green economy.

[32]   As of April 2011, there were 101 'species banks' in the United States (see http://www.speciesbanking.com).

ecosystem services. Second, there are distributional issues. This gives the rich and powerful even more influence over the life of the poor, as the rich, gradually, take ownership of more and more of these services. Once the ownership is endorsed, the moral right to criticize this imbalance is much weakened. The poor are also, in most cases, less competitive in the supply of ecosystem services as higher transaction costs are involved in dealing with many small suppliers than one large one. Third, the assignment of (monetary) value to these 'services' means that human relationship to nature is changing into a commodity as well.[33]

Surprisingly, there is not a very strong critique against the privatization that payment of ecosystem services actually entail, not even from normally anti-capitalist groups. On the contrary, many of them express sympathy and support programmes that pay farmers for ecosystem services. Probably they don't realize that payment of ecosystem services from taxes represents a giant privatization and a transformation of natural resources and commons into commodities. As we discussed, the total value of ecosystem services probably exceeds the total value of the current economy. And as more and more services are threatened, their prices will increase, fully in line with the principles of market economy. One of the main social battle lines in the future will certainly be drawn around the commoditization of nature.

---

[33]   Anthropologist Sian Sullivan (2010) has written a paper, 'The environmentality of "Earth Incorporated": on contemporary primitive accumulation and the financialisation of environmental conservation', which expands on these issues.

– 6 –
Energy

Energy is the currency of the future.

(CEO Nwabudike Morgan,
Sid Meier's Alpha Centauri)

There are only three primary energy sources on earth: solar energy, gravitation of the sun and the moon and geothermal energy. Of these, solar energy is predominant. These primary forms of energy are the origin of all secondary forms: biomass (from photosynthesis), circulation of water (which allows us to extract hydroelectricity), the wind, tidal flows and geological processes (volcano eruption, erosion, earthquakes, etc.). The primary and secondary energy sources that form oil — photosynthesis, geological movements and geothermal energy — are all diluted and difficult to use. But they have had a very long time to create a very concentrated product, which can be extracted for a low cost and which can be the basis for other energy carriers, such as electricity and petrol. It is the great 'work' of the bio-geosphere that has created the energy intensity of oil compared to the original biomass.

Energy has been central to the dramatic increase of productivity of human labour. The tremendously improved, and thereby cheaper, transport resulting from cheap oil is a key driver in the globalization of the economy and of human life. The economic importance and its implications largely surpass the importance of the World Trade Organization (WTO) and the European Union and other free-trade agreements, or possibly one could even say that they are a result of energy use.[34] Despite us living in an industrial society, or as some claim, an information society, agriculture is still very important and the interaction between energy and agriculture has major implications as oil reserves dwindle. It works both ways. It is largely through the

---

[34] The origin of the European Union was actually as a 'coal and steel union'.

use of external energy that labour productivity in farming has increased. On the other hand, historically and ecologically, agriculture was supposed to deliver more energy than it consumed. That was the whole point of farming in the first place. But agriculture today, and even worse our food system at large, is a huge net consumer of energy. At the same time, expectations are that we will produce more biomass and energy from farming. These are the reasons why this chapter discusses the energy and agriculture nexus thoroughly.

Energy costs represent some 3% of GDP. This fact can lure us into believing that energy is not so important for the economy, nothing could be more wrong. If the 3% of energy costs are taken away, most of the other 97% will vanish. Cheap energy, mainly in the form of fossil fuel, is the fuel of the whole economy — without it, all things will grind to halt like a car without petrol. This is also realized by many, even if most politicians try to do a balancing act of overplaying our ability to find other solutions and underplaying our extreme dependency on fossil fuel. The importance of energy is also reflected in the substantial subsidies that are directed to fossil fuel, which are calculated by the International Energy Agency (IEA) to be US$ 650 billion in 2008 (UNEP 2011).

Energy use and supply

Since the first oil crisis in the mid-1970s, the energy consumption per capita in high-income countries has stagnated. Transport energy is an exception and has almost doubled in high-income countries. Globally, the consumption of energy has doubled in the same period. The transportation sector accounts for over a quarter of total world energy use, and the proportion has increased since 1973. Cars, trucks and air transport use most of the transport energy; in the European Union, road transport represents 82% and air transport represents 14% of the transport energy. Transport of people both for work and for leisure increase dramatically as people become wealthier. A flight from England to Mallorca was a big thing and was very costly in the 1960s; now it is almost like catching a bus, and would be a lot more so if it weren't for all the security measures. Unless there is a major shift away from current patterns of energy use, world transportation energy use is expected to grow at 2% per year, with energy use reaching 80% above 2002 levels by 2030 (IEA 2009; UNEP 2009; European Commission 2010a).

With increasing awareness about problems with climate and with political and economic uncertainties about oil and gas supplies, there is a renewed interest in bioenergy and renewable fuels. It should be noted that the supply of energy from biomass is double the supply of energy from nuclear power, and that biomass represents a much higher proportion of energy in developing countries than in high-income countries. Still fossil fuel represents 80% of all supplied energy, as can be seen in Figure 6.1.

Figure 6.1 World total energy supply (1971–2006).

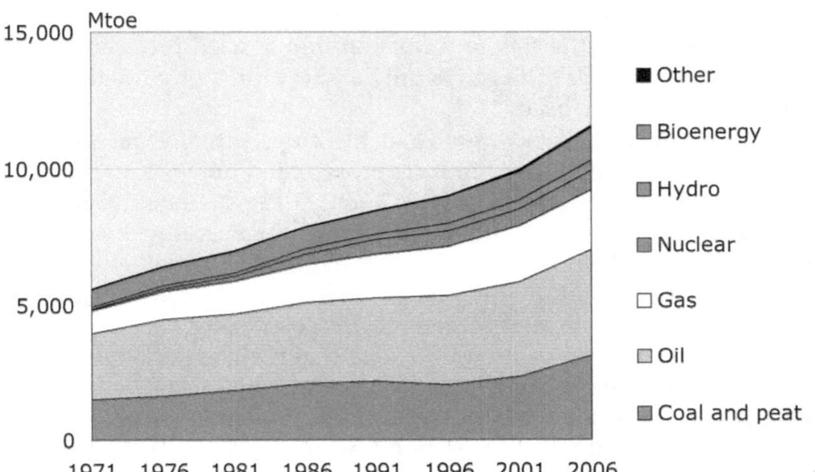

Source: IEA (2008).

The poorest people have very limited access to energy; it is largely limited to firewood that is collected by household members, often women and children, to manure or to other combustible waste products.[35] Approximately 2.6 billion people live with these as their main energy source (UNIDO 2008). There is a strict energy hierarchy for household energy in the sense that with increasing income people move from wood to coal (charcoal or mineral coal), kerosene, propane and, finally, electricity. Around 1.6 billion people have no access to electricity. Apart from the comfort and health advantages of electrical light over kerosene or firewood, electrification can lead to substantial economic development. In the Philippines, it is estimated that electrification of rural areas generates between US$ 81 and 150 per household per year (IAASTD 2009). More than any other factor, access to energy can explain the difference in productivity between the poor and the rich.

---

[35]   In the 1990s, I met farmers in China who used cotton straw as their source of energy for cooking.

We live off 1/400th of the total solar energy

Approximately 130 Joules[36] of energy per square centimetre reaches the earth as solar radiation. Some of this energy is absorbed by the atmosphere directly or is reflected. Of the 91 J reaching the surface of the planet, 18 J is reflected, 31 J is radiated as heat, 36 J is used for the evaporation of water, 6 J heats the soil and some 1 J is locked by plants as chemical energy (Bayliss-Smith 1982). It is this little part, less than 1% of the sunlight that reaches the surface, that is the plants' share of solar energy. And it is this tiny fraction that is used for food, fodder, fibre and biofuels. Or rather, it is only a part of this tiny fraction as the whole plant is rarely used.

If one calculates backwards from the crops actually harvested, we find that in 1993 harvested products represented only 0.4% of the solar energy reaching the fields. Of this 0.4%, only 61% was actually used; that is, real use was only around 0.25% of the solar energy reaching the ground (Uhlin 1997). A theoretical maximum for the efficiency of photosynthesis is around 3–4%, but in reality it is most likely hard to go above 1% for a total farming system. Of the solar energy not absorbed by plants, a small fraction is used in the form of hydroelectricity and wind power, along with biomass from forests and some of the solar energy that reaches the ocean in the form of fish, seafood and seaweed.

The simple agriculture equation has always been that one has to get substantially more energy out of the food than one put into its production. As long as in-energy is human labour, the equation is an iron law that can only be skipped for shorter periods. Obviously, human beings also need from food things other than energy, such as proteins, vitamins, etc.; however, without a positive energy balance at the core, these other components are not important as the farmers (or their families at least) fade away. The agriculture worker should not only feed herself or himself but also other family members who are too young, too old or too sick to work, as well as some others who supply services. Finally, in almost all societies, rulers have taken a great share of the production. In pure agrarian societies, around 80% of the population engages in farming; the other 20% lives off the farmers.

One can roughly divide the agriculture systems into three groups on the basis of energy use. This division is also reflected in the tools used, the degree of specialization, market orientation, etc.

---

[36]   A unit of energy (henceforth abbreviated as J). 1 J=1 Ws.

Pre-industrial farming, where the external energy is less than 10% of the total energy (i.e. human labour represents the totally predominant energy source);

Semi-industrial agriculture systems, where external energy is 10–95% of the total energy; and

Industrial systems, where internal energy supplies constitute less than 5% to completely negligible proportions.

Table 6.1 Energy efficiency in seven agriculture systems.

	Energy (MJ) use per hectare	Energy (MJ) harvest per hectare per year	Proportion fossil fuel (%)	Energy (MJ) productivity per person per day	Energy ratio
Pre-industrial					
New Guinea[37]	103	1460	0	10	14.2
England (1826)	183	7390	2	80	12–40
Semi-industrial					
Indonesia	1079	14,760	54	38	14.2
India (1955)	3255	42,280	58	49	10–12
India (1975)	6878	66,460	77	36	9.7
Industrial					
Soviet Union	6145	8060	96	59	1.3
England (1971)	21,870	44,980	99	2420	2.1

Source: Bayliss-Smith (1982).

Looking at these systems one can see that the total energy harvested per hectare can increase with increased use of ancillary energy; one can increase yield per hectare fivefold with the use of more energy. This energy can be in the form of better (and more timely) soil preparation, irrigation, fertilizers, etc. The ratio between energy output and energy input (i.e. efficiency in use of energy) seems to be fairly constant to a certain level after which it rapidly deteriorates. In industrial farming systems, the optimal use level has since long been passed (see Table 6.1). Harvested energy per labour unit increases dramatically with

---

[37] The example from New Guinea is of a very primitive farming that is based on swiddening. Energy in the swiddening itself doesn't seem to be included. As can be seen, the energy ratio appears to be maintained initially to be radically lower when external energy is becoming predominant.

increased input of energy with a factor of between 10 and 100, allowing the most advanced agriculture systems to have 1 farmer per 100 persons. Admittedly, a lot of support work is needed for that farmer. As the whole modern food industry employs a lot of people, the actual net efficiency increase is less, but still very substantial (Bayliss-Smith 1982).

According to FAO, 6000 MJ of fossil energy (corresponding to a barrel of oil) is used to produce a ton of maize through industrial farming methods, whereas only 180 MJ (corresponding to 4.8 litres of oil) is used to produce maize through traditional methods in Mexico. This calculation claims to include energy for chemical fertilizers, irrigation and machinery, but not 'shadow energy', that is, energy used for making machinery, for transporting products to and from the farm and for construction of farm buildings (FAO 2000). The energy ratio is below 1 (i.e. more energy is consumed than produced) for modern rice farming and just above 1 for modern maize farming, whereas traditional production of rice and maize gives a 60–70-fold return on energy used. The ratio is bad, and there is no doubt that it is worse in high-income countries than in low-income countries, and in industrial farming systems than in traditional farming systems, with the exception of swiddening farming.[38]

Table 6.2 Energy use and efficiency in developed and developing countries.

	Energy (MJ) per hectare	Energy (MJ) per ton grain	Energy (MJ) per farm worker
Developing countries	4019	2009	4144
Developed countries	13,062	4856	137,913

Source: FAO (2000).

FAO has also compiled average data for energy yields for developed and developing countries (see Table 6.2). It shows that, compared with developing countries, developed countries use more than twice the

---

[38]   I have not found any figures for energy use for swiddening, but if we assume that there is some 100 cubic metres of wood per hectare which is burnt and that the land is used for farming for three years, it would mean that some 35 cubic metres of wood is 'used' per year. That would correspond to the energy of some 3 cubic metres of oil, which would mean an appalling energy efficiency, even worse than most industrial systems.

energy to produce a ton of grain and three times as much per hectare (quantities are more per hectare because yields are a bit higher in developed countries). FAO notes that 'productivity is higher' when more energy is used (i.e. productivity per labour unit). One could of course inverse this and say that the productivity measured on used energy is very low. When we discuss bio-energy this discussion is suddenly very relevant.

Different kinds of agriculture production and different foods also have different energy ratios, as can be seen in Figure 6.2. The energy ratio is very low for deep-sea fishing, for meat production from feedlots and for vegetables in heated greenhouses.

Figure 6.2 Energy use per kilogramme for selected foods.

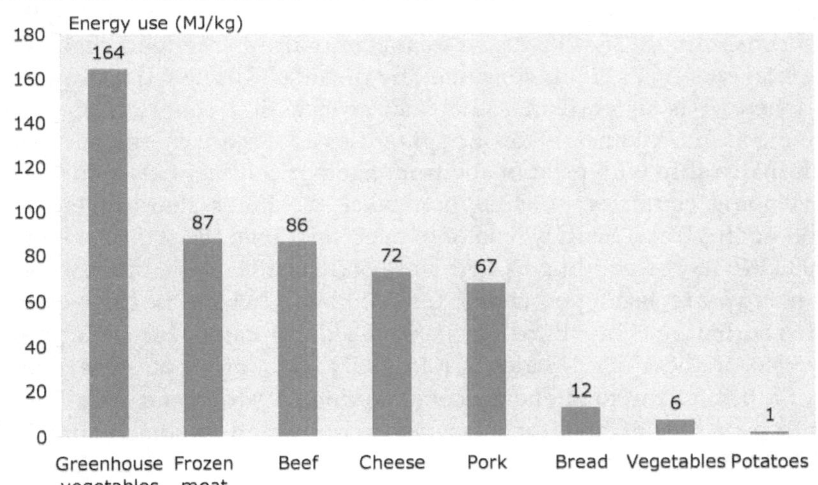

Source: Carley and Spapens (1998).

A big share of the energy use in farming is for the production of chemical fertilizers, in particular nitrogen fertilizers, and pesticides. In the production of bread wheat in the United Kingdom, 'over half of the energy used is in fertilization' and, globally, 30% of fossil energy used in maize is accounted for by nitrogen fertilizers (Woods *et al*. 2010: 2993) This also means that the contentious debate on organic versus conventional (nonorganic) farming has a strong element of the debate on energy dependency. If improved energy ratio is the primary goal of farming, skipping, or at least dramatically reducing, nitrogen fertilizers is a way to get there.

The increase of energy use in agriculture was in particular very rapid in the period after the Second World War and the first oil price chock in 1973; while labour force was reduced to half between 1952 and 1972 in England, energy use tripled. In the United States, energy use decreased from the mid-1970s to the mid-1980s as a response to increased oil prices; thereafter it has stabilized. Looking at the whole food chain, however, energy use has constantly increased. In pre-industrial and semi-industrial agriculture systems, most of the food was sold, eaten and prepared close to where it was produced, but modern food chains are highly centralized and globalized. And the energy ratio is just getting worse. In industrial countries, between 10 and 15 times more energy is used in the food system than in the food we end up eating[39] (Hendrickson 1994; Uhlin 1997).

A big part of energy consumption is caused by consumers buying, storing and preparing food. In Sweden in 1997, agriculture production represented 15–19%, processing 17–20%, distribution and retail 20–29% and consumption 38–45% of the total energy use in the food chain. Of the total energy, 7–11% is consumed by the much-discussed transports, and here it is in particular the final stretch that counts. A person driving a car 5 kilometres for shopping uses a lot more energy per food unit than a ship with meat or soy from another continent. Also in some developing countries, consumption takes the lion's share of energy use. In this case, mainly cooking over an open fire takes energy: 1500 kWh (corresponding to a bit more than a cubic metre of firewood) of energy are used per capita for cooking,[40] which is somewhere between half and one-third of energy used per capita for cooking in Sweden or the United States (Uhlin 1997). Cooking represents more than a fifth of the total energy consumption in Africa and Asia[41] and up to over 90% of household energy consumption in some countries (IEA 2006). Another observation is that energy used for cooking is more than the total energy in food. So while farming in developing countries and by traditional systems is energy efficient, cooking is not.

Energy ratios in agriculture give an interesting perspective but conclusions can't be taken to the extreme. First we are, at least not yet,

---

[39]   Here, many different figures are around.

[40]   As an interesting comparison, McDonald's Sweden states that they use 1 kWh per meal (McDonald's 2008).

[41]   The introduction of energy-saving stoves or the use of other fuels that are easier to regulate should be a prioritized measure. It is not only about conservation of forest and saving energy; soot and smoke indoors is a health hazard and one of the bigger killers. Between 1.5 million (WHO 2006) and 4 million (Pimentel *et al.* 1998) people are killed by this every year.

in a situation where we have to equalize energy in food and energy in oil, nuclear power and hydropower. At first glance, one might even consider such comparisons absurd. And they are absurd, if we think that there will be unlimited supplies of energy in the future. Not too many believe that anymore. All staple foods (i.e. the foods that provide the bulk of nutrition) are foods with a positive energy balance in their traditional way of production. If that were not the case, they would never have been staple foods in the first place. Second, calories alone don't give value to food. In such a case, it would be best keeping to sugar (and sugar cane is one of the most energy-efficient crops). Vegetables will always be inferior to grain when it comes to energy ratios. They contain a lot of minerals and vitamins, however. Meat is not primarily consumed for its energy content but for its protein content.[42] Finally, some food is eaten simply because it tastes good or for religious or cultural reasons.

Some countries have set targets for use of biofuels for road traffic. Biofuels are of several types such as ethanol from maize, wheat or sugar cane; biodiesel from rape seed, palm oil or copra; methanol from wood; biogas from fermentation of various waste materials or manure, etc. Some claim that it is unethical to produce biofuels. Others support it under the conditions that it is made in a sustainable way and some promote certification to safeguard this.[43] The debate is fairly simplistic and often disregards the economic realities of farmers. From the perspective of farmers, the food sector has been a buyer's market for most of the time. There is even a so-called economic law, Engel's law, according to which agricultural prices go down, and they do so more rapidly than other products. Thus, increased food prices and more alternative uses for farmland is a boon for farmers.

Like the rest of the agriculture sector, biofuels too are subject to large political interventions. Globally, biofuels received some US$ 11–12 billion in subsidies in 2006 (FAO 2008). During 2006/2007, one-fifth of the maize yield in the United States was used for biofuels, stimulated by heavy subsidies, and still the amount only corresponded to some 3% of petrol consumption (World Bank 2007). This shows that

---

[42]   The efficiency of meat production is discussed more later on.

[43]   For some reason, nobody promotes all fuels being subject to the same criteria and certification, in which case petrol also would have to be shown to be sustainable, which is obviously not possible.

biofuels present a very limited possibility for reducing society's dependency on oil. This doesn't have to be an argument against biofuels by itself, rather an argument for the need to totally redesign transport systems. Some biofuels are produced from waste products and don't compete at all with food production. A report for Friends of the Earth states:

> [A] realistic bioenergy potential on cropland and grazing land in the year 2050 may be around 70–100 EJ/yr,[44] with the lower number being environmentally considerably more favourable than the higher one. For comparison, we note that the global technical use of primary energy is currently around 550 EJ/yr (fossil energy use around 450 EJ/yr). This means that the bioenergy potential from cropland and grazing land is in the order of magnitude of 15–22% of current fossil energy use. (Erb *et al.* 2009: 25)

The results of Kersti Johansson *et al.* (2010) are more or less the same. One very disturbing aspect is that a lot of biofuel production has a bad energy ratio; some examples even show ratios below 1, implying that the production of biofuel uses more energy than the energy content of the fuel itself, something that is obviously only possible to achieve with massive political distortion. For grain-based biofuels, the energy ratio in a number of cases studied ranged from 0.7 to 2.8.[45] The energy ratio of biofuels from lignocellulose is normally higher as is the energy efficiency of sugar cane ethanol. The biggest problem with biofuel is the competition with other land use, as increased biofuel production is likely to either take place in now-low-on-production, but highly biodiverse, rangelands or expand into 'virgin'[46] lands, such as wetlands and primary and secondary forests.

Biomass is not used for fuel alone. First and foremost, it is used for food, of course, but increasingly it is also used for the production of other advanced modern products. Industrial biotechnologies use agriculture feedstock rather than petroleum-based ones to produce chemicals and plastics. Estimates are that this market will grow from around US$ 170 billion in 2008 to up to US$ 300 billion (*The Economist* 2010a). To use the sophisticated molecules made by photosynthesis for

---

[44]   Here 'EJ' stands for Exa Joule; 1 EJ=10^{18} J.

[45]   This also means that if you want to replace 100 EJ of oil with biofuel, you might need 200 EJ of biofuel if the ratio is 2.

[46]   I put virgin in quotation marks to indicate that there is no such thing. All landscapes todayare influenced by human activity.

this purpose is likely more interesting than to burn them. A bio-refinery industry built on farmed raw materials, where energy is *one of many outputs*, is not improbable, but it is primarily not an energy-producing unit.

Farming uses energy in many different forms: diesel for tractors and pumps; electricity for pumps, fans and indoor machinery such as milking machines, etc. Fertilizers represent a big energy use. Energy represents 90% of the production costs for nitrogen fertilizers, 30% for phosphorus fertilizers and 15% for potassium fertilizers. For production in the United States, energy costs represented 22–27% of the production costs for wheat, maize and cotton and 14% of the production costs for soybeans[47] (US CRS 2004). These figures do not include embedded costs in buildings, machinery, etc., so the actual share of the costs is substantially higher. In Argentina, energy costs were calculated to 43% of production costs in 2006 (Baltzer *et al.* 2008). In a situation with rising energy prices, agriculture prices will follow suit. This could also be seen in the case of food prices and in the case of the oil price hike in 2007–2008.[48] Increased energy prices influence food prices:

> by making the production more expensive;

> by making biofuel more interesting to produce and, therefore, reducing the production of food, leading to higher prices;

> through increased transport costs that directly reflect on food prices; and

> through reduced competition in the food sector (increased transport costs means that the pressure of global competition is reduced).

Our energy future

The radical change in energy supply would occur if we could harness solar energy more efficiently than we can today or if we could make our own sun (i.e. fusion energy). Already the little brother of fusion

---

[47] Soybeans can be grown without nitrogen fertilizers as they bring about natural nitrogen fixation.

[48] Other factors too were driving this, but increased oil price doubtless was one if not *the* main driver.

energy, fission energy or, popularly, nuclear power has so many problems related to security, safety and financial viability,[49] and it is frightening to envision a development of the other nuclear alternatives. But even without the risks, the practical and commercial applications of fusion power are far in the future. The proponents now say it is some 30–40 years away, but they said the same 40 years ago. There are reasons to be critical of scaremongers and doomsday prophets because they are often wrong, but there are at least as many good reasons to be sceptical and critical of the visions of the technological optimists who have promised many things that never materialize (landless food production is one such example of something that has been promised for ages). It is nice to have visions and fantasies, but when they want us to make strong bets on future technologies and take enormous risks they should be seen as dangerous.

Some promote hydrogen and hydrogen cars as if it were an independent, inexhaustible and clean energy source. But this is built on a total misunderstanding. Hydrogen is an energy carrier like electricity. It has to be produced from other energy sources. Currently, there are three commercial technologies for making hydrogen: a petrochemical process, gasification of coal and electrolytic division of water into hydrogen and oxygen. In the first two cases, fossil fuels are used, whereas for electrolysis, another energy carrier, electricity, is used; one will have to go further back to see how the electricity is produced to find the primary energy source.

Hydrogen is normally used in so-called fuel cells to produce electricity. This electricity in turn runs engines. So there is a very long chain of energy transformation; for example, gas/oil/coal/solar energy/hydropower/biogas ▶ electricity ▶ hydrogen ▶ electricity ▶ electrical engine ▶ mechanical work. In all these steps, there are energy losses, in the fuel cell itself perhaps up to 50%. Some estimate that only one-fourth of the energy used actually reaches the wheels (Trainer 2007).

---

[49] The nuclear industry has been and is still heavily subsidised by governments. The whole research foundation was paid by them and in most countries, such as the United States and Germany, the governments take most of the risks involved in a possible total failure. It is doubtful that the private sector ever would have built any nuclear plants or they would have to shoulder the risks (Radkau 2008). Another side of nuclear power is its dependency on oil for running diesel backup systems, for transportation of fuel and for construction of power plants. Energy return is also rapidly dwindling when lower-grade ores are taken into use. As such, it is not such an interesting solution either for peak oil or for climate change, as claimed by its proponents.

It is hard to understand how hydrogen can be so hyped. It might have the potential for storage of surplus solar or wind energy, but this technology is not here yet and it will probably not be commercial before 2050 (McKinsey & Company 2009; hereafter McKinsey).

Wind energy and, to some extent, wave energy certainly have the potential for expansion. A major problem with wind energy is its variability, which means that it can never supply more than a certain proportion of the needed electricity, at least not with the current structure of energy use and distribution, or it will have to use costly and energy-losing storages. Biomass will, inevitably, play an important role; its main limitation, which has already been discussed, is the competition for land, not necessarily in food production but in 'the wild'. A big proportion of biomass will in the future, just as throughout history, come from forests, but forests are also under great pressure in many places.

When discussing energy we need to look beyond the gross production of energy. Many aspects of energy are important. First, it is about forms of energy; for instance, one should not compare electricity with fuels. In a nuclear power plant around three units of nuclear energy are used to produce one unit of electricity. Second, it is about the price of energy compared to the price of labour and other resources. Third, it is about availability in time and place (see the discussion above about wind power). Finally, it is also about the efficiency of energy production, the energy returned on energy invested (EROEI). For EROEI too, it makes a difference, at least for short-term choices, if the energy needed to develop an energy source comes up front as energy investment in infrastructure or if it comes during the use of the source. For instance, for solar energy almost all costs in money and in energy are up front, whereas the energy needs for running a solar system are very low. Especially, during a rapid transition from the current energy system, this can play a big role. If availability of fossil fuel goes down and, at the same time, a lot of energy has to be used to develop new energy sources, we get a double squeeze, with potentially devastating effects, at least for the economy.

Another aspect of EROEI is also to consider the whole system. For instance, for solar systems, one can calculate the EROEI directly in the panel, but one should include the storage and distribution systems that are an integral part of the system. Finally, one should look at the EROEI all the way to consumption. So if the EROEI of petrol at the pump is 10:1, the EROEI of the use of petrol in the car is only 3:1, as only a smaller part of the fuel is converted into the movement of the car. An EROEI under 10:1 at the level of production (the mine/pump/-farm/factory gate) means that a very big part of the energy is used for production of energy. Similarly, a very big part of the economy is oriented to the production of energy — to drive the economy; an EROEI at 3:1 makes energy production more or less meaningless (Hall *et al.* 2009). Most biofuels have an EROEI in that range. And what is most alarming is that farming, our most basic energy production, has a negative energy ratio; it is totally dependent on subsidies of external energy.

If energy is going to be consumed in the current range, the only real alternative to fossil energy is solar energy. A few days of solar radiation to the surface of the earth represents all known reserves of oil, natural, gas, coal and uranium. It is a bit hard to understand why technological optimists so often are sceptic about solar power and enthusiastic over nuclear power. Currently, three kinds of solar energy technologies are available[50]: solar panels for heating a medium (mostly water, mainly for heating buildings or water itself), solar cells (photo-voltaic cells) for the generation of electricity and concentrated solar power (CSP). Concentrated solar power is based on the concentration of solar power through mirrors to one point, by which very high temperatures can be reached. This is used for generation of steam to run turbines for electricity generation. This is most likely the most interesting kind of solar power for large-scale projects and production on a level that can really replace fossil energy. For the time being, its energy efficiency is low (Hall *et al.* 2003). Another problem is that the natural place to put these constructions is in the desert, but firstly there are not many people in the desert, implying massive infrastructure for transmission and massive losses of energy in the transmission. Secondly, and possibly worse, there is a need for big quantities of water for cooling, which certainly is a scarce resource in deserts. This is overlooked in almost all projections for this option.

Solar panels can generate some 700 W per square metre and some 200–700 kWh per square metre per year depending on the system and where it is located. Solar panels for water heating are already today more or less commercially viable, but have limitations regarding seasonality and storage, and must therefore be combined with other

---

[50]   In addition, there is 'passive' solar energy, that is, constructing houses to make best use of the sun and to conserve as much energy as possible.

energy sources or with costly storage systems. Alternatively, one has to accept that availability of hot water fluctuates, something that is already the case in many simpler tourist places and homes in many countries.

Solar cells for electricity production generate some 150 W of electricity per square metre and some 50–150 kWh of electrical power per square metre per year depending on the system and location. It can be interesting to put this in relation to biomass production. Normal agriculture land produces some 50,000 MJ per hectare, which corresponds to 1.4 kWh per square metre per year. Biomass, produced only for maximizing energy yield, can perhaps produce some 5 kWh per square metre per year. The energy from solar cells is thus much more efficient than photosynthesis — if one counts energy as the only output. The real feat of photosynthesis is not the capturing of energy to be burnt but the production and reproduction of very complicated pieces of genetic information and a wealth of chemical substances ranging from fats, protein and vitamins to essential oils, in short life and all its support systems. Direct solar electricity is still expensive. Maintenance and operation costs are low, so investment costs and the lifespan of the equipment are the most critical factors for the profitability of solar electricity. Where there is no fixed grid, solar electricity can be profitable (as the costs for connecting to the net are substantial), especially in applications where the individual user doesn't use much, such as in simple holiday homes or homes in developing countries.

According to the Solar Energy Technologies Program of the United States, a solar cell needs to run for two years to recover the energy spent in making it.[51] The program estimates that both solar cells and concentrated solar power will be commercial by 2015 (EERE 2009). Obviously, the predominant factor here is the development of the oil price. There might also be some issues of the availability and toxicity of some of the minerals used for production of solar panels. The biggest solar power plant contract in 2009 was for a plant in Mongolia where cadmium telluride is used, which is very toxic and needs to be taken care of well (SvD 2010). But let us assume — together with the technological optimists — that solutions will be found.

---

[51]   Notably this is for direct use. For new installations for large-scale connection to the grid, energy costs for power lines and converters, etc. need to be added. Energy losses can also be substantial for large-distance transmission and even more for storage.

Still, the seasonality and variation of solar energy are real problems and that the sun, even when it shines, shines only half the day (for energy production, effectively even less time). Storage of heat is possible for daytime variations, but it is still very costly to take care of seasonal variations. Storage in batteries is only realistic in the smaller scale; storage as hydrogen made by electrolysis is promising. There are also various high-tech solutions such as space-based solar energy and a global system where energy production follows the rotation of the planet and electricity is transmitted to parts of the planet where it is night. But all this lies far ahead and it means that, in general, solar energy is not the short-term silver bullet that allows us to continue to consume and globally increase consumption. Even with breakneck expansion it will take decades before solar energy replaces fossil fuel. In the United States, in 2008, electricity generation from solar thermal and photovoltaic sources was up 41.2% from 2007, at 864,000 MWh, but that is just a fraction of a percent of the electricity use in the United States (EIA 2010). With the same annual growth for 25 years, *and no increase in consumption*, it would take 25 years to replace fossil fuel and nuclear electricity generation, and then we are only speaking about electricity. The United States uses substantially more energy in other forms, and a growth rate of 40% is very hard to maintain.

A problem with the debate regarding alternative energy sources such as solar, wind and biogas is that the debate is so confrontational. Those who favour these alternative sources tend to oversell them and overlook their weaknesses, and they avoid debating about the total energy use. The proponents of a continued blind march based on fossil fuel and nuclear power are, on the other hand, constantly undermining confidence in them and also avoid the real debate. Solar, wind and biogas sources are useful and good; they even have the potential to supply all the energy human beings need, *if* there are radical actions to curb energy use and *if* one effectuates less car driving and less air conditioning. These alternative sources will not do the trick if we continue with business as usual and middle-income countries continue along the same path.

Energy is central for the industrial society

In Part 1, we saw how important energy is for human development and how different sources of energy, and the technologies developed to harness that energy, shape societies in different ways. Energy consumption has reached levels where it threatens stability in the world and our future survival. Fossil fuel is the main source and also the main reason for the greenhouse effect. For high-income countries of the world, the first three measures should be to save, to save and to save. Still, for the poorest billions on the planet, more energy is a road to greatly improved conditions; we have to ensure that this can happen. Material wealth is almost a perfect linear function of energy use (Hall and Klitgaard 2006). Because of this, it is perhaps not realistic to get rid of society's oil and coal dependency very rapidly, even if it must be a goal. It can be radically reduced by substantial increase in energy from biomass, wind, water and the sun, combined with vigilant saving in rich countries. (Cheap) Energy is also one of the main components of the expansion of industrialism, capitalism and consumerism, all of which we will discuss more in Part 3. One could possibly go so far as to say that energy is the currency of the world today. More than money it is what keeps the wheels spinning. Or perhaps money and energy are today linked together in an inseparable union (also discussed in Part 3).

– 7 –
Global Warming

The climate change problem is at its heart an ethical problem. It's a problem of income distribution and it's a problem of income distribution with dimensions that we don't usually think about very much.

(Ross Garnaut,
'Foreword', in Climate Change and
Social Justice, 2009)

Global warming is so well known and there is so much written over the last years that there is no great need to go into detail here. The greenhouse function of the atmosphere, where so-called greenhouse gases let in sunlight but keep the radiation of heat from the planet within, is a natural phenomenon, which makes the planet habitable in the first place. Restless human activities have increased the proportion of greenhouse gases and thereby caused temperatures to rise. Before human beings started to burn coal and oil in a grand scale, the atmosphere had around 290 ppm carbon dioxide. In just a century, this increased to above 380 ppm. In the same period, temperature went up by 0.6°C. This might sound little, but already a few degrees change of average temperature makes a lot of difference. There is scientific agreement that a clear correlation exists between the increase in greenhouse gases and rising temperatures and there is also agreement about why the greenhouse gases are increasing (see more below).

The United Nations Intergovernmental Panel on Climate Change (IPCC) estimates that the temperature might increase by 6°C in the coming century, with a possible variation from 2.0 to 6.4°C. This will lead to dramatic changes in climate and the worst affected would be poor people and poor countries. The countries around the equator will be subject to droughts and shrinking food production. Other parts will be flooded by more rainfall, and ultimately low coastal areas will be flooded as an effect of melting glaciers, especially the huge ice caps of Greenland and Antarctica. The sea level could rise almost a metre in a century. A very big proportion of human beings live close to the sea, and the coasts are the richest parts of the world. Some countries, like the atoll Pacific Island states and Bangladesh, could be completely devastated.

There will also be direct effects on human beings; tropical disease, such as malaria, will spread to new areas where people are less resistant and don't know how to protect themselves. Higher temperatures also affect the human body in those parts of the world where temperatures are already limiting our ability to work. It is estimated that in India productivity of outdoor workers has reduced by 10% since 1980 because of higher temperatures and that a rise of another 2°C will lead to a 20% reduction in productivity (ICTSD 2009).

A worry, perhaps the worst of them all, is that some processes can be self-reinforcing; that is, they are first triggered by climate change and then the processes themselves add to the problem of climate change. One such an example is the melting of the Arctic tundra, the permafrost. If the tundra starts to melt, organic matter in vast marshlands will start to ferment or ferment more rapidly than they already do. This fermentation will release methane, a potent greenhouse gas, which will induce even higher temperatures leading to more melting. Similarly, if the ice caps melt, less sunlight will be reflected, because land or water reflects less than snow and ice, so it will also exacerbate the process of climate change. Drying out of the Amazon forest and methane emissions from the seabed are other such examples. The history of the planet shows that climate changes can be rather abrupt with dramatic effects on most species. It doesn't sound like a far-fetched idea that the changes can be even more abrupt when driven by a new phenomenon on the planet—one of the creatures suddenly start to feed on million-year-old storages of coal and oil.

Causes of global warming

The most important reason for global warming is the burning of fossil fuel, a fact that we don't have to expand upon here as I assume the reader is well aware. The second most important reason is change in land use. Around 13 million hectares of forest are cleared annually, mostly in developing countries. Logging for timber is a driver but the main driver is expansion for agriculture, both large-scale expansion—clearing of forests for palm oil or cattle ranching—and small-scale annual expansion by hundreds of thousands of smallholders. The transformation of forest to other lands has ceased in most industrial countries and several countries show increasing areas of forest.[52]

---

[52] As noted in Part 1, the forest resource was under extreme pressure in the early stages of industrialization, and huge tracts were completely

People often look too much at what is already visible, that is, what is above ground, and not enough at what is under the surface. The carbon stocks in the soil are many times bigger than all the carbon in green plants. The high drama, for greenhouse effect, is not cutting down of trees (this can of course be devastating from other perspectives such as biodiversity) but the change in land use, in particular the conversion of land to agriculture. The North American prairies have lost around half of all organic matter in 100 years. Even worse is the cultivation of organogenic soils (they are peat and marshlands made up of almost uniquely organic matter, i.e. carbon) in Northern Europe and South East Asia. Approximately one-third of the increase in carbon dioxide[53] in the atmosphere comes from the breakdown of organic matter, the humus, the life of the soils (Montgomery 2007).

Climate scientist William Ruddiman (2005) says that human beings started to change climate on a rather large scale some 5000–6000 years ago, by developing rice irrigation systems and by transforming forests, swamps and grasslands to farmlands. He shows that the areas affected by farming were already large thousands of years ago and that there were corresponding changes in methane and carbon dioxide levels. Because the underlying trend was one of cooling, of a new ice age, this went unnoticed until it was substantially augmented by the burning of fossil fuel. At the time of William the Conqueror, England below 1000 metres was largely deforested and most of this clearing took place perhaps 2000 years ago (Ruddiman 2005). This view is supported by others, such as the German historian Joachim Radkau, who note that the Alpine foothills were cleared by fire almost 6000 years ago: 'Discoveries of this kind [. . .] have revolutionized the picture of early history of the relationship between humans and the environment by revealing that large-scale changes of the landscape by human hand reach back far into prehistoric times' (Radkau 2008: 43).

Greenhouse gases and economic growth

There is a strong correlation between the emission of greenhouse gases and GDP. The higher the GDP the more the emissions. Largely, this is also correlated to the use of fossil fuel. Sometimes there are confusing (and confused) statements that emissions are reduced with higher

deforested.

[53]   Note that this is about carbon dioxide. Other greenhouse gases, primarily methane and nitrous oxide, are not increased to such a large extent by change in land use.

GDP, but these statements are not supported by data. What is actually meant is that there is less emission *for additional GDP* units in some high-income countries. But as GDP grows, even with small additional emissions, total emission goes up, even within the country itself. The situation looks worse when one counts emissions in other countries producing all the things the high-income countries import. As reported, the average North American causes emissions of 22-ton CO2e,[54] the average Briton 12 tons and the Swede 8 tons. But this is not the whole truth. The emissions for international travel and trade are not taken into account for any country (Björklund *et al.* 2008).

Table 7.1 Greenhouse gas emissions (million ton CO_2e from 2005 to 2030).

	2005	2030	Annual change (%)	Share of emissions 2030 (%)
World	26,620	41,905	1.80	—
OECD total	12,838	15,067	0.60	36.0
Developing countries	10,700	22,919	3.10	54.7
European Union	3944	4176	0.20	10.0
Japan	1210	1182	−0.10	2.8
United States	5789	6891	0.70	16.4
Russia	1528	1973	1.00	4.7
China	5101	11,448	3.30	27.3
India	1147	3314	4.30	0

Source: UNEP (2009).

In 2004, 23% of global CO2e emissions, or 6.2 gigatonnes CO2e, were embedded in products traded internationally,[55] primarily as exports from China and other emerging markets to consumers in developed countries. In some wealthy countries, including Switzerland, Sweden, Austria, the United Kingdom, and France, >30% of consumption-based emissions were imported, with net imports to many European countries of >4 tons CO2e per person in 2004. Net import of emissions to the United States in the same year was somewhat less: 10.8% of total

---

[54] CO_2e stands for 'carbon dioxide equivalent', the standard measure for emission of greenhouse gases whereby other gases are assigned values depending on their effect on the climate.

[55]   This had increased to 26% in 2008, according to Peters *et al.* (2011).

consumption-based emissions and 2.4 tons CO2e per person. (Davis and Caldeira 2010: 5687)

As a result, the actual emissions caused by people in high-income countries are a lot higher than official figures (as in Table 7.1). One question is whether the producer or the consumer should carry the burden of the effect. This is significant for the workshop of the world, China, whose record looks a lot better if the consumer countries take the responsibility for the emissions caused by their consumption. China has subsequently proposed in international negotiations that consumption should be the basis for any agreements on greenhouse gas emissions (ICTSD 2009).

Largely, global emissions need to go down from 45 Gt[56] CO_2e to 15 Gt CO_2e to avoid the worst scenarios, but it will still entail serious climate change. If a global population of 9 billion were estimated for 2050, some 1.5 tons CO_2e per person would be the global permissible average, which is the emission the average Indian caused in 2005. It is hardly realistic, or morally defensible, that North Americans or Swedes should ask Indians to reduce their emissions. This means that first they need to reduce their emissions dramatically to say the least, more than 10 times for the North Americans. In addition, people in low-income countries, with a certain moral right, also claim that the people in high-income countries should compensate for *historical* emissions. Even if this makes a lot of sense, realism says that it is already a very tough sell to reduce it to half of current emissions. So far the only dents in the curve for emissions coincide with economic recessions and the collapse of the Soviet Union.

Farming and climate change

Agriculture itself is directly responsible for some 14% CO_2e. Land-use changes (conversion of forest, savannah or wetland to farmland), land-degradation and deforestation linked to agriculture represent another 18% (FAO 2009). Agriculture plays four important roles in climate change:

Farming emits greenhouse gases, directly and indirectly.

Changes in agricultural practices have big potential to be carbon sinks.

---

[56]   1 gigaton (Gt) is a billion ton.

Changes in land use, caused by farming, have great impact on emissions.

Agriculture can produce energy, which can replace fossil fuels.

The two last points have been discussed above. The first two points are further discussed below. Agriculture is also very much affected by climate change. Climate change adaptation for the agriculture sector is at the centre of agriculture policies of many countries. We will not discuss that further here, even if it is a very interesting and pressing debate.

Farming contributes to emissions of greenhouse gases in many ways. Carbon dioxide is released from the direct and indirect use of fossil fuels in the agriculture sector, for instance, through diesel in the tractor and in the production of chemical fertilizers. Carbon dioxide is also released by all the processes that lead to a reduction in soil organic matter. The most dramatic change is in the clearing of new land, as discussed above. But it doesn't stop there. After an initial rapid deterioration of soil organic matter the process continues and a new balance is reached after many decades. At which level the new soil organic matter content will stabilize depends on many factors such as climate, soil type, precipitation, crop rotation, choice of crops, livestock, fertilizer use, etc. Cultivation of organogenic soils can lead to emissions as high as 20–40 tons CO_2e per hectare per year (Climsoil 2008). To convert organogenic soils back to the peat bogs, swamps and wetlands they were earlier is the single most important measure to reduce the contribution of agriculture to climate change, according to a report of McKinsey. If 1.1 million hectares of land were converted *annually*, this would lead to an annual reduction of 1.6 Gt CO_2e (McKinsey 2009).

Methane is emitted from ruminants, from paddy rice production and from manure. Knowledge is still scarce and incomplete for many aspects of methane emissions; for example, how fodder influences methane emissions or how one can reduce methane emissions from paddy rice. Rice production with no or less flooding (e.g. the SRI system[57]), releases less methane than normal practices. Methane emission is from the rice plant itself (up to 90%) and variety differences can be exploited for reduction of emissions (FAO 2003). Intensification

---

[57]   System of Rice Intensification; see more on http://ciifad.cornell.edu/sri/.

of livestock systems will lead to higher emissions from manure handling, but is claimed to lead to less methane per produced unit of meat or milk.[58] But we also need to relate the emissions from domesticated ruminants with those from wild fauna they have replaced. In some areas, cattle, sheep and goats have replaced herds of buffalo, deer and bison, animals that also are ruminants and emit methane.

On the American prairies there were 30 million bison 200 years ago (Chapman and Reiss 1999), and the number of cows today are more or less the same. Research has shown that permanent pastures, such as the Hungarian *puszta*, can accumulate organic matter for centuries and thereby be substantial carbon sinks and to some extent balance methane emissions from grazing cows (SvD 2008b). Methane emissions of a typical African cow are, according to researchers at the International Livestock Research Institute, normally offset by carbon sequestration in their pastures (Maarse 2010). In addition, if land cannot be used for pasture it can be converted into forest, left idle or converted into farmland. The last alternative leads, inevitably, to more carbon emissions; the average loss of soil carbon was 59% according to a meta-analysis of 80 reports on conversion of grassland to cropland (FAO 2009). Therefore, it is doubtful whether one can count the metabolism of ruminants like any other factor of global warming.

Agriculture represents two-thirds of the emissions of laughing gas (nitrous oxide), the third greenhouse gas of importance. These emissions are directly related to the nitrogen cycle (see Chapter 5). The increase in livestock and the use of chemical fertilizers are key drivers. The use of nitrogen fertilizers is extremely inefficient and a lot more nitrogen is added to the soil than taken away with the harvest. The rest of this nitrogen 'gets lost'; some of it as nitrate runoff and some of it as emissions of nitrous gases. In addition, nitrogen fertilizers also consume a lot of energy for their production. Reduced use of nitrogen fertilizers can thus be a main strategy for reducing greenhouse gas emissions from agriculture.

Both the FAO and the World Bank promote carbon emission trading encompassing carbon sequestration (sinks) in soils; that is, farmers should get paid for increasing soil organic matter. The carbon sink

---

[58]   The evidence for this seems weak and few. To measure actual emissions from the two ends of the cow in practical experiments is complicated and the empirical evidence is therefore very scarce.

capacity of the world's agriculture and degraded soils is said to be 50–66% of the historic carbon loss from soils or some 42–78 Gt of carbon (FAO 2009). Measures in the farm sector for mitigation of climate change are among the cheapest measures taken, and, in many cases, they also lead to other environmental improvements (reduced water pollution). Increased soil organic matter also leads to richer soils; that is, soils that can retain water and resist erosion longer.

A major hurdle is the large number of farmers and their diverse conditions; it is very hard to determine exactly how much carbon a farmer will capture if he or she does this or that. Measures at an individual level will be very costly. A new scenario emerges where farmers in developing countries can sell their services as carbon farmers to governments and companies in the developed countries. This could become a business replacement for aid, or a way to compensate farmers of developing countries for absurd subsidies to their competitors in developed countries. But it also means that a larger part of their activities are integrated into the world economy. It can also be seen as a new frontier of exploitation, where the rich countries will use the land in developing countries as 'dumping' ground for their waste, which is what this eventually amounts to.

It costs five times as much to do nothing

Before moaning over the costs for abating climate change, one should ask how much it costs to do nothing. According to the British *Stern Review on the Economics of Climate Change* (2006), costs for climate change in a business-as-usual scenario would amount to between 5% and 20% of the global GDP. That corresponds to the costs for the global depression in the 1930s and the two world wars together. In comparison, the cost for keeping temperature increase at an acceptable level would reach 1% of the global GDP (Porritt 2007); others say it can reach 2%. In that perspective, cost should clearly not be what stops us from action. McKinsey (2009) has compiled data to give a global overview of the measures that can be taken now, the extent of their abatement potential, the costs entailed and the investments needed. Some efforts are profitable in the longer view, but incur substantial investments, others are cheap to get going and very profitable, such as changing light bulbs to low-energy bulbs or LEDs, but their total effect is very marginal.

Abatement through change in land use and change in forestry and agriculture has great potential and is fairly cheap (McKinsey 2009). At the other end of the spectra we see *carbon capture and storage* (CCS), which will cost some €40–50 per ton CO_2e, substantially more than the current market price. The energy industry and the utilities are strongly promoting the CCS technique for several reasons. First, this is the only technology[59] that can cope with constant increase in energy supply, because while the rest of the world looks into ways to reduce energy consumption the energy sector looks into ways to continue increasing it. Second, the CCS technique is developed by the energy sector itself; by developing this 'service' the energy companies can ascertain that they will capture most of the carbon sequestration or emission reduction market that is under development. In other words, they will keep the profits in their pocket. Finally, like so often before, they think society should pay for this development.

> Shell's new chief executive has called on governments to intervene in carbon markets, the first time the Anglo-Dutch oil company has acknowledged that markets cannot be left to set the price of pollution. Peter Voser told the Guardian that action needed to be taken to make expensive green projects like carbon capture and storage (CCS) economically viable. He cited the example of Shell's CCS project in Australia where the government has introduced a carbon tax, or a minimum price of carbon. 'That is a way of making sure it gets the support,' he said. (*The Guardian* 2009a)

So far the EROEI for energy production from fossil fuels with CCS looks depressing; it will produce only 50% more energy than it consumes, making it one of the most inefficient ways of producing energy (Bull 2010), and therefore also increase energy consumption tremendously contributing to an escalating depletion of fuel resources.

McKinsey (2009) notes that there is great potential (3.5–5 Gt CO_2e) for abatement through changes in behaviour, for example, by driving fewer cars, accepting bigger variations in housing temperatures, consuming less, and so on; but this is dismissed as less likely to happen, despite these changes representing no costs but direct savings. The five sectors where reduction of emissions has the biggest potential are energy production (district heating, electricity generation, oil refining, etc.), forestry, agriculture, housing and transport. Within each sector, except for agriculture and forestry, energy use represents the

---

[59]   If it works that is; we still don't really know that, but it is a fair assumption that it *could* work.

biggest source of emission; that is, the greenhouse gas emissions from the transport, building and industry sectors are largely emissions from fossil fuels.

Notably, the *potential* for reduction in various sectors is not the same as which sectors cause most of the emissions. Some sectors cause a lot of emissions, but the assessment of McKinsey is that they can't reduce it very much; transport is one such sector that they think can only reduce 25% of emissions. Obviously, this is a question of choice and political will. In the McKinsey figures, it is just assumed that cars will run on the best possible technology. But if this is combined with a substantial reduction of the number of cars and trucks, the reduction of emissions can be much bigger (McKinsey 2009).

The investments needed to accomplish this level of reduction are estimated to be €530 billion per year by 2020 and €810 billion by 2030. The reductions in the transport and building sectors will ultimately lead to net savings (less use of energy through better efficiency) and one would assume that they would be implemented rapidly. They have, however, high investment costs and in some cases the investments fall on one party while the savings benefit another party. For instance, better insulation of buildings represents higher costs for the developers and lower costs for users of the building, which is an obstacle for their development (McKinsey 2009); which means that the market mechanism is not working properly for allocation of costs for mitigation of global warming. It also shows that the market mechanism doesn't work well for optimizing energy use.

The climate issue has, for rather good reasons, taken alarmist proportions. We are advised how to eat, travel, farm, perhaps even love 'climate-friendly'. Climate-friendly has in a short time become synonymous with environment-friendly. It is good that greenhouse gas emissions finally, after being ignored for more than 30 years, are getting attention. However, there are many other pressing environmental and social issues, and there is the risk of them simply getting lost and of decisions being made solely to address climate change, with potentially serious adverse effects on other important things. Big biofuel schemes can threaten valuable biotopes; the perception that meat is not climate-friendly can reduce the number of grazing cattle in Europe, with serious repercussions on landscape and biodiversity; meat consumption is geared towards (grain-fed) chicken instead of beef; that eating prefabricated foods from the supermarket is more energy efficient than cooking at home may lead to further industrialization of food; chemical pesticides and genetically modified organisms are promoted as alternatives to mechanical cultivation; banning campfires of scouts increases our distance with nature a bit more; nuclear power is promoted as a saviour. The list is long, and gets longer by the day. While we should take climate change seriously, very seriously, we should not use it as the ruler against which the whole world and all human actions are measured.

Which weather do you want to pay for?

It was always a human dream to control weather. Rain dances, sacrifices, prayers and other rituals have been used to call rain, stop rain and protect crops from frost or floods. And we know today that human beings can influence the weather. Cutting down forests, especially in the tropics, influences the climate so that was once a lush forest can become a dry steppe; draining of fields and wetlands leads to floods. At a very big scale, global warming shows that human beings can even influence the global climate, though not in the direction we wanted perhaps. And now the so-called international community will have to agree on which climate we should have. Should it be 5°C warmer or can we stop at 2°C? Can the Canadians get 5°C warmer temperatures, while temperature is reduced in the tropics?

Climate change has triggered ideas for large-scale engineering of global systems. Large-scale 'technological fixes' fall into two categories. Carbon dioxide removal (CDR) techniques are designed to extract carbon dioxide from the atmosphere. Solar radiation management (SRM) techniques are intended to reflect a portion of the sun's light back into space. CDR is based on biological, chemical or geological carbon sequestration. SRM is based on natural effects observed in the atmosphere following volcanic eruptions. A proposal for 'sunshade' geo-engineering consists of the installation of space-based sun shields, or reflective mirrors, to deflect a proportion of incoming solar radiation before it reaches the atmosphere. Sunlight deflectors would be placed in near-Earth orbits or near the Lagrange point, about 1.5 million kilometres above the planet, where the gravitational pull of the earth and the sun is equal. An array of sunshades in this position would pose less threat to orbiting satellites than would near-Earth objects (*The Economist* 2010a; UNEP 2010b).

If anything, these proposals underline the severity of the situation as well as how little we really know. They also give new perspectives on power. 'As Andre Matthews, an anthropologist at the University of California, Santa Cruz, puts it, it is not just a matter of constructing a switch, it is a matter of constructing a hand you trust to flip it' (*The*

Economist 2010a). We should also realize that the risks of such global large-scale engineering are huge and totally unpredictable. It is already a fact that human beings change many of the planet's systems enormously, but in the same way as most of these changes are a result of many small things (the car you drive, the meat you eat, etc.), the solutions too are to be found in the small things rather than in macro engineering.

Climate change has brought new credibility to perspectives such as ecological footprint and development space, and ultimately to the fact that there are biophysical limits for the economy. If we do agree that we can't increase carbon dioxide levels in the atmosphere, we agree that there is limited space. Negotiations also reveal that there is a real, and not only perceived, conflict between those who have used cheap fossil fuel to build their economies and those who are standing on the threshold of doing so. Climate change also puts consumption in the limelight again.

From an economic perspective, instead of seeing measures against global warming as costs, we can see it as production in the same way as parks, golf courses and other things we make for our pleasure. One sees, already, how the corporate sector wants to 'privatize' this service, and how the GDP can increase substantially. In the same vein as the earlier discussion about ecosystem services, this can be seen as a sign that capitalism works. By economic incentives, we can not only solve a problem but also turn the problem into a fantastic business opportunity, a new virgin market and more growth. The more far-sighted sections of the private sector have realized that huge opportunities lie in areas such as cap and trade; it takes away the government's right to collect pollution taxes and places it in the hands of the companies.

But one can also apply a completely contrary perspective, that we now start the process of final privatization of all resources on the planet. A privatization without parallel since the privatization of land; a process where markets take over all human relations with nature. In addition, carbon payment is another way to bring peasant farmers into the market economy for their management systems, something they have resisted so far. Once they have done this, the farmers and their resources will ultimately fall under the control of the capitalist market. Even if they nominally own their land, they will no longer own the organic matter, which is the basis of soil fertility.

Will We Run Out of Resources?

If you are a turkey being fed for a long period of time, you can either naively assume that the feeding confirms your safety or be shrewd and consider that it confirms the danger of being turned into supper.

(Nassim Nicholas Taleb,
The Black Swan: The Impact of the
Highly Improbable, 2007)

The early hunters and gatherers used fire and wood, both renewable resources. But they also used rocks, flint and obsidian for axes, knives and arrows. Stones were the first mineral resources human beings used, not only for tools but also for decoration. This was followed by metals, copper, tin, iron, silver and gold. So there is nothing new about digging, drilling and chopping up the soil to get non-renewable resources. What is new is the enormous scale and the many different kinds of uses. In this chapter, we look at our use of finite, non-renewable resources.

The long view

A hunter and gatherer society had a per-person consumption rate of natural resources (both renewable and non-renewable) of about 1 ton per year. This equals around 3 kilogrammes per person per day. In agrarian societies, consumption rose to around 4 tons per person per year or around 11 kilogrammes per day. A large percentage of this increase was attributable to the feed needed for animals that were kept for milk, meat production and as a power source (e.g. for ploughing fields). At the same time, larger buildings were erected and more metallic objects, such as ploughs, weapons and cooking pots, were produced. The Industrial Revolution in the eighteenth century launched the most important change in natural resource use to date. With the use of fossil fuels—at first coal, later also oil and gas—

suddenly much more energy was available to humankind. A resident of an industrialized country today uses 15–35 tons of raw materials and products annually—a multi-fold increase in comparison to agrarian societies (SERI 2009). The proportion between renewable and non-renewable resources has also shifted, and the most increase is for non-renewable resources (see Figure 15.1 in Chapter 15).

There is a lot of debate about 'peak oil', that is, the point where oil production culminates. Some say that we have already reached this point; others set the date quite far ahead; yet others seem to resent the whole notion of peak oil, in the first place. Obviously, no serious person can deny that sooner or later oil production will go down. What the debate really is about, I believe, is whether we need to do something or just continue with business as usual. Those who speak about peak oil normally also have some ideas about what to do, whereas those who deny peak oil mostly want to continue with business as usual, or put their faith in some not-yet-discovered future power source.

To assess how much oil there is isn't as easy as it might appear at first glance. How much 'there is' relates to the costs to take it out. If oil prices rise, there will be more oil, as sources of oil that were not economically viable suddenly become interesting, as is the case for the so-called tar sands[60] and shale gas and oil. Also, as long as there are known sources for some decades ahead, companies won't look around for more oil, in particular as most of the giant fields, and fields where extraction can be cheap, most likely have already been found.

Therefore, 'known reserves' tend to stabilize at 20–30 years ahead. As Jonathon Porritt (2007) mentions, according to Jeremy Legget in *Half Gone* (2005), the year 1980 was the last when newly discovered sources were more than consumption. On the other hand, newly discovered sources had then surpassed consumption for 50 years in a row. The production of many countries has peaked. The peak in the United States (excluding Alaska) occurred in 1972. Since then, a number of countries, including important oil-producing countries, have reached their peak: Canada in 1973, Iran in 1974, Nigeria in 1979, Soviet Union (excluding Kazakhstan) in 1987, Alaska in 1990, Egypt in

---

[60]   Tar sands are naturally occurring mixtures of sand or clay, water and an extremely dense and viscous form of petroleum called bitumen. They are found in large amounts in Canada and Venezuela.

1996, Venezuela in 1998, Argentina in 1998, Great Britain in 1999, Qatar in 1999, and Norway, Colombia, India, Mexico and China in 2000 (Helmfrid and Haden 2006). It is rather obvious that when so many countries have reached the peak the global peak is also reached. According to a study by Fredrik Robelius (2007), the global peak will be reached sometime between 2008 and 2018. This is based on analyses of the 507 giant oil fields, each of which will produce more than 500 million barrels of oil. They represent some 65% of all oil reserves and their status is therefore a good indicator of how production might develop. The study also includes tar sands and some other less accessible oil fields (Robelius 2007).

It is likely that oil prices will rise, sometimes as sudden hikes, but also as a longer-term trend. Meanwhile, we should be aware that oil and other energy sources so far have become cheaper relative to salaries in developed countries[61] for a very long period. If oil prices are constantly high, other technologies will be commercially interesting. The oil crisis of 1973 led to a shift away from oil in many countries. The Swedish industry used 75 TWh of oil in 1970 and only 20 TWh of oil in 2000 (Eklund 2000). The security costs[62] of oil that occur are another issue. Porritt (2007) claims that the costs for the United States to maintain safe oil supplies from the Persian Gulf are two to three times as high as the value of the oil. Higher price means that new sources, where extraction is more difficult, will be used. But extraction is not only more difficult and expensive; the energy return goes down as more energy is needed to get the oil out. The EROEI of oil went down from 100 in the 1930s to 17 in the early 2000s (Hall *et al.* 2003). The implication of this was discussed earlier and we will come back to the discussion in Chapter 32.

Oil is not only an important energy source it is also the basis of most of the chemical industry, being the basis for paraffin, waxes in cosmetics, adjuvant in paints, medicines and many other products. The plastic industry is also based on oil as well as nylon, elastane and most other modern fibres. Most chemicals, CD records, grease and propelling gases in hair spray can be traced to oil. In total, 567 Mtoe oil was used

---

[61]   This is obviously different for countries where salaries have not increased a lot.

[62]   To call massive wars for 'security costs' is perhaps an inexcusable euphemism.

for other purposes than energy in 2007, compared with the total 4107 Mtoe oil used; that is, 12% of all oil was used for industrial purposes (IEA 2008). Oil as a raw material will be more valuable for future generations than oil as an energy source.

The reserves of coal are substantial, many times bigger than those of oil, and coal can replace most of the needs for oil. It is dirtier and some of the uses will be more costly; to convert coal to a liquid fuel is not that difficult, but it will cost more than petrol. Still, it can keep things going for another 100 years. On the other hand, the energy return for coal is rapidly going down, and will be even worse if carbon capture and storage are used to mitigate carbon dioxide emissions.

Out of the rocks

In a similar way as for oil, there is a debate about whether we are running out of minerals. For many rare earth metals like indium and rhodium, more than 80% of known reserves have been mined in just the past three decades. Global demand for metals like copper and aluminium has doubled in the past 20 years (UNEP 2010a). The dramatic price hikes between 2006 and 2008 strengthened the view that there is indeed a real risk that we will run out of some minerals. According to some calculations, antimony, an important component of flame-retardants, will only last another 15 years and silver only 10 years. Indium and hafnium, both important for integrated circuits (for computers), could end in 2017, zinc in 2037 and terbium before 2012. There are, however, some indications of the stagnating demand for some metals; for instance, steel consumption in the United States has been flat since 1980. For some, however, demand is rapidly increasing, for example, for copper, despite fibre optic cable replacing copper for data and telephone (Cohen 2007).

Known reserves for most metal ores are enough for many decades according to the European Commission, but the reserves of copper, lead, zinc and silver represent less than 30 years of consumption. The European Commission (2003) says in 'Towards a thematic strategy on the sustainable use of natural resources' that even if a few decades sound little one has to keep in mind that the known reserves will always be a just a part of the total physical reserves. The reason is that, as for oil, there is little prospecting for new findings unless the known reserves are running out. There are several other reasons for why 'reserves of non-renewable resources tend not to decline.' First, process improvements lead to the extraction of a greater proportion of the resources present in reserves. Second, we can make more out of a ton of steel today than we could a century ago. This means resources are used at a slower rate than past or current consumption patterns suggest. Third, improvements in exploration and extraction techniques mean that today we can exploit reserves that were previously unknown or not considered viable. Fourth, some materials can be recycled to a very high degree, such as aluminium; increasing recycling rates mean that some resources are used at a slower rate than past or current consumption patterns would suggest (European Commission 2003).

Metals such as iron and steel, copper, aluminium, lead and tin enjoy recycling rates of between 25% and 75% globally (UNEP 2010a). In developing countries, formal documented recycling rates are lower, whereas informal waste collectors are very efficient in unearthing the last pound of valuables from waste. This informally collected, but not statistically documented, scrap finds its way into the formal sector (Hoffmann 1999). Even if we encounter shortage, there is also widespread substitution. When one raw material becomes too pricey because of scarcity or increased cost for mining or recycling, new technology is stimulated.

Apart from the aspects of shortage and price, many of the minerals are also dangerous, in particular the heavy metals. Asbestos, radium, arsenic, cadmium, mercury and uranium are among the most danger-ous substances let out in big volumes all around the globe and it needs to be ensured that they are not circulated in nature. It is rather easy to organize recycling for expensive metals that are used in big quantities in certain applications and a lot harder to organize it for minerals used in a low concentration in consumer products. To recycle copper from electrical cables and water pipes is very simple compared to recycling mercury from low-energy lamps, cadmium from plastic or anything from the flame-retardants that most consumer products are soaked in.

High prices and scarcity can also lead to conflicts or at least keep conflicts going. For many developing countries, it seems to be a curse rather than a blessing to have rich mineral resources. 'Blood diamonds'[63] are well known. Perhaps less known, but not less pro-blematic, is the role tantalum, an important mineral for cell phones, plays in the eternal conflict in Congo (Bond and Braeckman 2001).

---

[63]   'Blood diamond' refers to a diamond mined in a war zone and sold to finance an insurgency, invading the army's war efforts, or a warlord's activity, usually in Africa.

China has a unique position when it comes to rare earth minerals. The former leader Deng Xiaoping compared China's wealth of those rare minerals with the Middle East's unique role for oil. Of a total of 124,000 tons of rare earth minerals in 2008, China produced 120,000 tons (SvD 2009b). As with oil, this can have big consequences for security. Many of the rare earth metals and minerals are crucial for new technologies such as renewable energy and advanced batteries. For instance, indium is used in semiconductors, energy-efficient LEDs, advanced medical imaging and photovoltaic elements. But we don't have to go to exotic minerals to see how growth and exploitation create scarcity; even sand can become scarce. Some countries in South East Asia, for example, Singapore *imports sand for construction* and development. This has led to export restrictions in neighbouring countries and Singapore now imports sand from the United States (*The Economist* 2009b).

Recycling of metals has come far and is stimulated by increasing prices. Copper is recycled to a very large extent. Worldwide stocks of metals in society have grown such that there is copper 'above ground' equal to 50 kilogrammes per person. Since 1932, the amount of copper per person in the United States has grown from 73 to close to 240 kilogrammes. The lifetime of copper in buildings is 25–40 years whereas in personal computers and mobile phones, the in-service lifetime of the metal is less than 5 years. Existing buildings can become mines, where plumbing can be swapped for plastic tubes, to satisfy our ever-increasing hunger for copper. Serious proposals have been made to recycle platinum from road dust. Many metals can be produced from seawater; others can be brought from the moon. Of course, all those methods will lead to higher prices or, rather, they will only be commercially viable if prices go up (Bond and Braeckman 2001; UNEP 2010a).

Only when scarcity is a fact do prices go through the roof

All in all, we should not exaggerate the problem with scarcity of minerals or that we will run out of them. The reason to be worried is not that we will run out of a few of them, certainly we will, or at least they will become very scarce and thus prohibitively expensive; the reason not to be too worried is rather human ingenuity which is the real cornucopia. There is, however, a fundamental problem of how the market economy manages, or rather doesn't manage, prices of natural resources. It is first when scarcity is already here that prices increase, but then it is often too late to change consumption or find alternative solutions. When we discuss resources, it is also reasonable to discuss *whose resources*? We come back to that discussion in Chapter 22 in Part 3.

There is also a strong link between the availability of energy and other mineral resources. Often, there are technical possibilities to get some minerals from very diffuse sources, such as seawater, but the energy costs to do so are very high. Despite massive progress in technology the energy cost of producing copper has increased in the United States because the best ores are long gone (Hall and Klitgaard 2006). Similarly, one can see that energy return in nuclear power diminishes rapidly when uranium concentration in the ore goes down. When fossil fuel gets scarcer, minerals de facto also get scarcer, as it is simply too expensive to extract them. It is cheap energy that has been driving the raw material explosion. Meanwhile, it is also cheap energy that ultimately will drive up mineral prices (Johansson and Lönnroth 1975). Human use of minerals should aim at reduced consumption, substituting for dangerous minerals or those that risk becoming scarcer, reduce losses and increase recycling. We need to see those minerals as capital that we exhaust. First when we do that will we accept that it is inexcusable to use all resources in a century or to and just leave waste heaps to future generations. The capitalist market economy presents no viable or sustainable option for management of dead natural resources as little as for management of the living and the ecosystems.

– 9 –
The Chemical Society

What we have to face is not an occasional dose of poison which has accidentally got into some article of food, but a persistent and continuous poisoning of the whole human environment.

(Rachel Carson,
Silent Spring, 1962)

Chemicals and the chemical industry have contributed to a number of positive things in the development of human society. Sometimes they have contributed to considerable savings of other natural resources; sometimes they have contributed positively to human health (medicines) or made daily life easier (plastics). There are a number of problems with them, however, most of which are related to them simply not being 'natural'. One can wonder why it would matter whether or not something is 'natural'. But there are simple reasons for why natural is a safer bet.[64] All chemical components in nature are produced by processes that are part of nature and, therefore, there are also processes that break them down. What is waste from one organism is food for another; organisms have adapted themselves to the concentrations of the many substances in their environment, or made way for other organisms that are better adapted. No organism has the job of breaking down the synthetic chemicals human society releases. Some chemicals are so similar to natural products that they easily enter the circulation, and the human body, or the body of another organism, takes them up and confuses them with other natural substances. This is the case for endocrine-mimicking substances. There is also a group of toxic chemicals that do exist in nature, but normally in concentrations that life has adapted itself to. Among them are cadmium, mercury and radioactive substances such as caesium and plutonium.

---

[64]   A safer bet doesn't mean that natural things are always harmless; there are plenty of toxic natural substances, and some organisms have natural chemical warfare against others.

The chemical industry produces more than 100 million tons of chemicals per year; this corresponds to 17 kilogrammes per person (Vitousek *et al.* 1997). Most people seem to believe that the chemicals used in shampoos, clothes, building materials or toys are, in some way, tested for how they impact human beings and nature. The truth is they aren't. No country in the world requires the testing of chemicals before bringing them to the market. If a chemical is suspected to be harmful, it is instead the governments that have to prove, through costly and long-term experiments or studies, that it is dangerous for human beings or the environment. Considering the many chemicals that are spread and used, nobody can give a good summary of the situation or make a risk assessment. Anyone claiming to do so is most certainly an impostor. Below a number of examples of the chemical society are given; they represent only a ladle of a sample of lethal brew.

More than 100,000 chemicals

In the European market there are more than 100,000 chemicals. Around 30,000 are produced in quantities exceeding 1 ton per producer, and of these 2500 are high-volume chemicals produced in quantities more than 1000 tons per producer. The knowledge about these chemicals is fearfully little. Less than 5% of chemicals have been tested to any larger degree, which means we know nothing or very little about the other 95%, and still not enough about the 5% (SNF 2008). Only a few 100 of the more than 80,000 chemicals in use in the United States have been tested for safety (President's Cancer Panel 2010). Around 10% of all chemicals are suspected to be carcinogenic (Pimentel *et al.* 1998). Some 1000 chemicals are neurotoxic and 200 of them have caused direct damage on human beings. There are suspicions and fears that the combined effect, the so-called cocktail effect, of chemicals used in doses below damage thresholds can be the cause of several behavioural disturbances, such as attention deficit hyper-activity disorder, commonly known as ADHD (Rockström *et al.* 2009).

As of 1 June 2007, REACH,[65] the new EU directive for chemicals is in place. Over a 15-year period, producers will have to get data for the most used chemicals and seek permission for the marketing of them. The most dangerous ones will be taken out of the market. All chemi-

---

[65]   REACH stands for 'Registration, Evaluation, Authorization and Restriction of Chemicals'.

cals that are produced or imported in quantities surpassing 1 ton per company have to be registered and the industry has to supply basic information (SNF 2008). REACH will reduce risks but certainly not do away with them.

Already the planned use of chemicals is problematic, but they also appear where they shouldn't. For instance, it is a good idea to recycle food waste and human effluents from the cities to farms to close nutrient cycles. Sewage sludge is rather toxic today, however, both because sewage lines are not separated and because so many chemicals are used in daily life. In this way, a product that should be good for the land becomes a toxic waste and nutrient cycling is made impossible or undesirable.

DDT (dichlorodiphenyltrichloroethane), which has been (and still is) used as a pesticide, and PCB (polychlorinated biphenyl), which has been used in condensers and transformers, have been symbols for careless and irresponsible use of chemicals. Both are now spread all over the globe and their properties have made them accumulate at higher levels of food chains (e.g. in human beings, seals, and falcons). They were banned in most industrial countries in the 1970s and concentrations are slowly declining. Unfortunately, the World Health Organization (WHO) and the global campaign against malaria are reintroducing DDT as a component in malaria prophylaxis in Africa.[66]

New chemicals are introduced all the time. One such is PFOS (perfluorooctanesulphonic acid), a man-made fluorosurfactant. High concentrations of PFOS have been found in fish-eating mammals and birds in the northern hemisphere. PFOS is also a derivative from many other chemicals used in the electronic and mechanical industries and in detergents, impregnating agents and flame-retardants. The properties of PFOS are similar to those of DDT and PCB. It disturbs metabolism, impedes growth and damages the liver. In addition, several important cell functions are impeded, which in turn is suspected of causing cancer (OECD 2002).

Toxic products can reach the environment in mysterious ways. Nonylphenol is not easily degradable and is accumulated by plants

---

[66]   If the choice to weigh the loss of life from malaria with the potential danger of DDT was so easy, yes then it would most likely make sense to use DDT for malaria eradication. But the choice is not at all that simple; there are other substances and other methods to use and in the end the mosquitoes that carry malaria are likely to develop resistance to DDT.

and animals. It is very toxic for aquatic organisms and has long-term effects on the environment. It has endocrine-disrupting effects too; for example, male fish can get female properties. Within the European Union, the use of nonylphenol as well as the precursor nonylphenol ethoxylate is prohibited in any operation where there is a risk of them being released into nature. For almost 10 years, Swedish waterworks wondered why the levels of nonylphenol continued to be high. In 46 of 92 waterways, one could find increased levels of nonylphenol, despite the ban. Finally, the source was discovered: the washing of clothes imported from outside the European Union (SNF 2007; SvD 2007). Nonylphenol is also an adjuvant, a so-called inert ingredient, in a number of pesticides. The industry normally doesn't report the concentration in pesticides, but in California pesticide users must report the use of adjuvants. More than 1.5 million pounds of adjuvants that contain nonylphenol ethoxylate are used annually in California (Cox 2003).

The Hiroshima of the chemical industry
Death came out of a clear sky. From the factory a thin plume of white vapour began streaming from a high structure. Aziza Sultan, a survivor, remembers: 'At about 12.30 am I woke to the sound of my baby coughing badly. In the half light I saw that the room was filled with a white cloud. I heard a lot of people shouting. They were shouting "run, run". Then I started coughing with each breath seeming as if I was breathing in fire. My eyes were burning.' More than half a million people were exposed to Carbide's poison gases. When dawn broke over the city, thousands of bodies lay in heaps in the streets. The army dumped hundreds of bodies in the surrounding forests and the Betwa river was so choked with corpses that they formed log-jams against the arches of bridges. Families and entire communities were wiped out, leaving no one to identify them. How many thousands died, no one knows. Carbide says 3,800. Municipal workers, who picked up bodies with their own hands, loading them onto trucks for burial in mass graves or to be burned on mass pyres, reckon they shifted at least 15,000 bodies. Survivors, basing their estimates on the number of shrouds sold in the city, conservatively claim about 8,000 died in the first week. The official death toll to date stands at more than 20,000. (Bhopal.net 2010)

Chemicals don't have to be toxic to create problems. One such example is chlorofluorocarbons (CFCs). Many CFCs have been widely used as refrigerants, propellants (in aerosol applications) and solvents. The release in the atmosphere of such compounds contributes to ozone depletion in the stratosphere. In 1990, developed countries agreed to complete elimination of CFCs by the year 2000. By 2010, CFCs should be completely eliminated from developing countries as well. CFC elimination is one example of a fairly successful process to phase out a chemical, in a relatively rapid[67] and forceful manner. One reason it worked was that there were alternatives that were not dramatically costly or inefficient, and in that situation the interest of a few CFC producers weighted little in the balance, especially as the same producers could make the alternatives.

Another very disturbing class of chemicals are the endocrine-disrupting and endocrine-mimicking compounds. They are normally not toxic in the common meaning of the word, but they cause a lot of problems such as birth defects and fertility disorders. As the authors of *Our Stolen Future* write: 'Hormonally active synthetic chemicals are thugs on the biological information highway that sabotage vital communication. They mug the messengers or impersonate them. They scramble messages. They sow disinformation. They wreak all manner of havoc' (Colborn *et al.* 1997: 203).

Perhaps the hormone-mimicking compound *bisphenol A*, a ubiquitous compound in plastics, will be the next everyday chemical that will prove its nasty effects. First synthesized in 1891, the chemical has become a key building block of plastics from polycarbonate to polyester; in the United States alone more than £2.3 billion (1.04 million tons) of the chemical is manufactured annually. It is used in polycarbonate plastics used, for example, in bottles for baby instant formulas and in the lining of cans. Since 1936 at least, it has been known that bisphenol A mimics oestrogen, binding to the same receptors throughout the human body as natural female hormones. Tests have shown that the chemical can promote human breast cancer cell growth as well as decrease sperm count in rats. A host of studies show that it leaches from plastics and resins when they are exposed to hard use or high temperatures (as in microwaves or dishwashers). One hypothesis is that it can influence fat metabolism already at low doses. It could lead to obesity and thus cause diabetes and cardiovascular disease (*Scientific American* 2008; Arbets och Miljömedicin 2009).

There are many other candidates for the next scandal, such as short-chain chlorinated paraffin, which persists in the environment,

---

[67]   Ozone depletion was known already in the 1970s, so the reaction was not *that* rapid; but in the world of global policy, a 15-year period between seeing a problem and initiating actions for solving it is still quick.

accumulates in human breast milk, can kill small aquatic creatures and has spread over most parts of the globe. Polybrominated diphenyl ethers (PBDEs) used as flame-retardants, found in many different types of plastics and plastic-containing products, are powerful thyroid disruptors that are both persistent and bioaccumulative. It turns out that when they leach out of the plastics in which they are put, they quickly find their way into the biosphere. One prominent pathway is polyurethane foam. When it degrades, the most toxic form of PBDE (penta-BDE) escapes. In the United States, PBDEs are not only passively transported via water and air currents, but are also actively transported and dispersed because they are a prominent contaminant in sewage sludge, which spreads onto farmers' fields. Since the 1970s, PBDEs have increased more than 50-fold in breast milk (*Our Stolen Future* 2010). And there are of course pesticides, which are extra problematic as they are applied to human food (see more in Chapter 12), and medicines.

A medical cocktail in drinking water

Medicines, just as pesticides, are intentionally designed to influence organisms and biological functions. Many antibiotics pass through our body intact or as metabolites. Very little research has been conducted of the effects on the environment, because the authorities have not asked for this. Further, a lot of medicines are thrown away and drug manufacturers may cause substantial pollution. A study conducted in 90 companies in and around Hyderabad, India, in the period 2006 and 2007, found that levels of harmful emissions from most drug factories were extremely high. Analyses showed the presence of drugs including the antibiotic ciprofloxacin, at concentrations of up to 31,000 microgrammes per litre—a level 1000 times higher than the limit of toxicity for some bacteria, and the antihistamine cetirizine, at up to 1400 microgrammes per litre. Some of the factories released up to 45 kilogrammes of ciprofloxacin per day into the nearby river—the equivalent of 45,000 daily doses (MistraFarma 2010). The use of synthetic hormones is on the increase and they pass both the human body and the sewage plants and affect living organisms, often with fertility disturbances as a result.

There is a discussion that the medical industry should shift to compounds with short 'eco-shadow', for example, compounds that break down easily and don't create havoc in nature. But the industry is far from it. In Sweden, in 2005, only 27 out of 1200 active medical substances had been examined for their environmental effects by the

Swedish Medical Products Agency. There are, therefore, enormous knowledge gaps about the effects of drugs in nature (Region Skåne 2005). Most countries don't systematically follow drug residues in drinking water and there are also no maximum residue thresholds. In the United States, at least one contaminant was detected in 75% of the groundwater wells tested, in virtually all the streams and stream sediments tested, in about 80% of the estuarine sediments tested, in about 80% of the freshwater fish tested and in nearly all of the salt-water fish tested (H. John Heinz III Center 2008).

Of 62 big waterworks in the United States, only 28 had tested their water for drug residues by the mid-2000s. New York, Houston, Chicago and Miami had never tested their water. Those that did test found a disturbing reality. In Philadelphia, 56 different drugs were found in drinking water and 63 in the water source. In San Francisco sexual hormones were found in drinking water, and in Washington 6 different drugs (*USA Today* 2008). Among the human medications found in water in the United States are antidepressants, medications for high blood pressure and diabetes, anticonvulsants, steroid medications, oral contraceptives, hormone replacement therapy medications, codeine, non-prescription pain relievers, chemotherapy drugs, heart medications and antibiotics (President's Cancer Panel 2010). Purification of water to get rid of the drugs might cost in the range of US$ 200 per inhabitant per year (*UNT* 2008).

Many of the antibiotics used for animals are badly absorbed by the animals and go through them; by means of manure and urine they are spread in the environment. A study by the US Geological Survey found antibiotic residues in half of the surveyed waterways. Off the coast of Norway, 200 tons of chinolones were found in sediments originating from fish farming. These antibiotics are not readily biodegradable and will stay for years (Apoteket 2005). The carpet of antibiotics that we roll out into the environment, purposely, like human medication and medication of animals, and inadvertently, as in the drinking water, is most certainly a reason for the increased resistance of some bacteria to antibiotics, a resistance that threatens to nullify the gains of one the most important medicines ever found (see more in Chapter 22).

Russian roulette

It is simply not possible to know which of the 100,000 chemicals will be in focus for the next scandal. The only thing we can know for sure is that *there will be a next scandal*, and a next, and a next. As long as be have an industry that churns out new chemicals by the hour and markets them without comprehensive testing we are playing a dangerous game—a game with our health, our fertility and nature as stakes. But we don't know the stakes beforehand, and we don't know the risk of failure, a worse form of Russian roulette, where at least the risks are well known and statistically proven. We know even less about the combined effects of those chemicals.

That the chemical industry is driven by profit motives makes the whole equation even more worrying. Even the most adamant proponent of free markets must admit that government regulation is absolutely essential, to ensure that we are not poisoned by chemicals, and that market forces simply can't deal with long-term side-effects on consumers of products brought to the market. The idea that it should be possible to 'get the price' right by internalizing costs in the product doesn't work at all when it comes for chemicals. While chemicals themselves are the most explicit and extreme example, the problem is generic for new products brought into the market in a capitalist market economy; there are no self-correcting feedback loops to ensure that harmful products, or those whose use is harmful, in the long run are restricted.

– 10 –
Water and Land

We abuse land because we regard it as a commodity belonging to us. When we see land as a community to which we belong, we may begin to use it with love and respect.
(Aldo Leopold,
Sand County Almanac, and Sketches
Here and There, 1949)

In this chapter, we look into two fundamental resources: water and land. Both are, together with energy and ecosystems, absolutely fundamental for survival; the past and the future of humanity rest with the management of these four resources.

Water

Each human being needs a few litres of water for drinking and cooking, and an additional 25–50 litres for hygiene and laundry. In 1991, 3.4 billion people had to do with less than 50 litres per day and many with much less. I have seen women in Mozambique in the mid-1990s doing their laundry in potholes in the road after a rain. The average American was reported to use 350 litres per day and the Australian 570 litres, ironically so as Australia is the driest continent of all (Tansey and Worsley 1995). These figures don't tell the full truth as countries that import water-consuming raw materials are importing 'embedded' water from other places. For instance, direct consumption of water in the United Kingdom is only half of the quantity that has been used to produce things imported to the country (BBC 2010).

At the moment, human beings are directly using some 10% of all fresh water, and indirectly almost all of it. Certainly, the rainfall on forests, farms and rangelands is used to the benefit of society. In the Middle East and North Africa, 120% of all fresh water is used annually. That one can use more than 100% is explained by a substantial use of fossil water, water left from earlier epochs, which is not recharged by rain in present times, and by depletion of groundwater quicker than it is recharged. Agriculture scientists tend to believe that water use by human society, be it for households, industry or farming, of more than 40% is problematic, whereas ecologists tend to think that this is already far too much.

The tapping of water from rivers and lakes for irrigation and for city and industrial use was doubled between 1960 and 2000. Seventy percent of all water use is for farming. The irrigated area has increased from 210 million hectares to 277 million hectares between 1979 and 2003 (MEA 2005). The drawing of water leads to rapidly falling groundwater tables, in great parts of China and India up to 1–3 metres per year (FAO 2003). In Australia, the situation is precarious and the government's programme 'Water for the Future' allocates some A$ 12.9 billion to secure water resources for all Australians, but also for the environment. In the Murray-Darling basin alone, A$ 3.1 billion will be used to buy back water rights from farmers so that the water can be used for other purposes, mainly to maintain fragile and threatened ecosystems (Australian Government 2008).

Table 10.1 Countries with high dependency on water from outside.

Region	Countries getting 50–75% of their water from outside	Countries getting more than 75% of their water from the outside
Middle East and North Africa	Iraq, Israel, Syria	Bahrain, Egypt, Kuwait
East Asia and the Pacific	Kampuchea, Vietnam	
Latin America and the Caribbean	Argentina, Bolivia, Paraguay, Uruguay	
South Asia		Bangladesh, Pakistan
Africa South of the Sahara	Benin, Chad, Congo, Eritrea, Gambia, Mozambique, Namibia, Somalia, Sudan	Botswana, Mauritanian, Niger
Eastern Europe and Central Asia	Azerbaijan, Croatia, Latvia Slovak republic, Ukraine Uzbekistan	Hungary, Moldova, Montenegro, Romania, Serbia, Turkmenistan
The OECD countries	Luxembourg	Netherlands

Source: UNEP (2009).

The word 'rival' comes from Latin *rivalis*, meaning a person sharing the same stream (water). Thirty-nine countries currently get most of their water from sources outside their territory (UNEP 2009; see Table 10.1). In many cases, there are tensions between countries that share the same resources, and conflicts over water are likely to increase.

Irrigation plays a big role in food production. Probably no single investment in farming so clearly 'pays off' both economically and in terms of increased productivity. At the end of the 1990s, irrigated areas represented one-fifth of all agriculture areas, but produced around two-fifths of all crop and almost three-fifths of all grain. FAO estimates that 402 million hectares of land in developing countries can be irrigated, of which only half of it is irrigated today. In Africa South of the Sahara, only 4% of farmland is irrigated and the area has only increased by 4% in the last 40 years (World Bank 2007). Approximately, two-thirds of all rivers in the world are used for irrigation or hydroelectricity; in the United States, only 2% of rivers flow freely. Great rivers like the Colorado, the Nile and the Ganges have almost no water left when they reach the sea. Great lakes such as Lake Chad and the Aral Sea have shrunk tremendously. In the case of the Aral Sea, this affected both the fisheries and the whole climate around it, with sand storms and disease following (Vitousek 1997).

There are several limitations to the expansion of irrigation, however. One limitation is simply access to water; another one is the loss of biodiversity from the lands that can be irrigated, which today mostly are swamps or other wetlands. If wetlands are drained they will normally also be converted from carbon sinks to carbon emitters, thus driving climate change. In areas of high evaporation, salinization can be an additional problem. So there are reasons to believe that irrigation will not expand as much as theoretically possible. There are also many opportunities to improve water stewardship in existing irrigation schemes or to introduce small-scale technologies such as rainwater harvesting[68] in farming systems.

Pricing of water in most countries represents an actual subsidy of farming over other sectors. At the end of last century, global water subsidies were estimated to be US$ 33 billion, a fact that certainly reduces the incentive of farmers to save water (Worldwatch Institute 2000). Of 12 OECD countries, only the Netherlands and Austria priced water at a rate that could be called commercial.[69] Household and industries paid between US$ 0.5 and US$ 3 per cubic metre, whereas

---

[68]   Rainwater harvesting is the collection of rainwater to provide drinking water, water for livestock or water for irrigation.

[69]   Which still in this case only means that the costs for extraction and distribution is recovered, not that a price has been allocated to water itself.

farmers paid a few cents or even fractions of a cent. Now, irrigation water and drinking water are two different qualities of water, but the difference in price doesn't represent that quality difference. Even in a dry country like Australia, households paid almost 100 times as much as farmers did for their water (OECD 2001).

That water subsidies are sustainable neither for the environment nor for financial reasons is shown again and again. In the 1970s, Saudi Arabia supported massive expansion of grain production on the basis of subsidized irrigation. It was, by and large, very successful. In the 1990s, it became clear, however, that this wasn't sustainable. In some areas, groundwater tables dropped 5 metres per year and some farmers were drilling to 1000 metres to reach water. In response to this, production of wheat dropped from 4.5 million tons in 1992 to 1.8 million tons in 1993, and the production of barley dropped from 2.2 million tons to just 100,000 tons by 2002. In 2002, the Ministry of Agriculture announced that it would cease feed production altogether because of its negative effect on water resources (USDA 2002). In Santa Fe, New Mexico, the local government 'recently began requiring home builders to retrofit six existing houses for improved water economy to offset the added water use in each new house built' (Ruddiman 2005: 193). Las Vegans are paid to rip out their lawns and opt for (chic) desert landscaping (*The Economist* 2011a).

While a human being needs a few litres of water to drink, at least 1000 times as much water is used for production of food (Kijne *et al.* 2009). The water needed for foods varies tremendously and varies for the same product under different conditions. In Australia, the following quantities of water are needed to produce a kilogramme of a product:

1 kilogramme of wheat: 715–750 litres of water

1 kilogramme of corn: 540–630 litres of water

1 kilogramme soy: 1650–2200 litres of water

1 kilogramme rice: 1550 litres of water

1 kilogramme beef: 50,000–100,000 litres of water

1 kilogramme wool: 170,000 litres of water[70] (Meyer 1997)

---

[70] Notably the figures for meat and wool also include water that falls as rain on the grazing areas and the feed-producing areas for animals. Most of the

Other figures are reported elsewhere; in the United Kingdom, a kilogramme of beef is said to embed 15,000 litres of water (BBC 2010). That a certain food needs a lot of water is not an absolute reason to avoid that food. Nevertheless, a discussion on embedded water, or virtual water as it is also called, is obviously more and more relevant the more scarce water resources are seen to be. One weakness in the concept of embedded water is that it equates all sorts of water. There is 'blue' water — water in rivers and lakes, 'green' water — water in rainfall and in the soil, and 'grey' water — waste water; there is also fossil water in aquifers. Concentrated water in a lake is a more useful resource than rain.[71]

All in all global water supply is strained and, locally, water is already now a limiting factor for development. There are many ways to improve water management and the efficiency of water use, ranging from simple solutions such as rainwater harvesting to high-tech recirculation of irrigation water in greenhouses. To reduce wasteful consumption, subsidies should be abolished and perhaps water use itself should have a price tag.

'The fluid nature of water has always made it difficult to turn into private property,' writes Radkau (2008: 86), but then what increasingly is turned into property are water rights. The oil tycoon and investor T. Boone Pickens has bought up water rights of the Ogallala aquifer in the United States for some US$ 100 million and expects to sell water for some US$ 165 million per year once the thirst of Dallas is big enough (*Bloomberg Businessweek* 2008). Not surprisingly, this has triggered public outrage, regulatory responses and lawsuits. Parallel to private encroachment on water rights, maintenance of the water catchment areas is rapidly turned into a commercial service by means of 'watershed service payments'. In 2008, there were some 300 schemes with payments reaching a total of US$ 10 billion (Stanton *et al.* 2010). In the last decades, water utilities have increasingly been privatized as well.

The thin skin of the planet

Human existence is dependent on the thin skin of 20–50 centimetres of topsoil. This topsoil is one of the most important resources and it is

---

grazing areas are far too dry to be used for any farming.

[71]   A bit similar to discussions about energy where one can't equate heat energy to electricity.

under constant threat. It takes between 60 and 1500 years to create a centimetre of productive soil (Montgomery 2007). Stones that are eroded by wind, water and plant roots contribute with minerals; dead organisms provide food for the special life forms of the topsoil, such as worms. They in turn help decomposition and free nutrients for new growth. Soil formation is an example of a process with positive feedback loops, a self-reinforcing process, where a good soil will reproduce itself and actually improve over the years. The same goes for soil erosion; initially, erosion makes the soil vulnerable to more erosion until it might be completely washed away. Unfortunately, soil erosion is a process that can be rather quick compared to soil formation (Montgomery 2007).

Some erosion is unavoidable, it is a natural process, but human beings have increased the rate of erosion 10, 100, even up to 1000 times. With the introduction of the plough, soil erosion outpaced soil formation in many soils (Montgomery 2007). Depletion of agricultural soils has been a determining factor for the decline of many societies. As long as there is 'new' land available to farm, or there is forest that can be brought into periods of production and fallow, farmers' interest in caring for the land is normally low. In cultures and societies where land resources have been scarce for a long time, farming practices are mostly more sustainable. At least this was the case before the introduction of pesticides and chemical fertilizers. This might sound a bit contradictory for the layperson who would expect that extensive farming would be more sustainable. But it just follows the same logic as management of many other resources: when resources are scarce or expensive they are managed better. The United States provides a good case study of how plenty of land led to bad management. Since independence two centuries earlier, one-third of the American topsoil is lost. Up to 75% of the surface of New England was farmed, but land was depleted and large tracts were left again. In Georgia, agriculture area shrank by 38% from 1930 to 1970 as land was depleted and farming moved westwards (Cunfer 2005). This movement was arrested first by the lack of land to be ploughed and then by the availability of chemical fertilizers.

Agriculture represents the predominant form of human land use, corresponding to 37% of all land (FAO 2007; see Table 10.2). The real impact of agriculture on land use is bigger as at least 50% of all land now is within agricultural 'biomes' (see Table 10.3). Since 1860, more than a billion hectares of farmland have been brought under the plough (Montgomery 2007). Agriculture land area is still increasing in developing countries but the increase has slowed down lately. During the period 1960 to 2000 it grew by 5 million hectares per year, but today it is down to 3–4 million hectares per year (FAO 2003). Thus, agriculture land doesn't at all expand at the same rate as population. There are many reasons for this. One is simply that yield increase per area unit has been substantial and there has been no need, or commercial incentive, to increase farmed area. The other is that most of the good land is already occupied. The bringing in of new land is actually higher than the figures show, because a lot of agriculture land is lost in exploitation for human infrastructure (see more below). A lot of this lost land has been among the best farmlands as big cities normally are located in rich farming areas and when they grow, they grow on prime agriculture land. The other, quite different trend is that marginal lands are taken out of production (see the example of Turkey in Chapter 11). Some of that marginal land is converted to forests, some is just left idle and some of it has eroded to the extent that it is barren. To compensate for this, other land has been brought to use.

Table 10.2 Development of agriculture land from 1980 to 2000 (billion hectares).

	Arable land	Permanent crops	Pasture	Total farming	Hectare/ person
1980	1.345	0.102	3.244	4.691	1.05
1990	1.395	0.119	3.368	4.882	0.92
2000	1.397	0.135	3.442	4.974	0.81

Source: FAO (2007).

The rate of agriculture land per inhabitant is falling in Africa South of the Sahara and in South Asia. But hectares per person don't say too much. The 'natural' yield varies by almost a factor of 10 from the best to very low producing land. In addition, high-income countries 'use' land in other countries. Georg Borgström (1976) was most likely the first to raise this through the use of the concept of 'ghost areas' (now called footprint) and showed in 1962 that the Dutch used three times more land in other countries than in their own.

Changes in agriculture land use

The change in land use has many fronts: conversion of forests and wetland to agriculture land; loss or degradation of agriculture land through human infrastructure and erosion; transformation of pasture land either to forests or to other 'natural' lands; or the ploughing of pastures for agriculture often in combination with irrigation. In many developing countries forests are converted to farmland, whereas the process is the reverse in high-income countries and, increasingly, also in middle-income countries.[72] In Denmark, the proportion of land under forest cover has increased from a low of 4% in 1800 to 11% today. Much of this expansion has taken place on agriculture lands. In Sweden, forest coverage increased from 40% in 1875 to 53% today. In the same period, arable land increased first from 6.7% to 9.4% in 1920 but then dropped to 6.7% in the early 2000s. Pastures and meadows decreased markedly (SCB 1959, 2008). In West Africa, there is almost no natural forest left, whereas big parts of the forests of the Amazon, Congo and Borneo are still intact but under pressure. Countries like Thailand, China and India have, despite increasing population and strong economic growth, increased their forestland in the same way as Europe and Japan did before.

Draining of wetlands for farming has been very important for agriculture. The Po valley and the deltas of the Mississippi and the Nile are all examples of drained wetlands. In Denmark, some 200,000 hectares of lakes and wetlands were converted to farmland (Danish Ministry of Environment 2005). Since the mid-1950s to mid-2000s, the freshwater wetland area in the United States has declined by 9% and more than 160,000 hectares of vegetated wetlands on the Gulf and Atlantic coasts were lost, a decline also of about 9% (H. John Heinz III Center 2008). In high-income countries the draining of wetlands for farm expansion is no longer taking place; on the contrary, many wetlands are restored, for example, in the Netherlands and in the United States (Everglades). This is motivated by care for biodiversity, to reduce leaking of nutrients to the sea, to buffer floods and now also for carbon sequestration.

In developing countries, there are some 300 million hectares of wetland that could be converted to farmland. There are many obstacles however. One is the land itself, where draining can be difficult or the drained land would have too high concentrations of iron, aluminium or sulphuric acid all of which can be toxic to crops. There are also high environmental and economic values in these lands; for example, they might be breeding areas for fish. Wetlands act as buffers for sudden water flows and have therefore a big role to play in flood mitigation, bigger than forests in most cases. In China, draining of wetlands represents two-thirds of the reduced buffering capacity against floods. This

---

[72]   If this historical pattern holds, the Brazilian expansion into the Amazon will soon cease.

reduced capacity led to the floods of 1998, causing damages of US$ 20 billion above immense human suffering. Similar causality exists for the severe floods in the United States in 1993 (FAO 2003). We already discussed the role of wetlands as carbon sinks. We should therefore exercise caution in further draining of wetlands.

Soil organic matter is what makes the difference between a fertile soil and weathered rock or sand. Between 590 and 1180 million tons of carbon is fixed in the soil. Soil organic matter is essentially made up of carbon and plays many important roles. Reduction in organic matter leads to substantial carbon dioxide emissions and vice versa; increase in soil organic matter can result in a big carbon sink. Land with high organic matter content is, almost without exception, land that is more fertile. It has a better structure and is richer in nutrients. It also resists droughts better and is more resistant to water and wind erosion.

Some sources state that 5–10 million hectares are lost every year as a result of soil degradation (World Bank 2007). The loss of topsoil is calculated to a few tons per hectare per year. Soils in steep terrain are particularly vulnerable. In Nepal, the loss of soil in mountain areas goes from 20 tons to 200 tons per hectare per year (FAO 2003), and Rwanda reports an erosion rate of up to 557 tons per hectare per year (Minagri 2009). This can be compared to the formation of new topsoil that goes from 0.1 to 1.9 tons per hectare per year (Montgomery 2007). As a result of extractive farming systems,[73] the United States was affected by severe soil degradation during the so-called Dust Bowl in the 1930s. Extensive conservation programmes were initiated and have had effect; nevertheless, erosion is still rampant. Two billion tons of soil was lost per year in the early 2000s, an incredible number but still an improvement from the 1980s when erosion was estimated to three billion tons per year (Montgomery 2007).

The effect of soil degradation shows only slowly in many cases. This is particularly the case when other efforts (such as irrigation, use of fertilizers or new varieties) lead to higher yields. Despite that most Chinese soils were assessed to be degraded; wheat yields increased from 16 to 110 million tons and rice yields from 63 to 194 million tons from mid-1960 to mid-1990s. Soil degradation is often worse on marginal soils where yields are low already, which means that the effect on total yields is small.

---

[73]   George Washington said: 'Here it is more profitable to cultivate a lot of land poorly than a little land well' (quoted in Radkau 2008: 177).

The statistical basis for assessing how much land human beings have taken in direct use for buildings, roads, etc., is surprisingly weak. Literature quote figures from 1.5% up to 9%. In Sweden 3%, 1.29 million hectares of the land, is built on in some way. Land for housing was 29% of this; roads, railroads and airports 31%; industries, etc. 11%. Of course, Sweden is sparsely populated. In the county of Stockholm 15% of the land is built upon whereas in the northernmost county of Norrbotten only 0.6% of land is built upon. There were 550,000 kilometres of roads (some 60 metres per person) covering 345,000 hectares of land (SCB 2004). In the more densely populated Denmark, human infrastructure is calculated to cover almost 20% of the land area (Danish Ministry of Environment 2005). During the 1960s, 7% of European farm area was encroached upon by roads and 15% of the agriculture land of Great Britain was built upon (Montgomery 2007). By 2010, the United States lost almost 10 million hectares, more than 2.5% of farmland, to human infrastructure since 1950 (Talberth *et al.* 2007). Eastern United States has a larger proportion of its total area (4–5%) in urban and suburban land-scapes than other regions (H. John Heinz III Center 2008).

Although most ecologists classify the biomes (nature types or eco-systems) of the world largely as if they were not touched by human beings, the reality is that human beings now influence most parts of the world, even to the extent that some scientists speak about the *Anthropo-cene* as a geological era; that human beings actually change both the geology and climate. Erle C. Ellis and Navin Ramankutty (2008) identify 18 anthropogenic biomes and only 3 biomes that could be considered 'wildlands', most of them barren, permafrost or sparsely forested.

Table 10.3 Area, population and primary production of the world.

	Area (%)	Net primary production (%)	Population (billion)
Dense settlements	1.11	1.4	2.57
Villages	5.9	7.7	2.56
Croplands	20.8	32	0.93
Rangelands	30.4	15.5	0.28
Forests	19.4	32.8	0.04
Wildlands	22.5	10.7	0

Source: Adapted from Ellis and Ramankutty (2008).

The wildlands represent around 22% of the terrestrial land, but only 11% of the net primary production (the photosynthesis) because they are cold or dry, or both. All other nature types have been so heavily influenced by human beings that they can be called man-made landscapes. Of course, even in these landscapes there are patches of land that are less influenced, and even in a field there are weeds and wildlife, but they are all there under conditions created by human beings; 'human systems with natural ecosystems embedded within then'. It is assessed that human beings control at least one-third of all terrestrial primary productions (i.e. the biological production emerging from photosynthesis). Rangelands are in this classification the most common landscapes, covering nearly one-third of ice-free lands. As a result of arid conditions, the primary production is low, representing only 15% of the world's total production, and only 5% of the population lives in these landscapes (Ellis and Ramankutty 2008; see Table 10.3).

The naked soil and the unfaithful water

Society has not proven to be very good at managing either water or soil in a sustainable or equitable way. Pressure on the resources from increased population, material growth and bad management threatens the fundaments of the very existence of human beings. Very few countries have any systematic follow up of soil degradation and there are no criteria for how to assess the status or make measurements. That in itself is an indication of how little attention most countries pay to this critical basis for human survival. International conventions and treaties address all sorts of things, but not soil. Already human beings take a too high share of the land surface of the planet and should, therefore, not bring new land under the plough, but should also avoid covering more of the fertile farmland with concrete or asphalt. The loss of soil through erosion must be halted. The quality of topsoil is equally important; it represents a unique natural capital. When we evaluate different agricultural methods, the simple question should be whether the depth or the quality of the topsoil increases or decreases. With increased awareness of climate change, soil organic matter is receiving renewed attention, both as a problem and as part of the solution.

Unwavering efforts must be made to protect or increase soil organic matter and national and international protection of arable land resources should be put in place. Farming is mainly concerned with the annual return from the land. But it should be realized that soil organic matter is a 'capital', a capital that gives enormous dividends, but a capital that needs care, attention and replenishment. Thomas

Jefferson is supposed to have said[74]: 'While the farmer holds the title to the land, actually it belongs to all the people because civilization itself rests upon the soil.' Unfortunately, this is just a theoretical statement; in particular in his home country land resources are managed as private property. Increasingly, also water is privatized in various forms. Utilities are privatized, but also the water itself as well as the reproduction of water is turned into a commodity. Stealthily, capitalism is turning the basic fundaments of human life, for all life, into tradable commodities.

[74]   Quoted in the *Des Moines Register*, 8 July 1979.

– 11 –
Agriculture and Food Production

Sin maíz, no hay país.[75]

(Mexican proverb)

In agrarian civilizations, society was to a large extent defined and determined by agriculture. Even religions followed the season of the farm year. Power, status and wealth were built on ownership or control of land or livestock. People lived in villages where they took care of and were cared for by their kin. Life was simple, hard and unfair but mostly bearable. Most people had the same occupation and lived in the same place as their parents. With industrialism and capitalism, it all changed. Farming was gradually integrated in the market economy, and farmers in high-income countries became *entrepreneurs*. It is this development that caused the rural exodus — bigger and bigger farms and production for distant markets based on massive use of inputs.

Agriculture produces most of our food (a few people live mostly on products from the sea and even less people live on hunting and gathering) and approximately half of the world's population has agriculture as the main source of livelihood. Food attracts a lot of interest and how a growing population can be fed is a pressing question (see more in Chapter 13). Agriculture is also a very important element of many of the ecological challenges as it affects climate, energy consumption and biodiversity. Despite that, today, we live in an industrial society, agriculture remains fundamentally important for survival on this planet, not only because the products thereof are eaten, but also because it is a management system for the planet and the fundament for society. Industrialism has changed this management system, but has not given us any opportunities to actually leave it

---

[75]   'Without corn there is no country.'

behind. Synthetic food has been around as a vision for some time and this is the only way human beings could be radically independent of the most direct biological limits for continued growth of the species. For the time being, synthetic food remains a pipe dream.[76]

Agriculture needs to produce food for a growing number of people and to produce other raw materials such as fibre for the textile industry as well as biomass for energy. But it should also produce ecosystem services such as biodiversity and carbon sinks. It should do all this while also reducing its own negative impact on nature, biodiversity and ecosystem services and maintaining critical resources such as topsoil and water. For the individual farmer in high-income countries, agriculture is about economic survival as a farmer; for farmers in low-income countries, it is about survival as physical persons.

The big picture

Since the sowing of the first seed and the taming of the first animal, human beings have developed agriculture tremendously. Today (2012), agriculture has transformed half of the land surface and provides 7 billion people with food (well, almost a billion gets too little). Despite a doubling of population between 1960 and 2000, the nutritional status of the world's population has improved. The farmers of the world have reasons to be proud of their accomplishment.

The introduction of agriculture led to increased stratification in society. Not only animals were put in the service of men, but also men were put in the service of other men (the elite), and women were put in the service of men. Farmers in most places of the world and in most times since farming was introduced have mostly not been free in a modern sense; they were slaves, serfs, bonded labour or tenants paying exorbitant rates to landlords. Even the modern 'free family farmer' is mostly in the hand of the banks.

Human relation to other animals, to plants and to nature at large changed even more. In farming, a few species of plants and animals are chosen and favoured over others. Most radically, through sowing and weeding, one or a few crops cover large areas, a system quite

---

[76] There is indeed already synthetic food such as *quorn*, but as it is produced with farmed raw materials, it doesn't really change the fundamentals. It is more a competition to chicken, which is almost an industrial product in the first place, than to crops.

different from most natural environments where many species live together and occupy small niches in the environment for their survival. We nurture and protect them from other threats, at the price of ultimately eating them.[77] These species are evolutionary partners of the human species. Not only a few species are selected, but also within the species only the more valuable breeds are selected. Animals are bred to produce the most valuable product—meat, milk, fur or wool. Plants with bigger or more seeds or more of an edible root, or less bitter taste, are selected. Grains are bred and selected for bigger and more grains and less straw, which has less use. This in turn has made plants more dependent on human beings and the farmscape as they would not survive outside of its fences. Their weaknesses are compensated for with more intensive measures, lately with chemical pesticides and fertilizers.

From sickle to combine harvester

For centuries, the technological development of agriculture was slow, and it is first with the Industrial Revolution that farming became subject to a rapid technological development. Mechanization and, as important, the use of external energy resulted in a dramatic increase in productivity. But it doesn't mean that agriculture didn't develop at all earlier. Plant and livestock breeding has taken place over millennia and produced impressive results long before the laws of heredity and genetics were known. Crop rotations[78] were developed and many of the still used post-harvest technologies were introduced. Fermentation gave bread, cheese, yoghurt, wine and beer. Other technologies developed were pickling, drying, salting, freezing, etc. Ovens, wind and watermills were developed and refined.

Europe underwent a population boom in the period 1750–1900, and production from land was under serious pressure. In the nineteenth century, this was turned around with a great increase in agriculture production and productivity. The start of this was in the Flanders, parts of Germany and later in England. This was not a result of a quick

---

[77] One could turn the discussion upside down by stating that it was the wheat and the cow that selected the human being and through that strategy, they have reached global domination.

[78] The system whereby different crops are rotated in order for better use of the capacity of the land, reduction of disease, control of weeds and forming a fertile soil.

'revolution' but a rather slow process of agriculture intensification and modernization. Bernard Slicher van Bath (1976) says in his study on the history of agriculture in Western Europe 500–1850 that new methods spread some 50–70 kilometres in 30 years in France. There was a keen interest also from the ruling classes to develop agriculture and scientific methods were increasingly applied.

In earlier systems, land was fallow every second year or more; in some places, the land was actually fallow two years out of three. Fallowness allows the land to regenerate and yield decently in the coming years. In France (as it is now) on the eve of this first agriculture revolution, around the year 1700, fallow lands occupied 10 million out of 24 million hectares of arable land. At the end of the eighteenth century, 75% of these fallow lands had been cultivated. Many of the 'new' methods applied were actually known, already 1000 years earlier, but were applied only when population pressure made other methods less feasible. The new methods required more labour and employment rose; 'the average number of hours worked per year by a family of agricultural labourers and small peasants must have increased tremendously in the period from the agricultural revolution to the time when agricultural machinery became widespread,' says the Danish agronomist Ester Boserup (2005: 109–110).

In many parts of Europe, hay was grown on permanent meadows and cattle only grazed on permanent pastures. So there was a movement of nutrients from these meadows to the farmed land. The Swedish saying 'meadow—the mother of the field'[79] expresses this relationship. Another very land-demanding system was the North European *plaggen* system that was based on grass sod that was cut and applied to the land as fertilizer. This allowed continuous grain production without crop rotation. The system required an outfield area of at least five times the size of the cultivated fields (Radkau 2008). The result of this vastly extractive practice was a heath landscape, a landscape that today is adored in the same way as the permanent meadows that were subject to exploitation. In the new system, more grassland was brought under cultivation, more fodder was grown in the fields and more manure was taken care of and re-circulated to the crops. Instead of letting the land rest fallow, it was used for the production of fodder. Through the introduction of crop rotations of leguminous plants, plants that bind

---

[79]   Exists also in other languages in version such as 'The meadow is the mother of agriculture.'

nitrogen from the atmosphere, a lot of fodder could be produced. At the same time, productivity in grain production went up. Intensive row-cropping of potatoes, sugar beets and other root crops and cabbages also produced a lot more food per hectare than grain production.

All in all the intensification led to double or more yields. Grain acreage in France remained stable but yields increased 2.1 times in the period 1800–1900. Fodder crops were increasingly grown, and meat production increased by a factor of 3 and milk production more than doubled. The population increased from 27 million to 39 million and food intake rose from some 2000 to 3000 calories. All in all, production and consumption doubled and this was all before chemical fertilizers and tractors, which came with the second agriculture revolution (Mazoyer and Roudart 2006).

The second agriculture revolution came with specialization, mechanization and the use of chemical fertilizers and pesticides. In Europe, one horse was kept for every 1.5 hectares and perhaps up to 10 hectares (Bath 1976). The first tractors were heavy and very expensive and came into use only in large farms in areas with labour shortages. They were like a metal draught animal pulling basically the same machines that a set of two, four or eight horses could pull. The combination of increased labour costs, cheap oil, the innovation of the hydraulic system and the power transmission from the tractors to the implements (e.g. a harvester) paved the way for the breakthrough of tractors. This meant that draught animals rapidly disappeared.

The land used for grazing and growing food for horses[80] was now made available for the production of food and fibre for human consumption. Tractors also meant that the agrarian landscape was fundamentally reshaped and 'obstacles', such as trees, were removed at a rapid pace; open ditches were replaced by subsurface drainage systems. Employees needed more skills and salaries increased. Mechanization also drove further specialization, as each machine needed sufficient work to pay for it. When less fodder crops were needed for horses and specialization increased, crop rotations changed back to more grains, which increased disease and weed pressure. Fertilizers were introduced to compensate for bad crop rotations and in turn drove even more specialization. In one or two generations, in the high-income countries, there was a total system shift in farming

---

[80]   This could amount to between a half and one hectare per horse.

where the most prominent features were the tractor, market orientation, specialization, indebtedness, increased salaries and chemical fertilizers.

This development has also entailed a fantastic increase in productivity, measured in production per labour unit (see more in Chapter 14). Meanwhile, prices of agriculture products have fallen to half or even a fourth. In this development, less than 10%, down to 2% of the farmers in the high-income countries, have managed to stay on and develop their farms.

Globally, there are around 530 million farms of which 450 million are less than 2 hectares. Land ownership develops in different ways in different countries. In most of the high-income countries, units used for active farming are growing rapidly. In most developing countries (e.g. Bangladesh, India, Pakistan, Thailand, Malawi, Tanzania and Ecuador), the size of farm units is going down, mainly as a result of division in conjunction with generational shifts. In India, farm size has gone down from 2.2 hectares in 1970–1971 to just above 1 hectare in 2002–2003. Farm size has increased in a few developing countries (e.g. in Botswana, Brazil and Algeria). In Brazil, the average size masks the increase both for huge units and for very small ones, whereas the medium–sized units, family farms, are on the decrease (World Bank 2007; Braun 2010).

Countries in rapid transition

The high-income countries have already completed the transformation from an agrarian society to an industrial society. The middle-income countries are following in the same steps. China has to feed 20% of the world's population on just 7% of the land. In 2004, China became a net importer of food. Farmland is converted to other purposes at a rapid pace; around 600,000 hectares (double the cropland of New Jersey) is converted annually to roads, industries and houses (Porritt 2007). Simultaneously, China has afforestation programmes to reduce erosion, protect water resources and maintain biodiversity. It increased forest cover from 16.8% to 21.2% between 1990 and 2005 (UNESCAP 2008). Water resources are under enormous pressure, particularly in the north. Small farm sizes and plenty of manpower means that intensive crops such as flowers, fruits, vegetables and other speciality crops are prioritized over broad-acre crops.

The case of Turkey describes what already happened in Europe and in other countries during transition from agriculture. Farming has the longest history in Turkey, together with Iraq and Syria. Hundred years ago, farming in Turkey was much the same as it was 500 years earlier, which in turn was much the same as it was 1000 years earlier. But in the last 100 years, and in particular in the last 50 years, change has been rapid. The population of Turkey was 73 million in 2006, which was double the size of that in 1970 and five times as big as it was when people were first counted in 1927. Land use too has undergone rapid changes. Arable cropland increased from 8 million hectares to almost 27 million hectares in 1991. Pastures and grazed areas reduced from 44 million hectares to 12 million hectares, which means that the total area used by the agriculture sector actually decreased from 50 million to 40 million (representing 53% of all land in Turkey) despite a fivefold increase in population. This statistics hides that a lot of the grazing areas were very extensively managed, some of it like semi-natural areas (with amazing biodiversity), whereas today most land is intensively farmed. Irrigated land increased from 1.3 million in 1965 to 4.9 million in 2005,[81] and most pastures are now intensively grazed (Redman and Hemmami 2008).

Agriculture's contribution to Turkey's GDP shrank from 26% in 1980 to 9% in 2006, while still one-third of the population find their source of income in farming. The number of farms shrank from 4 million to 3 million just in the 10 years between 1991 and 2001. Half of the farmland is still in farms with less than 10 hectares. The self-sufficiency rate in the farms is rather high, even though few live solely on what is produced on farms. Most small farms are not part of any social security system, their economy is almost fully informal and they pay no taxes. The average income of rural population is 42% less than the income of city-dwellers. The increase in farmland, the decrease in permanent pastures and the decrease in the number of people involved in farming are combined with a rapid increase in the use of chemical fertilizers and pesticides, mechanization and irrigation. Annual synthetic nutrient supply to farmland increased from 70,000 tons to 2.1 million tons—30-fold between 1961 and 2000. The government intervened, as governments in most countries have intervened, to 'modernize farming', for example, with subsidies for irrigation and chemical fertilizers. The subsidy for chemical fertilizers took up 45% of the government's agriculture budget in 1997 but has declined since (Redman and Hemmami 2008).

---

[81]   This also has political implications. Turkey now uses a lot of the water that earlier flowed to Iraq in Euphrates and Tigris.

Above the social, cultural and economic effects of this transformation, natural resources have been put under pressure; soil erosion is rampant. Of Turkish soils, 59% are estimated as seriously or very seriously eroded; 1 billion ton of soil and 87 million tons of nutrients (compare that with the use of chemical fertilizers above) is lost every year. Erosion is caused by deforestation, over-grazing and bad arable field management. Through intensification of pasture land management the pressure of animals is now unsustainable. Another troubling development is that the whole traditional farmscape is disappearing. This landscape had both an impressive variety of farmed crops and animals and a lot of wildlife that had found its ecological niches in this relatively extensively managed landscape. As a result of this, 6 wild plants, 14 of 20 cow breeds, 2 of 19 sheep breeds and 2 of 5 goat breeds have been lost. How much diversity of agriculture crops is lost is impossible to say as it was badly documented earlier. Both the traditional farmers and the landscape that was their creation are rapidly disappearing (Redman and Hemmami 2008).

The visible hand in farming

Many governments of developing countries invest little resources in the farm sector. Despite this the farm sector is the most important sector in many of these countries. Only 4% of government budgets in African countries is spent on things related to the farm sector. There is a political commitment that this shall increase to 10%, but little happens. And the money that does get spent is often spent on in-efficient and/or harmful subsidies, especially subsidies for chemical fertilizers. This kind of subsidy amounted to 37% of the farm budget in Argentina (2003), 43% in Indonesia (2006) and 75% in India (2002) and in Ukraine (2005). Of the farm budget in Zambia (2003/2004), 80% was used for fertilizers and price support to maize. Most of this support went to the farms that were best-off to start with (World Bank 2007).

In India, subsidies reached new record heights in 2009. The Food Corporation of India used US$ 10 billion for food subsidies and the Ministry of Fertilizers (the only country with a ministry for chemical fertilizers?) used US$ 20 billion dollars for subsidies of chemical fertilizers. To this one can add federal support for food security and crop insurance as well as state programmes for energy, seeds, seed-lings, livestock, tractors, pumps, irrigation, etc. Despite all support measures, Indian farmers are under huge stress. Just in 2007, 16,600 suicides were reported among Indian farmers (*Tehelka* 2009). Ironically, they often use pesticides for finishing their lives.

Earlier, most high-income countries had policies that primarily stimulated higher production and structural rationalization (i.e. making farms bigger). Today, in most high-income countries farm policies are made up of a jumble of measures and programmes, often with contradictory purposes. A country that supports the promotion of genetically modified organisms might also support organic farming; a country that pays farmers to leave farms to allow for 'structural rationalization' might also promote farmers' markets to give small farmers a market outlet; and a country that promotes intensification of farming might at the same time pay farmers to maintain traditional farming landscapes. Some measures were developed in order to supply an increasing industrial population with cheap food, others to promote quality small-scale production.

The policies of high-income countries — and the ever-increasing productivity among farmers — have led to huge surpluses that have been dumped in global markets, thus pressing down already low global prices (and inducing even more subsidies from their government). Public support to the farm sector has for the last 20 years oscillated between 30% and 40% of total farm income in OECD countries. The value of public support to farms was US$ 238 billion, whereas the total income was US$ 673 billion on average in 2001–2003 (OECD 2009). The trend is that the support is (slowly) reduced and it is also shifting from production-driven support mechanisms (e.g. price support or subsidies for fertilizers) to support for ecosystem services, rural development and farmers' income. Agriculture policies in high-income countries have not been very efficient in delivering their promised results while being directly harmful for farmers outside the countries.

Market-based farming is problematic

Great differences exist between agriculture practices say in Thailand, Brazil, France and the United States, but a number of trends are clearly global. In most parts of the world, farmers are linked to a market and are affected by price fluctuations in global markets. In industrialized countries, farming for self-sufficiency hardly exists at all, except in a few lifestyle farms. In developing countries, a large share of the harvest is consumed directly by the farm family, but also here farmers sell or have to sell substantial parts of the harvest to pay for implements, clothes, medicine, school books, school fees and taxes.

Farms are integrated in global markets not only for what they sell but also for what they need, the inputs, which range from seeds to jerry cans to diesel and fertilizers. Even the most basic farm tool, the hoe, is often imported; for example, in East Africa the hoes often come from India or China. The use of external inputs, in particular chemical fertilizers and pesticides, has increased enormously. Farmers are also increasingly relying on credits that make them dependent on markets to generate income to pay their debt. Communal and state-owned land is converted into privately held land, which can be sold and used as collateral for loans.

In all countries, the *share of* (not the numbers though) farmers in the population is decreasing and farmers and farm workers have lower income than people employed in industry. In the richest countries, farmers and farm workers are less than two percent. In the poorest countries, farmers and farm workers exceed two-thirds of the population. To a large extent, they are part of the informal economy.

The share of agriculture land is going down in almost all high-income countries and is rapidly increasing in countries that are in the early stages of demographic transition (see more in Chapter 16). Extensive pastures are shrinking in most parts of the world. In Europe, this land is now subject to conservation measures (because of high biodiversity).

Intensive animal keeping, especially of species with quick turn-over, such as pigs and poultry, is rapidly increasing (pigs limited to non-Muslim cultures). Government intervention in farming is big in most countries, much heavier than in most other sectors. Very few countries have a truly liberalized farm sector.[82] Many countries have subsidies for fertilizers and generous schemes for water and electricity supply to farms.

In countries with high labour costs, mechanization also plays a big role in shaping agriculture. The process has also shaped a 'food industry' of what was earlier taking place on the farms (e.g. cheese- and sausage-making) or in the households (cooking).

The key in farm development has been to *transform the farmers into market actors* who buy and sell produce, need financial capital or credits, services, etc. Ironically, once farmers go there, compelled by the state, the interventions of society are geared to mitigating the negative effects of a commercial farm sector. One reason for this is that farming doesn't lend itself easily to the same logic as industrial

---

[82]   Of OECD countries, New Zealand is the most liberalized.

production. Farming is an intricate livelihood system that carries culture, religion and social belonging. It involves a complicated management of natural resources, as a matter of fact the very basis for human civilization and is the key component of our planetary stewardship. And it is here that stewardship of the planet and markets collide. To sell your surplus plums in a marketplace is one thing, but to subject the whole farming system to the market is quite a different thing. By and large, market-based farming focuses on production but has lost the aspect of reproduction of the system and the maintenance of the ecosystems of which agriculture is a part. The modern farming system doesn't reproduce or regenerate fertility of the soil; it buys it in bags. It doesn't reproduce genetic resources; it buys them in sacks. It doesn't reproduce the tools; it buys the energy and gradually loses the people. It is not the management system for the earth that it has to be, as it shapes more than half of the terrestrial landscape. Still, the market paradigm is promoted as the best way to shape agriculture for the future.

– 12 –
Environmental Problems in Farming

In farming, the Polluter-Gets-Paid-Principle applies.

Author

There are two different kinds of environmental problems in farming. The first kind is related to the fact *that* we farm, that we use the land in the first place, that is, that we expand farmland at the cost of other ecosystems. The other is caused by *how* we farm, in particular the use of chemical fertilizers and pesticides. The former we have already discussed, and here we discuss the latter.

Loading farms with nitrogen and phosphorus

In many farming systems, the availability of nutrients is a major limiting factor, in particular the availability of nitrogen and phosphorus. Human beings have loaded farming systems with more and more nitrogen and phosphorus. Initially, the soil could take care of it, but now it leaks into the water and the air everywhere. Eutrophication[83] is a substantial problem in all countries where modern, industrial farming is practised. Global annual use of nitrogen fertilizers increased from 11 million tons in 1960 to 85 million tons in 2003 (MEA 2005; see Figure 5.2). At the same time farming lost 95 million tons of nitrogen in 1990; that is, most nitrogen is simply wasted through denitrification, leaking, erosion and volatilization of ammonia. In Rwanda, erosion causes loss of almost 1 million tons of organic matter, some 40,000 tons of nitrogen, 280 tons of phosphorus and 3000 tons of potassium—more than the total use of chemical fertilizers (Minagri 2009). In the United States, half of the fertilizer used is used to compensate the nutrient losses caused by erosion (Montgomery 2007).

---

[83]   Over-abundance of nutrients leading to algal blooms, etc.

These losses represent not only an economic and agronomic problem but, to an even higher extent, also an environmental problem. The Mississippi, the Columbia, and the Susquehanna rivers together discharge approximately 1 million tons of nitrogen in the form of nitrate per year to coastal waters (H. John Heinz III Center 2008). Nitrogen losses will continue, and increase at the same rate as the use of nitrogen fertilizers. Through denitrification, nitrogen is lost as nitrogen gas (which in essence is harmless) but some 5–7% is emitted as laughing gas, which is a potent green house gas (Ayres 1998). Mono-cropping of grain, which is rather closely associated to the use of chemical fertilizers,[84] leads to a reduction of carbon in soils and, thereby, an increase in the greenhouse gases.

In high-income countries, a lot more nitrogen is used than what is taken out from farming in the form of products. At the end of the 1990s, countries like South Korea and the Netherlands, with very intensive farming systems, used more than 250 kilogrammes of nitrogen per hectare more than they took out, and these figures do not even include all nitrogen sources (OECD 2001). The International Assessment of Agricultural Knowledge, Science and Technology for Development[85] (IAASTD 2009) assessed the fertilizer uptake efficiency to be less than 30% for rice production in South and South East Asia. Globally, the nitrogen efficiency in grain production has deteriorated drastically and rapidly. Around 1960, each ton of chemical fertilizer resulted in an increase in grain yield of 75 tons, whereas at the end of 1990 this resulted in just 25 tons, an evident example of decreasing marginal utility, as nitrogen fertilizer use increased tremendously in the same period.

Over and above the use of chemical fertilizers, there is a substantial supply of nitrogen through biological nitrogen fixation. This is partly done by bacteria living in symbiosis with (mainly) leguminous plants and partly by bacteria and algae that fix nitrogen independently. Biological nitrogen fixation represents one third of the nitrogen brought to farming (Vitousek *et al.* 1997). Even if nitrogen in chemical fertilizers and in biological nitrogen fixation is from the same source

---

[84]   It is nitrogen fertilizers that have enabled farmers to skip sound crop rotations and opt mono-cropping, so it is not so far-fetched to 'blame' nitrogen fertilizers for being a major factor for increase in mono-cropping.

[85]   The IAASTD was an intergovernmental process with the representation of UN agencies, the World Bank and international and regional NGOs. Its main report was published in 2008.

(i.e. the air), one can't treat the source as equal in both instances, especially with regard to the effects in the soil. In theory, there could be substantial nitrogen leakage caused by biological nitrogen fixation; in practice, this is difficult to substantiate. One reason is that the process, as most natural processes, to a large extent is self-regulating; if there is a lot of free nitrogen in the soil, nitrogen fixation from the air ceases, as it is easier (it 'costs' less) for the organisms to take it from the soil than from the atmosphere.

We have mainly looked at nitrogen, but there are similar problems with phosphorus. A main difference is that phosphorus leaks primarily through erosion. Phosphorus is mined and is thus a limited resource, and there are indications that we approach 'peak phosphorus', that is, the point at which less phosphorus can be produced than previously because of limited supply. A complication with phosphorus fertilizers is that they often contain cadmium, a highly toxic heavy metal of which the load in our food is already alarmingly high. Rich countries can choose the cleaner phosphates, or even purify contaminated ones, whereas low-income countries are left with the contaminated fertilizers.

The biological war

From a bigger perspective, the use of chemical pesticides is a logical result of trying to take a larger share of the production of the photosynthesis per area unit, by killing all the weeds and competitors that might eat the crop. Our farming system is more and more sensitive for each step taken in that direction, and problems have to be corrected with sophisticated technologies. But this comes at a very high price. The World Bank (2007), which certainly is no anti-pesticide crusader, estimated the number of persons killed by pesticide use to be 350,000 annually. Most of the direct and visible harm from pesticides is caused by its use in developing countries, where toxic substances, many of them banned in high-income countries, are used by untrained people without protective clothing. I have seen, in North Uganda at the end of the 1990s, how pesticides were stirred in a bucket (the same which is used for carrying water to people or animals) and then spread by dipping with bare hands a bunch of grass into the liquid. Thereafter, the pesticide was shaken out over the crop, while a big proportion ended up on the skin of the person doing it.

Pesticides are spread everywhere. Most countries do no systematic follow up of pesticides in nature and no country monitors all active substances, yet what is found is frightening enough. In the United

States, 80% of all rivers and 60% of all wells have been found to contain pesticide residues; the proportion of contaminated wells in urbanized areas was almost as high that in rural areas, owing to use in home gardens, gravel or stone paths, golf courses, etc. In France, pesticides have been found in all rivers and half of all water sources had at least traces of them. Of the 50 active substances monitored in the Netherlands, two-thirds have been found in ground water (OECD 2001).

Nearly 1,400 pesticides have been registered (i.e. approved) by the US Environmental Protection Agency for agricultural and non-agricultural use (President's Cancer Panel 2010: 45). The effect on human health by the residues of the pesticides in food is very difficult to ascertain. The food and agriculture establishment, agri-business and the associated scientists, go to great lengths to convince consumers that it is safe. According to a report of the President's Cancer Panel:

> Exposure to these chemicals has been linked to brain/central nervous system (CNS), breast, colon, lung, ovarian (female spouses), pancreatic, kidney, testicular, and stomach cancers, as well as Hodgkin and non-Hodgkin lymphoma, multiple myeloma, and soft tissue sarcoma. Pesticide-exposed farmers, pesticide applicators, crop duster pilots, and manufacturers also have been found to have elevated rates of prostate cancer, melanoma, other skin cancers, and cancer of the lip. (2010: 45)

In addition to the effect on human beings, damage on biodiversity is the main effect of pesticide use. An insecticide typically harms not only the target insect, the pest, but also several others, often also those that regulate the pest. The target insect might be valuable food for other organisms, which then are threatened. Herbicides typically reduce the amount of food for birds and other small animals living in the agriculture landscape. If a fungicide is used it normally affects various fungi, some of which compete with the fungal pest one tries to control. Bees are sensitive to pesticides and there are many examples of bee deaths caused by pesticides. This is harmful not only for the beekeeper but also for society at large. Some estimate that the value of pollination of agriculture crops alone is worth around €150 billion, corresponding to some 10% of total agricultural value (Gallai *et al.* 2009).

All in all the soil, the plants and the agriculture landscape make up an intricate web and we have really no clue of the total effect of the use of one single pesticide on this web, let alone the effect of hundreds of different ones in combination. In the individual case, there are often very good reasons for a farmer to use a pesticide, and if used at the 'right' time and place, it will mostly be economically viable and rational. But the end result of economically justifiable and rational decisions by half a billion farmers can still be — and certainly is in the case of pesticides — negative. The agriculture systems that are built around the use of pesticides and its twin chemical fertilizers[86] create new needs for more pesticides. Scientists are speaking about the pesticide paradox: by applying pesticide to a pest, one may in fact increase its abundance or stimulate the emergence of a new pest. Or conversely, when reducing pesticides, pest may also wane.[87]

Most of our food comes from three crops

The biodiversity in the farming systems has been impoverished dramatically over the last century. This is the case for the system's 'own' biodiversity, that is, the plants and animals that are grown or bred in the farms, and it is also the case for the 'wild' biodiversity in the farming system itself and its surroundings.

Over and above human labour, land and water, animal and plant material are the most important resources in farming. From the first small steps towards farming, farmers all over the globe have created a fantastic diversity, but it has declined rapidly in the last century during which 75 percent of the biodiversity has been lost according to some assessments. Fewer numbers of species, and fewer varieties of species, are grown. Only 150 species are cultivated, of which 3 (!) supply 60% of the calories human beings consume (Pretty 1995). A survey of 75 crops in the United States shows that 97% of the varieties in early public lists of seeds have disappeared. Paddy rice production, which earlier showed a tremendous genetic variety, has through the so-called Green Revolution[88] become very monotonous. In Sri Lanka

---

[86]   Those two are companions in most farming systems and they reinforce each other; using more fertilizers increases the use of pesticides. Both lead, or lure, farmers into more specialized and less diverse systems that in turn motivate their use.

[87]   The International Rice Research Institute has shown that, if pesticides are used less frequently, then nature itself, in the forms of predators and parasitoids, will join the fight on the farmers' side. 'In 1993, a new scheme for spraying pesticides was introduced [. . .] After 14 years of the program, pesticide use on the farm has decreased by a staggering 87.5%. Insecticides, which are the main type of pesticides used on the farm, have fallen in use by 95.8%' (Hamilton 2008: 32). This coincides with observations made by the more than 2 million organic farmers in the world, including myself.

[88]   The Green Revolution refers to a process of rather rapid transformation of agriculture in some developing countries. Key features were

75% of the rice varieties come from one mother line, in Bangladesh 62% and in Indonesia 74% (Stolton 2002). Of all wheat grown in developing countries, 80% are modern varieties.

The original varieties of wheat, rice and other food crops have, in principle, disappeared from their centre of development. Varieties show varying degree of sensitivity or resistance to disease and pests. If all farmers grow the same variety there is a bigger risk that a disease rapidly can wipe out the production in large tracts of lands. An outbreak of wheat stem rust in the 1950s reduced wheat yields by 40% in the United States and resulted in a loss of income of US$ 3 billion apart from its effect on the food supply (World Bank 2007). A new strain of the rust, *ug99*, is about to spread. Norman Borlaug (2008), legendary proponent of the Green Revolution, says: 'Today's lush, high-yielding wheat fields on vast irrigated tracts are ideal environments for the fungus to multiply, so the potential for crop loss is greater than ever.'

Farmers and breeders who demand 100% uniformity (mostly a result of demands from the food industry and supermarkets) of their crops are in fact throwing away most of the genetic possibilities accumulated through generations. To preserve a big share of the giant common heritage of human beings in the form of old varieties and races has a tremendous value. Genetic diversity (=information) is a source of wealth, a high-quality capital that would be stupid to waste. Realistically, however, we can't have the ambition to save all varieties and races that human beings have developed over millennia. As little as we can take care of every single species in nature, we can take care of every variety of cabbage or breed of sheep. Most of the things humankind has created are not very competitive in nature and some of them are functionally defective, for example, turkeys that can't reproduce or cows (like the Belgian blue) that can't give birth naturally. And now genetically modified crops have the 'evolutionary advantage' of being able to survive a spray with a certain herbicide, glyphosate (better known by one of its trade names, Roundup). These crops are linked to an 'ecosystem' where this herbicide is the main selective force. There is certainly no value in keeping those genes alive! Many non-farm plants, insects and animals live in the farmed landscape. Many species also live in the border zones between the fields and 'nature'; border zones are often very rich in diversity. As agricul-

---

expansion of irrigation infrastructure, modernization of management techniques, and distribution of hybridized seeds, chemical fertilizers, and pesticides to farmers.

ture undergoes rapid change, this affects all species interacting with the farmed landscape. This, in combination with farming now occupying between 37% and 51%[89] of the planet's surface, makes the farmed landscape a centre of attention for biodiversity conservation. Much of the effects are a result of the system as a whole taking into consideration habitat and feed, while some species are directly harmed by something very specific, such as a pesticide or destruction of their nesting areas. According to the list of threatened species (the 'red list') of the International Union for Conservation of Nature, agriculture directly affects 70% of all threatened birds and 49% of all threatened plants (Stolton 2002). Like in the example of Turkey in Chapter 11, the diversity is strongly related to the traditional farming systems and in particular to grazing areas and border zones.

Where are the self-correcting mechanisms?

Environmental problems in farming are rampant and represent most types of problems: pollution, deterioration of external resources, degradation of the system itself and the destruction of other ecosystems and, thereby, of their services. Many of the environmental problems associated with farming are caused by the specialization and break up of the nutrient cycles. This was mainly triggered by the increasing market orientation and was made possible by the introduction of chemical fertilizers (i.e. external energy). One could increase the use of environmental fees on pesticides and fertilizers; some countries do it to a limited extent, whereas most countries still subsidize them. The fees must be set very high to actually change usage patterns. To have fees and payment for all external, negative and positive, aspects of farming is simply not feasible, it would lead to a control apparatus and planning system that would surpass the Soviet regime.

To tackle biodiversity issues and ecosystem services associated with farming only through market-based measures seems mind-boggling. Instead of fiddling with hundreds of different subsidies or fees regulating what the farmer should or should not do, we need to look at the whole system. With this perspective, a farming system

---

[89]   The lower figure is from FAO (2007), the higher from Ellis and Ramankutty (2008). The latter includes areas that are part of the agriculture landscapes, or biomes, so even if they are not farmed, they are heavily influenced by farming.

closer to what today is called organic farming emerges. Still, organic farming and other supposedly sustainable farming systems are pushed by 'market forces' to specialize, to overuse certain resources (such as fossil fuel), and have problems to close nutrient cycles as a result of broken links between production and consumption.

Food Consumption and Markets

The coffee in your cup, for which you pay one, two or three euros or dollars, costs less than five cents, of which in turn the farmer only gets some two or three cents, that is, one-hundredth of what you pay for it.

Author

Not only has farming itself undergone rapid changes, the consumption of farm produce has gone from being a local business, sometimes even mainly self-consumption, to a big international trading system, where food is handled and traded as any other commodity. Transnational corporations take over not only the food trade but also the retail, and gradually the leverage in the food chain is moved to the retailers, those closest to the consumers, but farthest away from the producer.

Human diets and cooking habits also change. The increase in meat consumption is largely an effect of increased income, whereas the increase in the consumption of industrial food and ready-made food is more an effect of how time is getting more and more precious, how women have been emancipated and how food industries have rationalized to make food cheap (the taste and the nutrition is another story). Commercial actors can now earn money from activities that earlier were out of reach of the market (as they were done within the family), including cooking, preparation, processing of food, feeding infants and brewing. All in all, the consumption pattern follows—not surprisingly—the same market logic as the production pattern; they are after all just a reflection of each other.

More burgers and sodas

The farmers' share of the GDP, farm produce prices and peoples' proportional expenses for food are all reduced over time. Around 2000, consumers in high-income countries spent an average of 13% of their total household expenditures on food, whereas consumers in low-income countries spent an average of 43%. These shares ranged from a high of 55% of total household expenditures in Indonesia to 7% in the United States (Regmi and Gehlhar 2005b). In the European Union, the share was between 7.5% (United Kingdom) and 15.6% (Greece) in 2002. Compared to other sectors farm produce prices are falling, and this is also the case even if one compensates for yield increases, which is also why farms grow bigger and bigger. Despite impressive yield increase per hectare, price developments are not enough to give the farmer a decent income. Even to maintain a stable income he or she has to expand.

The share of income spent on food has certainly shrunk, but in absolute numbers people spend a lot more money on food today than they did before. Although food raw materials, agriculture produce, have become much cheaper, consumption of those in their un-processed form has decreased. As the share of processed and ready-made food increases, the cost of food also increases. And when people earn more money, they also spend more money on food.

The share of 'processed' food was 80% of the value of the global food market in 2002 and the share increases rapidly in developing countries. Eating 'out' as opposed to at home is also increasing across the board; 22% of the food budget in Brazil and Indonesia is spent on food services. Eating habits are equalized at a rather rapid pace. The Consumption Similarity Index,[90] which measures the similarity between a certain diet and the average American diet, for Japan increased from 0.45 in 1961 to 0.7 in 1995, where '1' means an identical diet (World Bank 2007). Even if shops are loaded with products, consumption is mainly constituted of rather few products and an increasing proportion of 'empty calories', food that just has energy (sugar, fat, starch and other hydrocarbons). In the United States, the number of people working in restaurants and bars to serve patrons is almost five times as many as the number of people engaged in farming (Bartsch 2009).

The markets for fruits and vegetables have grown very rapidly. It is also one of the few production lines where developing countries are rather competitive, largely because it is hard to fully mechanize the

---

[90]   It is the choice of the World Bank to measure the consumption of other countries against that of the United States. This obviously has nothing to do with whether the American diet is desirable.

markets for fruits and vegetables. Even in high-income countries, a big share of the work in the markets for fruits and vegetables is done by workers (mostly illegal) from developing countries. Fruits and vegetables earn a lot of money per area unit, perhaps 10 times as much as grain production, and their production employs much more people both in farming and in post-harvesting operations (grading, washing, packaging, marketing). Fruits and vegetables use 28% of all pesticides and huge quantities of chemical fertilizers, and they are almost always irrigated; this means that the environmental footprint of fruits and vegetables is bigger than that for most other crops (World Bank 2007). Although fruits and vegetables have little energy and little proteins, they are important for nutrition because of the content of vitamins and other important substances.

Fish is a valuable protein source for several people; almost half of the world's population eats fish regularly. Fish contributes to more than 15% of the protein intake of 3 billion people. Because of the decline in fisheries, aquaculture is in rapid rise. Soon, most fish will come from fish farms; the last century aquaculture's share of fish rose from 16% to 44% in 2007. Most spectacular is the development in China. In phase with increase in income, the Chinese don't only eat more meat; they also increase fish consumption, especially farmed fish. The average Chinese eats 24 kilogrammes of farmed fish and 6 kilogrammes of wild fish, compared to the average for the rest of the world of 3.5 kilogrammes of farmed fish and 10 kilogrammes of wild fish (Ababouch 2009).

Global meat consumption per person doubled between 1983 and 2005, which meant that total production increased even more as population also grew by almost 2 billion. Consumption of meat is estimated to double again between 2000 and 2050, the biggest increase in developing countries. China is today both the biggest producer and consumer (FAO 2006). Meat consumption is fairly strongly correlated to income. For an average Brazilian an increase in income by 10% leads to an increase in meat consumption by 7%, but for the average American, who already eats so much meat, the same increase results in only 1% increase in meat consumption (Regmi *et al.* 2008). Livestock provides 17% of all energy and 33% of all protein for human nutrition. Meat consumption ranges from 5 kilogrammes per person per year in India to 123 kilogrammes in the United States (FAO 2006).

At any given time, around 4.3 billion animals are in the service of human beings. Not only has the number of animals increased but also their role in farms has changed. In traditional European farming, livestock was an integrated part of farming; the animals were a source of power and they occupied different ecological niches from human beings; they transformed nutrients, both to the field and directly to human beings, from lands that were not suitable for arable farming and they supplied meat, milk, eggs, fibre, feather, fats and natural remedies. In many cultures, especially the pastoral cultures, livestock was and is a source of status, pride and social position. The changes in the livestock sector influences also the arable farming as more meat is nowadays grain fed. This is particularly the case for pigs and poultry.

Of all grain in the world, 38% is used for animal feed; in high-income countries this is as much as 60%. It takes 3 kilogrammes of grain to produce 1 kilogramme of chicken and slightly more for pork. Ruminants, cattle, sheep and goats, are increasingly also fed grain, something they are not well adapted to (CAST 1999; FAO 2006). Their main diet should be grass, which human beings can't eat. This has led some to conclude that animal products should not be consumed at all. Things are not so simple though. Globally, it is assessed that it takes 1.3–1.4 kilogrammes of protein in grain and other sources to produce 1 kilogramme of animal protein, but animal proteins are mostly of higher quality than vegetable proteins.[91] Livestock is an important buffer of food and provides livelihoods for people who otherwise had no income; no evidence suggests that livestock production detracts food from those who currently go hungry (CAST 1999; FAO 2006). Today, 600 million people are engaged in small-scale livestock production and 200 million people live on pastoralism. This kind of animal production uses ecological niches that would not be suitable for arable farming.

Ruminants graze pastures that are no good for arable farming, for example, too steep, too dry, too cold. Small-scale pig or poultry rearing is based on waste products from the field or kitchen, sometimes from food processing (e.g. pigs getting whey or distiller's wash). Therefore, that kind of livestock production doesn't really compete with production of food for human beings. In addition, 400 million heads of livestock are used as draught animals, and their role is mainly to help human beings to grow more food for themselves, even if they also, mostly, also end up on our plates (Worldwatch Institute 2006; FAO 2007; Erb *et al.* 2009). Plants and animals are complementary and use and need each other. There are no natural ecosystems without animals, and there never

---

[91]   Increasing consumption of animal protein is one of the easiest (admittedly not the only one) methods to rapidly improve nutrition and avoid long-term damage caused by malnutrition of children, who may never develop to their full bodily or intellectual potential.

were any sustainable intensive agriculture system without some degree of livestock production. We have seen in the case of the first agriculture revolution in Europe that integration of livestock in farm production, including the growing of fodder on arable land, increases production tremendously, both of feed and food. It also increased consumption of livestock products.

Industrial livestock production is, however, the opposite of traditional systems; it needs a lot of capital and natural resources, water consumption is big and it is largely grain-based. Energy use is also high; a calorie of beef requires 33 times as much energy to produce as one calorie of potatoes. Large-scale ranching might be better in many of these regards, but instead poses threats to other valuable ecosystems for example, from forests that are cleared for grazing. A big share of the increased meat production in Brazil is from the rainforest zone (Worldwatch Institute 2006; FAO 2007).

From this brief overview, it should be clear that one can't make general statements on the effect of eating meat. Most of the criticism of how meat is produced targets two different production systems: grazing of cattle in the rainforests zones and the industrial grain-based feeding in richer countries, but this has almost no relevance to large parts of the livestock production, and certainly not for the pastoralist systems. Still, the total increased pressure on natural resources caused by the combination of growth of population and growth of meat consumption is worrying.

While yields have been constantly increasing over the last century, the nutritional composition of agriculture products has not. There are many indications that the food harvested today is less rich in valuable things, and contains more water and carbohydrates. According to American studies, there are statistically significant reductions in nutritional value in the period 1955–1999 for 6 nutrients: 6% for protein, 38% for riboflavin, 16% for calcium, 15% for iron and vitamin C and 9% for phosphorus (Davis *et al.* 2004). The differences are mainly explained with changes in varieties and, subsequently, indicate that modern varieties have not been bred with sufficient attention to their nutritional value. How we grow also plays a big role. Irrigation and use of chemical fertilizer have a big impact, mostly negative. Studies comparing organic and non-organic (conventional) products show that, on average, organic food has a higher content of 'good' things such as vitamin C, dry matter and omega-3 and a lower content of 'bad' things such as nitrates and — quite obviously — pesticides (QLIF 2009).

Supermarkets take over

The changes in trade and food processing have been almost as dramatic as the decline in the number of farms. Multiple retailers have taken over the food market not only in the high-income countries but also in developing countries. This causes direct and indirect effects on all parties from farmers to food shoppers, and there is immense pressure on small independent food shops. They have largely disappeared for mainstream food shopping in high-income countries. Also in a country like Argentina, the number of food shops dropped from 209,000 to 145,000 in just 10 years between 1984 and 1993. While supermarkets accounted for 15–30% of the national food retail sales before the 1980s, they account for 50–70% of the retail sales in many Latin American countries in the mid-2000s, registering in only one decade the level of growth experienced in the United States in five decades. Multiple retailers are also on a rapid expansion in China, India and Africa (Regmi and Gehlhar 2005b; World Bank 2007).

The strict quality requirements and sharp business conditions in terms of supply security, penalties for failure and slow payment instead of the earlier cash-on-delivery affect farmers and small food producers a lot. In some cases, the demands from buyers lead also to efficiencies of producers, with less wastage or less exposure to dangerous pesticides. Mostly, it is a pressure that favours the more resourced as they are the ones who can climb the ladder and invest in their production, in cooling facilities, in trucks, in toilets, in bore holes, etc. Farmers who don't want to or can't (more often the case) are marginalized and will have to sell in traditional markets, but as they are waning so are their suppliers, the smallholders. In Guatemala farmers in modern supply chains have double the land area; 40% are more educated, more often have access to irrigation, four times as often have access to a truck and have 2.5 times more employees than the average farmer. Studies from Kenya and Indonesia show the same trends (World Bank 2007). Earlier, export markets were seen as the markets that most clearly discriminated against small farmers, but with the expansion of supermarkets, this discrimination has moved 'home' as well. This pressure on small farmers is the same on small farmers in high-income countries as well. Their numbers are, however, already so reduced that the few remaining ones can occupy some market niche (speciality food, farm shop, local processing for demanding clients).

The winner takes it all

Concentration is also high in the production and trade of inputs to farms. The market share of the four biggest[92] seed companies went from 23% (1997) to 33% (2004) and for pesticides from 47% to 60%. Many companies (Monsanto, DuPont, Syngenta and Bayer) work in both segments. One company, Monsanto, has 91% of the market for genetically modified soybeans. There are, very roughly, 25 million coffee farmers in the world, while 40% of the trade and 45% of the roasting is controlled by the four biggest companies. Simultaneously as coffee consumption doubled during the 1990s, the price share of the coffee-producing countries decreased from one-third to 10%. The trend is the same for cocoa and tea (World Bank 2007). This pattern is not unique for produce from developing countries; it is basically the same all over the planet. In the United States, four companies control 80% of the meat market, three companies control 80% of maize export and 65% of soy export and four companies control 60% of the domestic grain market. Many companies integrate production both 'upstream' and 'downstream',[93] for example, by contract farming. Most of the transnational companies in the food sector are from the United States or Western Europe (Regmi and Gehlhar 2005a).

One could say that the power over food moved first from farmers and landowners to raw material traders such as Cargill, thereafter to food processing giants such as Nestlé and Unilever, and now finally to multiple retailers. This is also expressed by the spread of supermarket-owned brands and private labels. The retail share of private labels among food products has reached 50–60% in Switzerland and 20–40% in most other Western European countries (Regmi and Gehlhar 2005b). In 2005, the revenue of the world's four biggest multiple retailers was US$ 339 billion for Walmart (USA), US$ 117 billion for Carrefour (France), US$ 80 billion for Ahold (Netherlands) and US$ 72 billion for Tesco (UK). Walmart was the biggest company of all categories in the world measured in sales, with sales of US$ 408 billion in 2009 (Forbes 2010). The dominance of the retailers is almost total. That the power in this way has moved further away from farmers should mean — at least in theory — that it has moved closer to consumers. But consumers are

---

[92]   The market share of the four biggest (CR_4) is a common measure of concentration in a given sector.

[93]   This terminology is used regarding value-chains and supply-chains, where 'upstream' denotes something 'earlier' in the process (e.g. a supplier) and 'downstream' refers to those 'after' a process (e.g. a buyer).

easily manipulated by advertising, so the much-heralded 'consumer power' and 'voting with the wallet' is to a large extent one of the modern myths of our world.

Global trade in food

Since the 1960s, the value of global food trade has increased 10-fold. Most of the growth has been in high-value products such as animal products, fruit, fish, vegetables and flowers. Despite the impressive growth, the growth of food trade has been slower than that of other trade, so its share fell from 25% to 10%. The total value of food exports was estimated to be US$ 442 billion in 2000. Only 10% of US$ 3.2 trillion global processed food sales were traded products in 2002.

Compared to industrial goods, prices of food dropped by 2% annually. It is worrisome that the money flow for agriculture products has shifted direction. At the beginning of the 1960s developing countries sold agriculture products for US$ 7 billion more than they bought for, but at the beginning of the 2000s that balance had shifted to a negative of US$ 20 billion, excluding Brazil this value was US$ 27 billion. And this development is despite developing countries having engaged themselves more in high-value crops (flowers, fruit and vegetables in particular), which now constitute 47% of the exports (Regmi and Gehlhar 2005b; FAO 2007).

In the period before the Second World War, trade in grain rarely exceeded 30 million tons; in 1975 it was 175 million tons and in 1990 it was around 200 million tons (notably the population tripled in the same period). Of the total production of rice, only 6% was traded across the borders in the early 2000s. In the case of wheat, the world's largest export crop, some 17% was traded globally. Less than 10% of the meat was sold across the borders (Ploeg 2009).

Despite globalization and cheap transports, there are still substantial advantages of local food processing. This has to do with customs tariffs, technical barriers to trade and the need to use local raw materials, but also to adapt products to local preferences. Therefore, direct investments are common; American food producers sell five times as much from their production in the country of consumption as through direct exports (Regmi and Gehlhar 2005a). Trade between developing countries has also increased. From 1994 to 2004, trade in soybeans doubled. Seventy percent of the increase was import to China, mainly from Brazil and Argentina. Population growth and growth in agriculture production is increasingly de-coupled, which means that global trade in food is likely to grow.

Growing hunger

In total, the number of malnourished people has been fairly constant over the last 20 years and has fluctuated around 850 million (see Table 13.1). The optimist notes that, taking into account population growth, the proportion of malnourished people has decreased from 16% in 1990 to 13% in 2005. In the countries South of the Sahara, the situation is rather gloomy. It is worth noting that India alone has more hungry people than the whole of Africa despite rapid economic growth in the last 10 years. This also shows that solving food security with the Green Revolution in agriculture is not working; India is after all a poster child for the Green Revolution. In addition to lack of food, badly composed food is also a big problem. It delays the development of children and can lead to permanent damages in the eyes or in the brain. Every year a million children die from faulty diet. This could be addressed by better education, especially of women, reduced poverty and a more diverse farming system.

Table 13.1 Number of undernourished people (million).

	1990 -1992	1995 -1997	2003 -2005	2010
World	842	832	848	925
High-income countries	19	21	16	19
Developing countries	823	810	832	906
Asia	582	535	542	578
Latin America and Caribbean	53	52	45	53
North Africa and Middle East	19	30	33	37
Africa South of the Sahara	169	194	212	239

Source: FAOstat (2011).

In 2006, the world spent US$ 1200 billion on weapons; the value of food waste was estimated to be US$ 100 and excess consumption by the world's obese amounted to US$ 20 billion.

> Noting that the time for talk was over and that action was urgently needed, FAO Director-General Jacques Diouf today appealed to world leaders for US$30 billion a year to re-launch agriculture and avert future threats of conflicts over food. [. . .] 'Against that backdrop, how can we explain to people of good sense and good faith that it was not possible to find US$30 billion a year to enable 862 million hungry

people to enjoy the most fundamental of human rights: the right to food and thus the right to life?' Dr Diouf asked. 'It is resources of this order of magnitude that would make it possible definitely to lay to rest the specter of conflicts over food that are looming on the horizon,' he added. (FAO 2008)

Don't we have to increase production to supply the hungry with food? This kind of question obscures the real causes for hunger and malnutrition. The real causes are mainly war and civil unrest, poverty and inequality, and bad agriculture and trade policies. It is mainly in low-income countries where people starve or are undernourished. And except for in very short periods, rich people have never starved even in countries devastated by famine. There are also many low-income countries where hardly ever any person starves, mainly as a result of good distribution and relative equality. Many countries that have hungry populations export food or other agriculture crops at the same time as people starve to death. For centuries mass starvation caused by food shortages have been predicted, but globally this has not happened. Obesity is gradually becoming a problem of the same magnitude as undernourishment.

This snapshot makes clear that food security is not mainly about food production, but rather about social and economic factors. Or as expressed in a report from the United Nations Conference on Trade and Development (UNCTAD):

At the country level, food security does not depend on whether countries are able to cover domestic food consumption through domestic food production, but whether they are able to generate sufficient financial resources to finance necessary food imports. Similarly, at the household level, food security is determined by household income more than anything else. (Herrmann 2009: 1)

In India, some 40% of all fruits and vegetables are spoiled after harvest. In the United States, 40 million tons of food is wasted by households, retailers and food services each year. The United Kingdom, United States and Europe have nearly twice as much food as is required by the nutritional needs of their populations. Up to half the entire food supply is wasted between the farm and the fork. If crops wastefully fed to livestock are included, European countries have more than three times more food than they need, whereas the United States has around four times more food than is needed, and up to three-quarters of the nutritional value is lost before it reaches people's mouths. Households in the United Kingdom waste 25% of all the food they buy (*ATL* 2009; Oxfam 2009; Stuart 2010).

Globally, there is simply no shortage of food; the poor just can't buy it. Or seen from the other perspective, rich people don't starve. Even countries that import the bulk of their food, such as Japan, are food secure, because they have money to buy food.[94] Most of the undernourished children live in countries with food exports. Increasing production when there is no demand leads to falling prices, which paradoxically can lead to an even worse situation for the rural poor,[95] most of whom are dependent on farm income, either as smallholders or landless individuals seeking employment by farmers. A surplus of food, with falling prices, is a bigger problem as it drives small farmers off the market so that they can't buy the things they need for production or for their families. Better prices enable farmers to respond with an expansion in production, something that was very visible in the period 2004–2009. That low prices are no cure is evident from the food prices falling almost constantly for a century but people still starving, or perhaps they are starving exactly because of those low prices.[96]

Overall, it is a diversion from the real problem to discuss agriculture methods as a solution to hunger, be it fertilizers, pesticides or genetically modified organisms. The solution to the problem of hunger lies in economic and social conditions, in securing the right and access of the poor to critical resources such as land and water.[97] Even if the discussion about hunger and food security should not be confused with the discussion on agriculture methods, more food will have to be

---

[94]   At least they are food secure as long as the international trading system is working. The tsunami and earthquakes of 2010–2011 also showed the vulnerability of such dependency. On the other hand, it is already clear that rich Japan could cope much better than poor Haiti, even if Haiti produced a bigger share of its food before the earthquake than did Japan. Not a single person in Japan died for want of food, whereas Haiti experienced mass starvation after the 2010 earthquake.

[95]   The rural poor constitute the majority of the starving.

[96]   The effect of prices on food production and distribution is a complicated matter and the effects are different for the urban poor and the rural poor, different among different groups of rural poor and different in the short and the long term. Higher prices will undoubtedly lead to higher food production, but higher prices themselves are most likely an effect of a shortage, unless they are a result of government support.

[97]   Or as expressed in a recent report from UNCTAD: 'In fact, only few problems in agriculture are mainly caused by a lack of technology, many are related to social, economic and cultural issues that require structural changes, not techno-fixes. It is therefore critical to first of all define what problems are best solved by changing legal frameworks, trade policies, incentive structures or human behaviour and, second, what contribution technology could make within this very context' (Hoffmann 2011).

produced because of increasing population, increasing consumption per capita and increasing meat consumption with rising incomes.

Production must increase in developing countries, where most people, and almost all hungry people, live. In order to do so, should these farms use chemical fertilizers, pesticides and genetically modified organisms? No. The reasons are similar as to why they shouldn't be used in developed countries, but the arguments against them are even stronger in developing countries: the number of deaths by pesticides is already too high; poor people can't afford inputs, even less when they have to pay an interest rate of 25% or more; most tropical soils are very sensitive for soil degradation and just a few years of monocultures with chemical fertilizers can be devastating; and finally low income countries can't implement the necessary controls for registration and monitoring of pesticides or genetically modified organisms. The impact of these methods and tools on yields is in any case overstated. Luckily, plenty of other opportunities exist to increase agriculture productivity in developing countries, on the basis of improved traditional systems (IAASTD 2009; Hoffmann 2011).

Beyond control

In many countries, there is low trust in the food industry's commitment to healthy and sound food; there is a desire to support local farms. Repeated food scandals and absurd transportation of food adds to this. There are valid concerns for what happens to our food if it is allowed to be fully subject to the logic of the market (i.e. that profit and unlimited competition rule). Interestingly, the confidence in that the global trading system will work well for food is weak on many sides, also among countries that otherwise are committed to global free trade. At the time of the food price hike of 2008, several food-importing countries made bilateral agreements with food-producing countries to safeguard their supplies of food. Some went a step further and initiated large-scale projects for production of food in other countries.

I am not raising these points to argue in favour of customs barriers or the continuation of absurd agriculture subsidies, but the development of the food industry and food trade is another aspect that, if left only to market logic, develops in a direction which most people don't like or desire. To make the basis for human survival a thing for the market to sort out and to place it in the hands of a handful of short-sighted companies with almost no accountability is a risky strategy. This is the reason for most agriculture policies in the first place. Having said that, having the state run farms, control the prices or plan food production has historically had very bad results.

Food production is still not a limiting factor for human expansion. Starvation and malnutrition are symptoms of an unfair and unequal society rather than signs of overpopulation or the result of not using modern technology. Nevertheless, there are biological limits for food production. The space used for farming can't be expanded a lot; however, the yields per area unit can be increased and production systems (and diets) can be changed to produce more food. Meanwhile, emerging energy scarcity is mounting pressure. An equitable world will have the potential to feed more people, as a result of equalization of diets as well as access to resources. It will certainly ensure that the food is distributed more fairly among the world's population. The market based food-production and distribution system has in no way contributed to a more equitable world.

– 14 –
Productivity in Farming

Productivity is about what we measure and how we measure it, what we believe has value and what we believe does not.
Author

We have the idea that we have a very productive farming system. But what does that actually mean? Productivity in farming can be measured in many ways.

Per area unit: how much useful resources can be produced from a certain area

Per person-hour: how much useful resources can be produced per worked hour

Per capital: how much useful resources can be produced per deployed capital

Per energy: how much useful resources can be produced per energy unit

Per water: how much useful resources can be produced per water unit

Comparisons can be made with 'nature'. For example, the extent of the total biological production (mass, energy or protein) in a farmed system can be compared with that in the natural system before farming. Comparisons can consider total biological production or only what is directly useful to human beings (and thus not to other creatures) in the form of food, fibre, fuel or ecosystems services. And the results, the useful resources, can also be expressed in different ways, for example, as tons, as calories, as kilogramme-proteins or simply as money. Let us discuss these different ways of looking at productivity.

A basic tenet is that, in the long term, it should be more interesting to increase the productivity relative to a resource that is limited rather than relative to a resource that is abundant. From this perspective, human obsession with labour productivity is mystical. In the chapter on energy (Chapter 6), we have already studied energy efficiency in depth. With clearly targeted uses of energy, such as driving pumps for irrigation, energy efficiency can increase a lot. But on average, the energy efficiency in modern industrial farming is embarrassingly low.

More from your plot

Yield per area unit has increased considerably (see Figure 14.1), both in developed and developing countries, and this doesn't really support the scenario of a looming food crisis.

Figure 14.1 Global yield per hectare of selected grains (1979–1999).

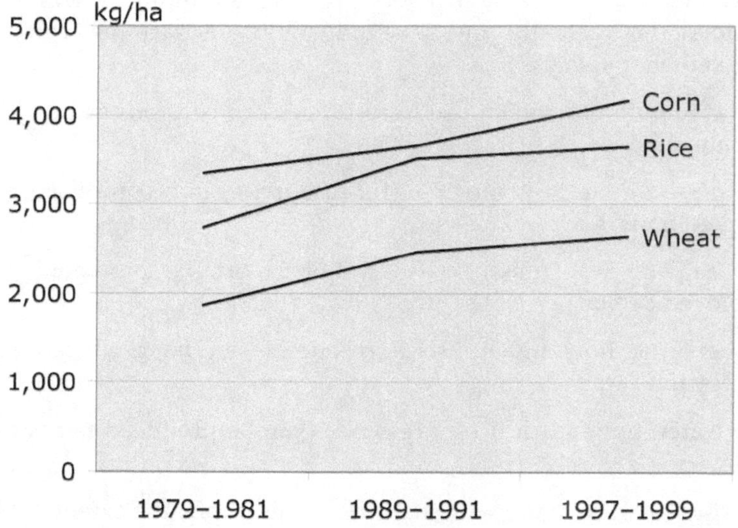

Source: FAO (2003).

The reason for increase in yield is always disputed, with different parts of the agriculture sector claiming credit. According to the World Bank, seed breeding is central. Hybrid rice is supposed to explain half of the increase in yield in China from 1975 to 1990 and improved varieties are said to explain 53% of the increased productivity in Punjab (Pakistan). Roads (!) are said to have improved farm productivity in India by 25%. A study from Africa shows that better food means a lot for productivity for African farmers, shedding light on the vicious circle of low yields, low income, little food, much illness, low labour availability, low yields, etc. (World Bank 2007)

Twenty percent of the increase in yield is said to be attributable to use of chemical fertilizers, perhaps surprising compared to how often it is said to be so important and even more surprising in the light of the tremendous increase in the use of chemical fertilizers in the same period (World Bank 2007). Equally surprising is that the normally very market-orientated World Bank doesn't explain any of the increase in yield with market factors, except for the Indian roads perhaps. The reality is, of course, that there are many factors cooperating in a technological and socioeconomic complex where energy access and the integration of farming into markets are key drivers.

One can study the objective conditions for yields (soil, rainfall, temperature, etc.) in different parts of the world and compare them to the real yields, and thereby get a fair idea of how much yields can increase without dramatic changes in the conditions (such as the introduction of irrigation or covered structures). There was, at the end of the century, a substantial potential for increased yields in Argentina, Australia, Canada, Hungary, Italy, Poland, Romania, Turkey, Ukraine and the United States. If these countries increased their yields by half the gap between the current and the theoretical optimal, it would correspond to a 23% increase in the world's wheat harvest. Studies of wheat in India show a similar potential in many Indian states (FAO 2003). Comparisons between real yields and yields at research stations in many developing countries show that farmers often harvest one-fourth of the yield registered at research stations (Kijne *et al.* 2009). Obviously, in real life, farmers will inevitably harvest less than a research station, but it should be possible to reach at least half the yield.

Without doubt, yields can still be increased substantially, but it has a price environmentally as well as economically. One should not forget that in many cases yield per hectare is low just because it doesn't pay to increase it. Human beings have never maximized yield per hectare as such, but yield per labour input. Most dramatic increases can be accomplished in highly regulated systems, such greenhouses with climate and water control where productivity per hectare easily is double and even up to 10 times the productivity in an open field.[98] Land price is a factor that influences productivity per area unit the

---

[98]   The more advanced producers in the Netherlands take more than 60 kilogrammes of tomatoes per square metre, whereas an open-field production may reach some 5 kilogrammes per square metre. Interestingly enough, the very high yields doesn't lead to increased profits.

most. To some extent, it works both ways, that is, land with high productivity potential has a higher price than land with low productivity potential, and when you pay a lot for land *you have to* get a high yield per area unit. In many countries and places, the correlation between land productivity and land prices is weak, however, and land prices are influenced also by alternative uses (conversion to forest, industries, housing), status and culture.

More bang for the buck

The return on invested capital in farming is mostly low. The real return is often realized first at the sale of a farm. Farmers 'live poor and die rich' as the saying goes and most profits are reinvested in the farms. For most farms, especially family farms, the concept of return on capital is somewhat theoretical. Families invest their capital and their labour on the same farm and there is no distinction in their way of thinking or in their books (if they have any) between the return on capital and the return on labour. One way to assess return on capital is to look at land rents. But land rent is currently only a proportion of the total investment as modern tenants also bring their own capital for purchase of machinery, etc. In addition, land rents are often regulated in ways to protect the tenants. For instance, in the very commercial Netherlands, land rents are administratively defined on a level of 2% of the calculated production value of the land. If this was not the case, if the farms had to render a commercial rent of say 4% of the land price, this would make farming unprofitable (Ploeg 2009).

A lot of the support to farms in high-income countries is capitalized in the form of higher land prices and recently also in quotas such as milk quotas. In 2006, milk quotas in the Netherlands were valued at an astonishing €20 billion, whereas the gross annual value of milk production was around €3 billion. The right to produce milk has thus become a commodity, more important than actual milk production (Ploeg 2009). Contrary to what many believe, return on capital in developing country farming is rather high. Very little capital exists and even less is invested in farming, but that which is invested is often borrowed at rates of 20–40%, a return on capital which few investors in high-income countries can dream of. Despite this, capital is not rushed to the farming sectors of developing countries because the risks involved are also very high and farmers are unreliable borrowers who often can get away with not paying their debts, or at least delaying payment for years.

Can rain be productive?

Different foods need different quantities of water, which we discussed earlier (in Chapter 10). It is well known that meat needs a lot more water than plants, but we have also discussed that meat may be the only food possible to produce from arid rangeland and therefore meat production would have higher water productivity than any crop. But what is it like within a certain production system? Seen on a larger scale, attempts to increase productivity per hectare mostly also increase water productivity. A hectare of wheat that is not irrigated uses more or less the same quantity of water when it produces 1500 kilogrammes as when it produces 5000 kilogrammes. Notably this applies to situations where water is not the limiting factor for the yield, which is typically the case in most parts of Europe.

If irrigation is introduced into the systems, the effect is, quite surprisingly, that water productivity increases; that is, the increased yield resulting from irrigation uses less water per kilogramme than the original yield based on rainwater. Of course, if one takes it to the extreme, that is, to a starting point of a desert where no yield is possible, this is quite apparent. For instance, in a situation where the annual rainfall is some 400 millimetres, water stress poses a real limitation to crops, and with some 100 millimetres of water applied at critical moments yields can easily double (Renault 2002). One problem with this kind of calculation is that one equates rain water to water in the ground or in a lake. The latter is concentrated and can have alternative uses, whereas rainwater is dispersed and largely just falls on the ground. There is a link between energy and irrigation as many irrigation systems are based on pumps.[99] In India, between 10 billion and 11 billion litres of diesel is used to run irrigation pumps (EENGO 2008). The potential for improving water productivity is great.

> There is a large potential to improve water productivity through improved and known water management practices. Globally, the water requirement to feed the world in 2050 would be an increase of ~4500 km3/yr from the current ~7000 km3/yr. Our estimates suggest that water productivity improvements could save up to 2200 km3/yr reducing the future additional needs to ~2300 km3/yr. This saving is larger than the world's current total consumption of water in irrigated agriculture. (Kijne *et al.* 2009: 1)

---

[99]   Conversely, in many places water in dams can be used either for generation of electricity or for irrigation.

That water is available for free and that many governments subsidize the use of water for agriculture are reasons for waste of water and obviously limit improvements in water productivity, something we discussed in Chapter 10.

More per hour

In 2012, I visited Susan, a small farmer in Zambia, dependent on the manual labour of herself and her husband. She grew 0.5 hectare of corn and got 900 kilogrammes from it. A bit later, I visited Bob, a big farmer in Illinois. Seven workers managed 3800 hectares and harvested some 38,000 tons of corn, i.e. more than 5000 times as much per person and year[100]. Because labour costs have represented a very high share of farm costs, productivity per person-hour or per person-day has been central in the mechanization of farming. Therefore, there is reliable data available for this. Quite naturally, productivity, measured as added value per worker, is much higher in high-income countries than in developing countries (see Table 14.1).

Table 14.1 Agricultural labour productivity (US$ value added per person-year).

	1990–1992	2001–2003	Agriculture as share of GDP (%)
Low-income countries	315	363	20
Middle-income countries	530	708	9
High-income countries	14,997	24,438	2
France	22,234	39,220	2
United Kingdom	22,506	25,876	1
United States	20,797	36,216	1
Brazil	1507	2790	5
India	332	381	4
China	254	368	12
Malawi	72	130	36

Source: World Bank (2007).

---

[100]   Susan, like all poor smallholders were engaged in a number of other things, managing the household, raising chicken and five children.

As with land prices it works both ways; if workers are productive they can get a good salary which in turn means that there are incentives for saving labour, which in turn means that income per worker will increase. In addition, if wages in other sectors are high, agriculture wages will follow (even if they mostly will stay comparatively low). A worrying fact is that the difference in labour productivity between rich and low-income countries is increasing.

Considering the romantic view of French farmers (herding sheep in the mountains, making exclusive cheeses or wines) the productivity of the French farming sector is perhaps surprising, but the reality is quite different from the idyllic image. One person in the United States occupied in farming can produce food for some 300 people, whereas one farm worker in some developing countries can only feed a few. Another way of looking at labour productivity is to see how much grain[101] can be produced per person-year. In the areas with lowest productivity, one person can produce not even 1 ton of grain per year, whereas the most productive farms exceed 2000 tons per person-year (Mazoyer and Roudart 2006), well up to 5000 tons as in the example from Illinois.

These comparisons mask the fact that a modern farmer doesn't even 'feed' him or herself or the family or their animals. A modern farm is a big company where other people's goods and services are bought and sold. There are accountants, repairmen, consultants, machine operators, computer service technicians and meteorologists, all in the service of the modern farmer. All this work is embedded in the farm produce. Earlier generations of farmers, and still most farmers in developing countries, produced not only food and fodder for themselves and their animals, but also fibre for clothing, leather for shoes, most of their own tools and medicines from their farms. But even considering that, the difference in labour productivity between farms in high-income and low-income countries is staggering. As we discussed earlier, energy use is the single most determining factor here. There is a very strong correlation between the energy resources commanded by one person and that person's productivity.

---

[101]   Because of the critical role of grain for our food supply as well as its character of global commodity, grains often serve as global proxy for crops in discussions of agriculture production and productivity.

Perhaps we are not as brilliant as we believe

It is often believed that the biological production is high in farmed systems compared with nature. But that is a truth with modifications; often it is the other way round. Compare a pasture that is established in place of a rainforest or a field that forms when a swamp is drained. In both cases, the biological production is likely to be higher in the original, natural, system than in the farmed system. The biological production is concentrated into plants that human beings find useful. One or a few crops are selected, for human use, and allowed to dominate an area. In this way, we can use a higher proportion of the biological production from a farmed system than from a 'wild' system. Biological production really increases mainly when external resources are brought into the system; for example, when irrigation is introduced to drylands or when greenhouses are heated in cool climate or when nutrients are brought from outside the system, productivity can increase a lot.

In many 'marginal' areas, farmed systems are often not very competitive and in many cases directly harmful. Carl Linnaeus (1929) noted that the 'Swedish' settlers had a miserable life — and couldn't pay any taxes — in the mountains whereas the non-farming, reindeer-keeping Sami people did well under the same conditions. Diamond (2005) observes a similar situation for the Viking settlers in Greenland. When the climate changed, their farming system was inferior to the economy of the Greenland Inuits that was based on extraction of wild resources (Diamond 2005). And this is not restricted only to cold areas. The livestock expert R. N. B. Kay writes:

> From remote historical times up to the present day, clumsy attempts to introduce domestic species of plants or animals into marginally arid regions have led to catastrophic collapse of the ecosystem and the appearance of dust-bowls and deserts. [. . .] East African acacia savannah and bushland is able to carry roughly a five times greater biomass of wild ungulates than of domestic animals. Moreover, the restricted feeding habits of the domestic species may make poor use of the vegetation and at the same time cause the range to deteriorate. (1970: 271–72)

Around 30% of the land mass is too cold or too dry for farming to work well and here wild systems are mostly — even with a narrow human utilitarian perspective — more productive. 'Productivity' would have a completely different meaning if one counted the impact on ecosystems and the external costs caused by farming. Figure 14.2 shows four different situations; the white bars show the value per hectare generated from the now farmed systems, and the black bars show the value per hectare from a primary or secondary ecosystem that existed before. Both include ecosystem services, but not external costs caused by the system. In all the cases, the value of the original system is substantially higher. But, importantly, the benefits of the systems accrue to different people. Choices are made on the basis of the profitability for the person or entity that has managed to, with whatever means, control that piece of nature[102].

Figure 14.2 Values of different production systems.

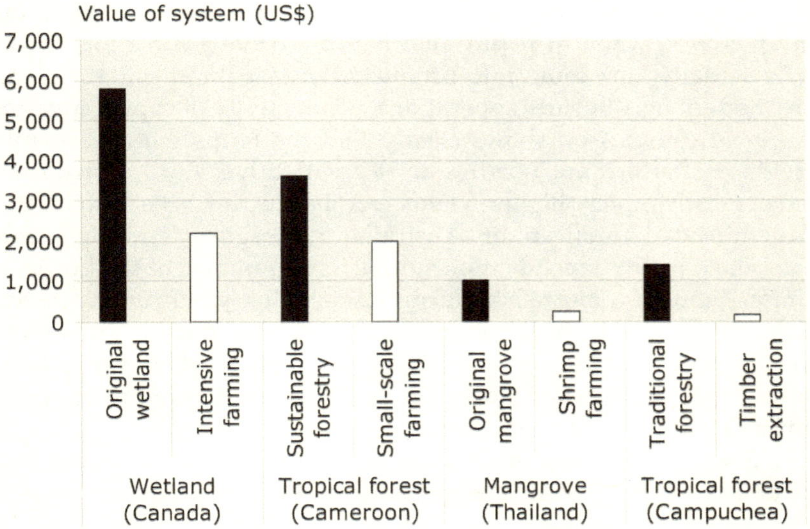

Source: MEA (2005).

If we also include external costs that arise from modern farming, it looks even worse. External costs for farming in Great Britain reached some £1.5 billion, corresponding to £265 per hectare (Pretty *et al.* 2005). The calculations are conservative and only include proven costs; for example, it includes no costs for health effects of pesticides in food. There are similar calculations from the United States and Germany. The external effects of pesticides alone has been estimated to amount to between US$ 8 and 47 per hectare of arable land, or an average of US$ 4.28 per kilogramme of pesticide active ingredient applied in Germany, the United

---

[102]   The cases have obviously been selected to show this, so this doesn't prove that farming is always futile.

Kingdom, United States and China (rice only). In the case of China, these external costs exceeded the market value of the pesticides—for every US$ 1.0 worth of pesticide applied, costs to society in the form of health and environmental damage averaged US$ 1.86 (Pretty and Waibel 2005).

How shall we measure?

It is a bit puzzling why most agronomists and institutions such as the World Bank, in capitalist society, to such large extent focus on yield of food crop per hectare as the main measure of agriculture productivity, when in reality that is not a driving force for farmers who actually look more into productivity per labour unit, or if they are modern agri-business operations, productivity of capital invested. A global comparison shows clearly that the farms with the highest yield per hectare are not the most competitive. European farmers have mostly much higher yields per hectare of wheat than their Argentinean, American or Australian colleagues, still they can't compete and are dependent on support programmes of the European Union. Similarly, most small farms have higher yield *per hectare* than large farms, still large farms gradually squeeze smaller farms out of the market because of market access, possibilities for rational specialization, economies of scale, better access to credits or simply governmental policy distortions.

As we can see from this discussion, there are many different ways to look at farm productivity and, depending on what we measure and how we measure, we may draw different conclusions. In principle, it is the factor of which there is a shortage that will, and should, determine which factor is the most important. Farms in high-income countries are shaped by high input of energy and low input of human labour (energy). In high-income countries, there is no shortage of labour but it has been costly and therefore productivity per work-hour has been the strongest driver of change, whereas farms close to the cities are shaped differently by high land prices.

To farm with high land productivity, water productivity and energy productivity is important if one takes into account natural resources. If the market 'worked', there would be prices on all resources and their comparative values would be 'right'; the profit per invested capital would ensure correct resource allocation. But it is very apparent that such a market doesn't exist in farming and will never exist.

Farming shows clearly the limitation of markets for getting the costing right. There is, at least, the need for a whole battery of other mechanisms (fees, subsidies, pricing of resources, taxes, regulation) to have the correcting functions that a sustainable system will need. The link between human labour (energy) and external energy is very strong and that is also the point of intervention in the farming system if one were to change it fundamentally. Humankind will soon be 9 billion strong and nature's resources are more and more stressed; it is hard to see the logic of increasing production per person-hour being the most interesting for farming in developed countries.

– 15 –
Green Growth and Green Capitalism

To quibble with this kind of system [capitalism] about its values, to frighten it with visions about the consequences of growth, is to quarrel with its very metabolism.

(Murray Bookchin,
Toward an Ecological Society, 1980)

Commerce and industry appear to have taken environmental issues to heart. Many environmental organizations are either oriented to big business, such as the Natural Step or Forum for the Future, or formed by big business, such as the World Sustainable Business Forum. Organizations such as the World Wildlife Fund, Sierra Club and Rainforest Alliance increasingly orient themselves towards the industry. Environmental considerations are regarded as profitable; examples of saving energy, saving materials or saving water are repeated again and again to show how human beings, and the industry, without any sacrifice can have a smaller environmental footprint. The signs about saving water, detergent and energy are in hotel rooms all over the world. Eco-efficiency, including Factor 4 or perhaps Factor 10,[103] is talked about.

The United Nations launched a new initiative in 2009, when the financial crisis surfaced and climate change still was on everybody's lips, the 'Green Economy'. The message in all this is quite consistent, and has its origin in the report of the Brundtland Commission, *Our Common Future* (United Nations 1987), which was the starting point for the widespread usage of the term 'sustainable development' in 1987. 'What is needed now is a new era of economic growth — growth that is

---

[103]   *Factor 10* refers to the possibility of creating products and services that have a massively lower resource intensity than the conventional alternative. It evolved from the concept of *Factor 4*, as developed at the Wuppertal Institute for Climate, Environment and Energy.

forceful and at the same time socially and environmentally sustainable, writes Gro Harlem Brundtland in the foreword to the report (United Nations 1987: xii). One learns that it is possible to combine economic growth with social and environmental development; not only that, the message is actually that *more* economic growth is needed to deal with these challenges, and that a free market is best situated to deal with the issues. But as we will see this is an illusion, a delusion or just wishful thinking.

Eco-efficiency and de-coupling

Those who argue that the market economy can deal with the environment and resource use in a good way are very fond of terms such as 'eco-efficiency' and 'de-coupling'.[104] They refer to inverted u-curves or Kuznets curves that are supposed to show that, at a certain stage, increased growth will reduce environmental problems instead of increasing them. It is shown that energy per GDP unit goes down in high-income countries or that use of other raw materials per GDP unit is less. In the period 1975–1993, the total material need per GDP unit decreased in Germany, Japan, the Netherlands and the United States (WRI 1997). Deforestation is often rampant in countries in early stages of their growth, but tends to go down with increased income and finally reverses in high-income countries; no nation where annual per capita GDP exceeded US\$ 4600 had a negative rate of forest volume (Kauppi *et al.* 2006).

Meanwhile, the average American will, during his or her lifetime, consume some 450 tons of construction materials, 18 tons of paper, 23 tons of wood, 16 tons of metals and 32 tons of chemicals. Consumption of raw materials in the United States increased 17-fold whereas population trebled between 1900 and 1990 (Carley and Spapens 1998). The global GDP grew from US\$ 19 trillion (constant 1995 dollars) in 1980 to US\$ 35 trillion in 2002, a growth of 83%. The materials use, meanwhile, increased from 40 billion tons to 55 billion tons, a growth of 36%. Expressed as kilogramme per dollar it means a reduction from 2.09 to 1.55 kilogrammes per dollar, but still the total use increased substantially (Giljum and Hinterberger 2004). The first long-term global study says:

---

[104]   De-coupling refers to economic growth that does not lead to increased use of resources.

Humanity currently uses almost 60 billion tons (Gt) of materials per year. In particular, the period after WWII was characterized by rapid physical growth, driven by both population and economic growth. Within this period, there was a shift from the dominance of renewable biomass towards mineral materials. Materials use increased at a slower pace than the global economy, but faster than world population. As a consequence, material intensity (i.e. the amount of materials required per unit of GDP) declined, while materials use per capita doubled from 4.6 to 10.3 t/cap/yr. (Krausmann et al. 2009: 2697)

Between 1900 and 2005, the only periods of absolute de-materialization coincided with economic recessions and the two world wars (see Figure 15.1). Across the whole period, materials use grew much faster than the population but slower than the GDP. Use per person thus doubled in the period whereas use per GDP unit decreased to only 40% of what it was earlier. Biomass, which is among other sources for human nutrition, seems to be linked primarily to population growth, whereas the use of non-renewables is much closer linked to economic growth (Krausmann *et al.* 2009). And the other way round as well, the energy expert Robert Hirsch (2008) estimates that a decrease of oil supply by 1% percent will shrink the GDP also by 1%.

Figure 15.1 Materials use by material type (1900–2005).

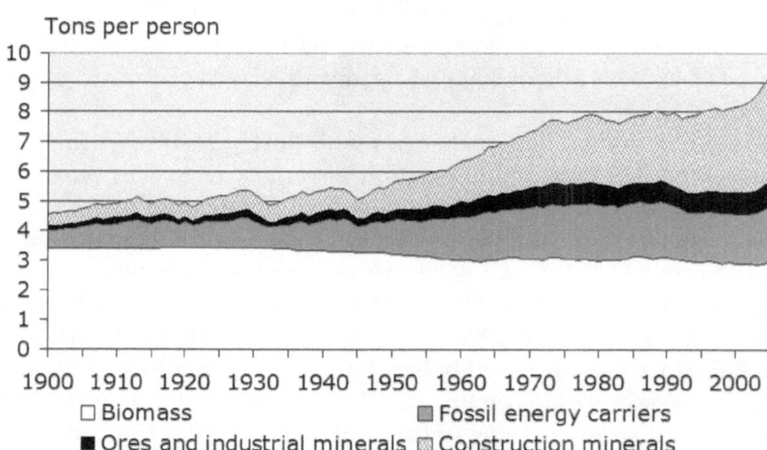

Source: Krausmann *et al.* (2009).

Emissions and resource use in some rapidly developing countries, such as China, are to a large extent the result of export to high-income countries. Most high-income countries are net importers of emissions and pollution that take place in the exporting countries (see earlier discussion about greenhouse gas emissions) and they also 'use' a lot of resources in exporting countries. In sync with increased consumption, transports are increasing, and for them there seems to rather be the opposite of de-coupling. Between 1990 and 1999, the growth of transport within the European Union was higher than the growth of GDP (SCB 2003). Global maritime transports are also increasing at a pace quicker than the GDP. The International Maritime Organization predicts that, in the absence of correcting policies, emissions from ships may grow by 150–250% between 2007 and 2050.

Aviation is also growing exponentially. Technological improvements have not at all sufficed to compensate for the emissions from the growing transport sector. New aircraft are 70% more fuel-efficient than those designed 40 years ago. A further 20% gain by 2015 over 1997 levels seems attainable, and perhaps a 40–50% gain by 2050. However, such improvements are insufficient to keep emissions and resource use at reasonable levels in view of aviation's rapid expansion (UNEP 2008, 2011).

There are certainly areas where industries have been able to grow and total emissions and resource use have decreased. The paper and pulp industry in Sweden has, in 25 years, gone from a big waster of resources to an almost closed system. Use of water, sulphur dioxide emissions and emissions of organochlorides have all been reduced by more than 90% (Eklund 2000). But these are exceptions. Materials use seems to flatten out in more mature economies; their economic growth is also moderate. For humanity at large it is now more interesting to see what happens in the growth economies of today. A middle-income country like Mexico shows a very strong correlation between materials use and GDP, even if GDP grows at a slightly quicker pace (Gonzalez-Martinez and Schandl 2007).

A similar discussion as the one comparing different countries can be held with regard to households in the same country, and perhaps that is a better comparison as they exist under the same conditions. Data from relatively equal Sweden show clearly that the richest 10% emit thrice as much carbon dioxide (which is a rather good proxy for general resource use) as the poorest 10%. Meanwhile, the rich emit less carbon dioxide per Krona; the pattern is quite the same as for a comparison between countries, and is in a way rather obvious: money is largely used to consume things, buy bigger houses, travel more, etc.

Modern technology is said to be cleaner and more efficient than the old industries, and, for sure, that is often true. Mobile communication is a fantastic technology and it appears to be a 'light' technology, despite people changing phones every year. But no evidence suggests that it has released pressure on any other resource or service apart from—possibly—the fixed telephone lines. People move around as much as before, or perhaps even more, as the mobile phones have taken away important reasons for planning and coordination. In Western Europe, despite all the talk about paperless offices, paper consumption quadrupled between 1960 and 2000, and the only time paper consumption went down was during economic recessions (Solberg, undated). Modern communication technology made it possible to use video conferencing and thus reduce travel. Meanwhile, travel is one of the real growth sectors, so if video conferencing increases it is probably because it is one more technology added to the existing ones, perhaps mainly replacing a less demanding technology, a non-video conference call. It seems like it is only when nature makes it impossible, for example, with the Eyafjallajökull volcano ash cloud in Europe in 2010, or when there are security disruptions, such as after the attack on the World Trade Center in New York in 2001, that one travels less.

Information technology is often referred to as a 'clean' and resource-saving technology, which can replace other technologies with more resource uses. Although there are a few such examples, evidence is not at all compelling that information technology leads to reduced resource use in general. Of 8 million new servers each year, 50% are bound for data centres. That's 4 million additional servers per year that will be used to store and deliver computing programs and services over the Internet, or 'the cloud', services that generally are taken as a good example of lean and clean. Information technology companies are quickly developing cloud-based offerings and building data centres to house servers that store all of the data the cloud delivers—email, photos, videos, entertainment, news and computing programs. More and more electricity is required to operate these data centres and run the servers. According to a report from Greenpeace (2010), 623 TWh of electricity is used to drive the servers and the Internet globally. To put this into perspective: the United States produced some 800 TWh from all its nuclear plants and the total electricity consumption of the country was just above 4000 TWh (EIA 2010); the total electricity consumption in the United Kingdom was 344 TWh in 2008 (EIA 2011). Even more worrying is the forecast that 1964 TWh of electricity will be used to drive the Internet by the year 2020.

One example of an unrealistic expectation from technological efficiency is given by the pioneering American environmentalists Paul Hawken, Amory Lovins and Hunter Lovins in their book *Natural Capitalism: Creating the Next Industrial Revolution*, where they describe the magic Hypercar powered by fuel cells:

> When you are not using it, rather than plugging your car into the elec-
> tric grid to recharge it—as battery cars require—you plug it in as a
> generating asset. While you sit at your desk, your power-plant-on-
> wheels is sending 20-plus kilowatts of electricity back to the grid [. . .]
> it wouldn't require many people's taking advantage of this deal to put
> all coal and nuclear power plants out of business, because ultimately
> the U.S. Hypercar fleet could have five to ten times the generating
> capacity of the national grid. (Hawken et al. 1999: 35)

Many similar unrealistic propositions abound; the odd thing with this one was that it came from the environment camp.[105] Certainly, there is a lot of efficiency to gain in the automobile industry, but here Hawken *et al.* (1999) forget that the hydrogen in the fuel cells has to be produced in some way—by electricity that is (see earlier discussion about energy—Chapter 6). *Perpetuum mobile* is simply an impossibility. All in all, the proponents of eco-efficiency and de-coupling promise much more than they can deliver. Increased efficiency and clean up of pollution is mainly driven by higher prices of raw materials or energy, costs for effluents, local community action or governmental regula-tions.

While people in high-income countries may have the impression that the problems with smokestack industries are already solved, by seeing how industrial areas such as the Ruhr, Birmingham or Pitts-burgh don't smoke or smell any more, this is jumping to conclusions. Many of those dirty industries are only relocated somewhere else, where they cause the same kinds of problems they did before, but to poorer and powerless people. Even if one doesn't want to admit it, both the standard of living and the clean environment in the West is built on the polluted shores of India, the backs of Chinese workers, the mining of rocks in Bolivia and the erosion of African soils.

Good and green consumption

Another version of green capitalism is green consumption; that is, consumers through their choices select environmentally friendly, or at least less harmful, products. The most successful such concept is organic products. The global market value of these products was around US\$ 50 billion in 2009. A couple of million farmers grow crops

---

[105]   It is perhaps these kinds of 'innovations' that are needed to make 'natural capitalism' work. We will come back to this discussion later on.

organically, not only in high-income countries, but also in developing countries such as India which has the largest number of organic farmers in the world, 600,000. Organic products mostly command premium prices of between 10% and 100%, with most products in the range of 15–25%. Price premiums are not guaranteed, however, and occasionally (rarely) organic products are sold for conventional prices. Five countries or territories (French Guiana, Sweden, Switzerland, Austria and Lichtenstein) have more than 10% of their land under organic management (Willer and Kilcher 2011).

Examples of other such labelling schemes are the Forest Steward-ship Council (for sustainable forestry) and the Marine Stewardship Council (for sustainable fisheries). The Forest Stewardship Council system certified 90 million hectares of forest by 2009, which corre-sponds to 10% of the forests that are commercially exploited. The highest share is in Europe, with Sweden alone having 10 million hectares (FSC 2009). Eco-labelling schemes are thus fairly successful. It is nevertheless clear that they have quite some limitations. Mostly, their consumption is not harmless for the environment; they are just better than the 'normal'. Consumers may think that as long as they buy an eco-labelled product, there are, or should be, no limitations to consumption, and thus they can lead to increased consumption threatening to offset the gains. Some schemes try to counter this by a constant increase in the threshold for rewarding the label.

There are similar so-called rebound effects on energy-saving devices. Although they do save energy, the total energy consumed tends to remain similar as people buy more or bigger gadgets; one swaps the small energy-wasting television to a triple-the-size energy-saving appliance. There is also an important limitation in eco-labelling as a tool for transformation. The attraction of it, for businesses and consumers alike, is to differentiate a product from other products that are without the label. But the more successful the label, the less value it has in a competitive market.

There are also more principle objections to eco-labelling; why should those consumers who behave most responsible pay for the failure of the market or governments to solve the problems. It is like institutionalizing free riders. With this view, eco-labels are more a sign of market failure than a sign that the market is working. Shouldn't society, the governments, ensure that products sold are as little damaging as possible? Should it not be the responsibility of all companies to take enough consideration of the environment, without claiming to be something special? One could compare with the social conditions in a factory. In high-income countries, today, no one thinks there should be social labels for products from their own country (possibly with the exception of the United States where such schemes are launched for the agriculture sector). This is not because they don't care about social issues, but because a combination of government regulation and shifts in attitudes to a large extent guarantees that working conditions are reasonable.[106] It would almost be embarrassing for a company to advertise that they provide living wages and a decent work environment.

These objections to eco-labelling have merit, but if one is a bit more pragmatic it is clear that eco-labels can make a difference and that they can drive technical development, until it becomes 'normal' to be environmentally friendly.

Good capitalism

Not only products but also companies can be certified for being environmentally friendly or socially responsible. The layman might believe that companies that have a certified environmental management system (ISO 14000 or EMAS) perform better in the environmental arena than those companies that have no such system. This impression is enhanced by the certified companies, the consultants helping them and the certification bodies selling certificates stating compliance. They use expressions like 'environment certified' or 'eco certified' (83,000 and 423,000 hits, respectively, on Google in Dec 2012). These certifications mean not much more than that national legislation is followed, that the company has a system to control its environmental effects and that the environmental impact is monitored. It is also, typically, the worst industries that are keen on letting themselves be certified to these standards. There is less direct value in terms of improved performance from these systems than from eco-labels. Companies, nevertheless, or perhaps exactly because of this, are more keen on these systems.

It is not negative as such, they still, mostly, perform better than they would without these systems, but it is misleading when they are promoted as making a big difference. In all fairness, there are companies that try. McDonald's in Sweden reuses 90% of all packaging: it uses up to 90% renewable electricity, serves organic coffee and milk

---

[106]   There are certainly exceptions in the same way as there are thieves or fraudsters even when theft or fraud is forbidden.

and has a ventilation system that adjusts itself to the numbers of visitors thereby saving a lot of energy — and money (McDonald's 2012). Coca Cola aims to be 'water neutral' by 2020 and Rio Tinto, one of the world's largest mining companies, launched its biodiversity strategy in 2004 with a voluntary commitment to achieve 'net positive impact' on biodiversity. To fulfil this commitment, the company first aims to reduce its impacts on biodiversity through avoidance, minimization and rehabilitation activities, and then aims for a positive impact through the use of biodiversity offsets and additional conservation actions (TEEB 2010). What this means in reality remains to be seen.

'Corporate social responsibility' (CSR) or just corporate responsibility is another concept that has been adopted by large parts of business, cheered by governments and the United Nations. The then-Secretary General of the United Nations, Kofi Annan, launched a CSR initiative called 'Global Compact' in 1999. In the same manner as companies that produce rubbish get quality certification (ISO 9000) and the worst polluters are certified for their environmental management systems (ISO 14000), companies committed to CSR are often involved in questionable business. The military contractor BAE shows responsibility by using lead-free bullets and less poison in missiles and grenades, thereby causing less pollution (Porritt 2007). Scandal companies such as Enron and many of the banks that were caught red-handed in the financial crisis of 2008/2009 are among the pioneers in CSR.

Green capitalism—a pie in the sky

Some almost totally deny any conflict between growth and environmental considerations or between profit-seeking and social or environmental sustainability, for example, Julian L. Simons, Johan Norberg and Bjørn Lomborg. I believe that this chapter has made quite clear that they are wrong. Others, however, agree that it is very problematic, almost desperate, but that capitalism and the market economy can take care of the problems, given the right framework. Hawken *et al.* (1999) coined the term 'natural capitalism'. They claim that this new capitalism will (and has to) be driven by values different from those of the current system.

On the other side of the Atlantic, Porritt (2007) follows a similar discussion in *Capitalism as if the World Matters*. Tim Jackson (2009) repeats it with his *Prosperity without Growth: Economics for a Finite Planet*, and the UNEP (2011) published the 600-page *Green Economy Report: Pathways to Sustainable Development and Poverty Eradication*. With all due respect for the prominent, intelligent and serious environmental pioneers and for the UNEP, not much in these books explain what mechanisms and incentive structures should be put in place to support this new kind of capitalism and the values belonging to it. We return to this discussion in Part 3.

A risk of green consumerism and green capitalism is the individualization of politics, or the reduction of politics to consumer and market choice. It is not society that shall choose by means of politics (regulations, incentives, etc.) how, for instance, farming shall be conducted. If consumers (not citizens!) want good animal welfare, they will choose products certified to some animal welfare standard; if they want farming to be organic, they will buy organic products. And if consumers don't like the unfair global trade, they will buy fair trade products. But this reduction of politics to individual consumer choice has the tendency to steer people away from political action and away from challenging the underlying mechanisms that created the situation in the first place. The effect of consumer choice is certainly very little for many of the big developments in the world today.

– 16 –
Population and Migration

It's no coincidence that most of those who are obsessed with population growth are post-reproductive wealthy white men: it's about the only environmental issue for which they can't be blamed.

(George Monbiot,
in *The Guardian*, September 2009)

The world population has increased from 1 billion at the beginning of the nineteenth century to 6 billion at the end of the twentieth century. Such a growth certainly justifies the expression 'population bomb' (as broadcast by Paul Ehrlich in his 1968 book). Many believe that population size is the most critical issue for the survival of human beings on the planet. Porritt (2007) gives the following example. The total carbon dioxide emissions are 30 billion tons from 6.5 billion people. That corresponds to 4.6 tons per person. Total emissions should go down to 18 billion tons by 2050. But the population is expected to be 9 billion then, which means that only 2 tons per person can be emitted, a reduction to less than half of current emissions. If, instead, the population stabilizes at 6 billion,[107] 3 tons per person can be emitted, a reduction of only two-thirds. Porritt (2007) believes that few people want to take a grip on the population issue because they are afraid of being seen as elitist, intolerant or right-wing extremists.

To regulate the reproduction of the human species has always been a tricky thing, and to do so at a time when the freedom of the individual is worshipped is even worse. Limits to population are simply not wired into human genes, rather the opposite. However, history shows that human beings have curbed population in earlier periods, not only through external factors but also by norms of society. Regulation of population, directly and indirectly, has been the norm all through

---

[107]   Things move rapidly and this option is already gone as the population is approaching 7 billion.

human history. It is the unfettered growth that is the exception. The immediate problem is not *how* people will get food but *how many* people the ecosystems can 'carry' before they start to collapse. In this perspective, *which* people become more plays a big role. For the planet as a whole, each new American represents a much bigger burden than an Indian, even if the United States is not yet so densely populated.

Grey hairs are good for growth

It is possible to influence population growth without force. Iran reduced its growth to half between 1987 and 1994 using a broad range of programmes including education of women and television broadcasts of family planning information. In Ethiopia, radio dramas have included information on family planning, gender roles, reproductive health and HIV. A study two years later showed that demand of condoms had increased by 157% (Brown 2008). Birth rate in Bangladesh fell from 6.1% to 3.4% between 1980 and 1996 through massive family planning efforts (Sen 1999).

Amartya Sen (1999) argues in favour of social action and, in particular, empowerment and education of women as most successful in population control. Some poorer states in India, such as Kerala, have lower birth rates than richer ones, a difference Sen explains with better education and literacy among women. He argues against a policy by force, as in China. The Chinese government has undoubtedly been successful in its efforts to limit population growth, but it has come at a high price, for example, selective abortions and even, allegedly, killings of small girls.[108] China has also had good education of women, good access to contraceptives and a high proportion of women in the work force, all things coinciding with lower birth rates, and it is not possible to disentangle which is the effect of these and which is the effect of the one-child policy (Sen 1999). It can be profitable to invest in population control. If the government of Bangladesh invests US$ 62 to prevent one birth it can save US$ 615 in social costs (Brown 2008). In the transitional phase, the country will benefit from the so-called demographic bonus.

Studies of the demographic transition in the high-income countries show a strikingly similar pattern in most countries. First death rates

---

[108]   This seems to be at least as spread in India.

decline rapidly — and it is not primarily the old who live longer but more children who survive. This leads to very big cohorts of youth, reaching 35–40% of the population. When birth rates go down after a while, a big wave of people will reach the stages of young adults, mature adults and old people. A reduction in fertility immediately leads to fewer burdens to take care of infants and children. Meanwhile an increase of burden for old people will not occur until those born before the reduction of fertility become old. Between those two situations, the start and the end of the demographic transition, the burden of taking care of non-productive persons is less than it was before and less than it will be in the future. This can be sufficient to initiate a positive loop of investment and productivity (Brown 2008; Malmberg and Sommestad 2000).

There is strong correlation between a number of society developments and this transition. The great, uncontrolled expansion of agriculture land coincides with large cohorts of young adults, clearing new lands (or like in the European case, emigrating to other countries for clearing land). Economic growth is greatest in countries where there are many people in the range of 40–60 years; they have savings; they can invest and they still produce. Finally, when the old become a big share of the population, growth inevitably slows down (Malmberg and Sommestad 2000). This is already shown by Japan and Europe, where fewer people are working and more pensioners are to be taken care of; on the other hand, fewer children are to be taken care of. Statistics from Sweden, one of the countries that first stabilized its population, project that the number of people in the workforce will be maintained at over 50%, actually never falling as low as it was at the beginning of the twentieth century, when Sweden experienced its population explosion. More women have entered the work force and people can also work longer.

The population follows our production system

Human beings, like any other species that conquers a new habitat, show an exponential population growth. What is unique is that human beings have been able to radically expand the ability to tame nature and other organisms and to direct production to the stomach and the body. Instead of following a typical S-curve of growth, humankind has twice restarted population growth. The first time was with the introduction of agriculture, the so-called Neolithic or agrarian revolution. Before that, human population grew mainly through expansion into larger and larger areas and into more and more ecological niches. The role of human beings was more or less the same, living on the surpluses from other species, which were collected, hunted or fished depending on the niche. Population growth was very low, perhaps just 0.001% per year (to be compared with 1.8% in 1990).

Human population was probably less than 10 million 12,000 years ago, and, most likely, the ecological limits for further growth were reached. With the transition to agriculture, many more could be fed from a defined area, but not everywhere, as some parts of the world are not suitable for farming. A period of rapid growth ensued. In 10,000 years, the population grew 100 times. At the time of the Industrial Revolution, the population had reached new ecological limits in many places in the world. Not so densely exploited areas were rapidly filling up with settlers from Europe, representing one-third of the annual population growth of Europe. Through the deployment of huge amounts of fossil fuel, first coal followed by oil, it was possible to, once again, take a giant leap to new levels (Livi-Bacci 1992; Chapman and Reiss 1999).

Improvement in technology could, in each kind of society, lead to some possibilities for expansion; for example, spears and digging tools could allow hunters and gatherers to harvest a larger share of the surplus of their habitat; the introduction of crop rotations and other improved agriculture methods could perhaps double or triple the yields, but it was only through employment of industrial technology that yields could increase 10-fold. In the same way, improvements in technology have constantly expanded the ability to feed more and more people on the planet also during this industrial phase, but limitations are now arising from all sides. It is hard to see what could be a similar radical change of conditions that would allow humankind to make another giant leap in population.

While we can agree that the size of the population is a problem, many examples show how high population density forces a more sustainable use of the landscape than a sparse population. For example, in Germany, Japan and Sweden it was first the exhaustion of, or the threat of exhaustion of, forest resources that compelled people to manage the forests properly and it was the increasing population that made farmers abandon the extractive practice of sod cutting to fertilize their soil (Radkau 2008). Similarly, regardless of how much environ-mentalists and foresters condemn swiddening agriculture, it will continue to be the norm until population density reaches a certain level, simply because it is the most comfortable, secure and sustainable way of getting a good return on labour input.

It is pointless to try to make poor farmers switch to other cultivation systems before the productivity of the agro-ecosystems is threatened. The Egyptians, who operated a limited but fertile land resource, knew that it was beneficial to alternate cereals and fodder crops such as clover or alfalfa (Mazoyer and Roudart 2006), but it was only under the pressure of population that crop rotations integrating fodder production and grains became widespread in Europe. Similarly, it is the population density of China over a long period that made the Chinese utilize human excreta in an efficient way, whereas in sparsely populated Africa it is rarely in use, even taboo in many places. '[Soil] fertility may be a result of the use of intensive methods of land utilisation and not *vice versa*' (Boserup 2005: 19). So while a big population in general is problematic, there are situations where it is not that bad or at least where society has not managed well with a low population.

Some of the ecosystems created are dependent on human beings, and with continued urbanization, despite population growth, more and more areas are depopulated. What these depopulated landscapes shall look like and how they shall work have not been planned, and so societies end up paying farmers or entrepreneurs to 'maintain' the landscapes or preserve the 'cultural heritage'. And this problem is not confined to high-income countries alone. Some areas with traditional terraced rice paddies experience labour shortages to maintain the infrastructure. In 2007, I visited traditional farms in Bali and their main problem was that the youth preferred to work in the tourism industry rather than on farms. Thus, the truly amazing landscape and culture were no longer sustainable as people voted with their feet for another life. If the terraces are not maintained, they will collapse from erosion, with potential devastating landslides as side effects.

Numbers are stabilizing

The global growth rate of population has been decreasing since its peak of 2% in 1965–1970 (UNFPA 2011). In 2010, 43 countries, including Japan, Russia, Germany and Italy, have populations that are stagnant or even decreasing. A larger group of countries, including China and the United States, has reached the stage where new families will be smaller. When the next generation reaches fertility, the population will stabilize. The third group will have doubled its population from now to 2050. This group includes many of the African countries such as Ethiopia, Congo and Uganda. The United Nations predicts three alternatives of population size for 2050: 10.8 billion, 9.2 billion or just below 8 billion. Most seem to bet on the middle alternative. That Malthus' horror scenario of mass starvation has kept its appeal over 200 years coincides with the start of the demographic transition of England following an unprecedented population explosion. This pattern has been repeated in country after country, and in the same way that Ireland was the frightening example 150 years ago, Nigeria or Ethiopia are now.

Why is population stabilizing? It is perhaps a strange question, but it is much more remarkable that the demographic transition has taken place and that populations *are not* growing than that there has been a population explosion. After all, we have been taught that the strongest driver of them all is to reproduce, to spread one's genes. And why are populations stabilizing now and not a hundred years earlier or later? Why is it not stabilizing in the countries where population is still growing? Is the reason technical, such as contraceptives; economic, such as increased wealth; human, such as improved education of women; or social, such as perhaps a result of shift in values? It is interesting to understand what drivers make individuals change their reproduction. There are actually very few reasons for why a European woman (or her possible partner) today only wants two children, while she wanted four some 100 years ago. It appears to completely contradict the sociobiological ideas that human beings only act with the purpose of spreading their genes. If that were the case, voluntary birth control would rarely occur.

For a long period, it was well known that smoking causes cancer and other diseases. Still, it was a long and slow process until this insight led to a changed view on smoking and subsequent regulations. It is interesting that the view on smoking is now changing in all countries, also in countries where smoking is perhaps not a primary health problem (because people don't live long enough to die from cancer). In the same way it appears that the regulation of populations has little to do with whether or not the country is overpopulated. Sweden and the Netherlands stabilized their population more or less at the same time; despite that the actual population density is very different, and large tracts of Sweden are rather 'underpopulated' if there is such a thing. So-called soft factors, values and education, seem to play a big role here. At the same time, on a global level, it is hardly a coincidence that populations level off at the same time that we see more and more limits to growth. It seems as if collective human behaviour re-affirms those limits.

Moving around

In 1960, 20% of the population in developing countries lived in cities; by 2000, it was 40%; and by 2030 the proportion is estimated to be 56%. The pace of urbanization is quickest in Asia and Africa. In Latin America, urbanization is already very high; Mexico City, Sao Paolo and Buenos Aires already have populations larger than 10 million, with Lima following suit. Migration is not proportional within the population. Those with higher education move more often and women move more than men. Education is, in some countries, a forceful driver of migration; youth move to the cities for education and don't return. Some parents move to cities just to ensure a better education for their children. Although urbanization in developing countries is rapid, most of them still have *increasing* populations also in rural areas, contrary to what most people seem to believe.[109]

We tend to believe that our time is unique, even if it often isn't. People are moving around the globe at a rapid pace, but this is nothing new. During the decline of the Roman Empire, Europe entered the 'Age of Migrations'. There were numerous invasions and migrations amongst the Ostrogoths, Visigoths, Goths, Vandals, Huns, Franks, Angles, Saxons, Slavs, Avars, and Bulgars. Similar movements occurred in many other parts of the world, for example, the Bantu expansion in Africa. At the close of the nineteenth century, the fraction of foreign-born residents in many countries was higher than today (UNDP 2009) During the first modern wave of globalization from 1870 to 1910, almost 10% of the world's population moved from one country to another. The biggest migration ever was the 50 million people who migrated from Europe.

During the latest wave of globalization, some 150 million people have moved, but they represented only 2.3% of the global population (FAO 2003). The common view is often that the poorest are the ones migrating, but that picture is not really true. To move you need money and contacts and often some kind of education (World Bank 2007). Another common misconception is that most migration is from developing countries to high-income countries, whereas actually the biggest flow is between developing countries (Collodi and M'Cormack 2009).

---

[109]   In countries with rapid increase in population it is possible to have rapid urbanization and increasing populations in the countryside simultaneously.

According to FAO, migration plays a bigger role in equalization of income than trade or capital movements. Migration plays that role in at least four ways.

First, the millions moving away from Europe between 1850 and 1910 created a shortage of labour in Europe and, therefore, a press upwards on wages. In the receiving countries, such as North America, it was the opposite. All in all, differences around the globe shrank owing to migration. As a result of migration, salaries are estimated to have increased by 32% in Ireland, by 28% in Italy and by 10% in Norway, whereas salaries shrank by 22% in Argentina and by 8% in Australia (FAO 2003). Second, foreign remittances, that is, what migrants send to their home countries, amounted to US$ 338 billion in 2009 (ILO 2010), a figure three times as high as the total development aid from high-income countries! In some countries, in Moldova, Central America, the Pacific and the Caribbean, it amounts to more than 10% of the GDP. Third, some emigrants go back to their home countries. Sometimes they come back having failed, but more often they have saved some money and come back with new ideas as well, ready to invest. Fourth, many of those who don't return or send money to relatives still keep contacts in the old country, and they sometimes invest in it or start trade with it.[110]

There is little or no evidence for the popular perceptions that migrants displace local workers, increase crime or exploit welfare systems (ILO 2010). There is, however, some evidence (as pointed out above) that it does lead to a somewhat more even level of salaries globally. But isn't that just another reflection that global inequalities need to be reduced? Quite obviously, this will not happen by everybody reaching the income of the richest, even if this message may not be too popular. Despite all these advantages of migration very few advocate free movement of people, and the issue is hardly on the international agenda.

Two important, globally valid, problems are linked to large-scale migration. One problem is that the social cohesion is easily disturbed when 'strangers' enter a population. This is not because they are bad; it is just because they are different. Societies are built on predictability, that one knows and understands what other people are doing and will do, and that one understands the signals. But in populations develop-

---

[110]   For instance, Armenian and Bosnian diaspora are major investors in their respective home countries.

ing apart from each other, these signals and behaviours are different, and it will always create some stress when different populations are mixed. The other problem is linked to human interaction with eco-systems. Each culture has co-developed with nature, influenced nature and been influenced by nature. This means that human relationship to nature and what is actually done, that is how one farms, how one manages forests, how one builds, etc. represent a kind of ecological adaptation. But when one moves, this doesn't work well. Several examples show how migrating populations have not understood the ecosystems they moved to and therefore have created havoc, for example, the effort to introduce English systems of farming in Australia or the US mid-west, or high-altitude Peruvians or Bolivians moving to the Amazon for farming.

We are reaching stability—is it too late?

World population has exploded, literally. Still it now moves towards stabilization, perhaps not rapidly enough, but quicker than many thought 20 years ago. Many would argue that the human population is too many already, and this can be discussed for eternity. The large numbers are a reality that we have to cater to. The problem is not primarily the numbers but how much 'space' each person takes, how much of the biological production, how much of finite resources. More and more migration will occur.

To make migration free seems to be a most human and morally just thing to do. It also makes sense for economic development. It is a mystery why most proponents of more globalization don't take this up in their propaganda. Or perhaps it isn't; perhaps, they are just promoting what is in their narrow interest and what supports their privileged position by just promoting the liberties and rights that they profit from themselves, while denying others the same? After all, for the decision-makers and the international jet set free movement is largely assured. We will discuss this more in Part 4.

– 17 –
Man and Nature

We live then in a world different from that of a mountain lion—he is neither bothered nor blessed by having ideas about ecology. We are.

(Gregory Bateson,
Steps to an Ecology of the Mind,
2000)

Human beings have a troubled relationship with nature. On the one hand, the other creatures in nature are admired and it is mostly agreed that they have a right to exist. We fight for the survival of whales or eagles, while slowly squeezing out species after species. Is the human species not just following the same logic as all other species, to expand at the expense of others? To see human beings as just one of many species, with no unique right, is tempting in its simplicity, but where does that leave us? To see human beings as the overlords is to make virtue out of necessity. We de facto control the survival of an increasing number of species.

Of all species, 99% have been extinct

Of 10,000 bird species, 90 have been extinct since 1600 and another 1000 are considered to be threatened. Of the extinct species, most were endemic, that is, existed only in an isolated environment, mostly an island. A third of the extinct birds were on Hawaii. Also so-called indigenous people exterminated other species. Before arrival of the Europeans, the Maori of New Zealand exterminated 12 of the 90 bird species (NRM 2008). The role of human beings in the great death of big mammals in North and South America is still disputed[111], but the

---

[111]   Some claim that the reason for why this happened in the Americas but not in Africa was that in Africa human beings and the big mammals developed side-by-side for a million years, and they had adapted themselves

mammoth, the mastodon, camels, horses, bears and big cats disappeared within some 1000 years after the first appearance of human beings in the Americas. And there is no other, better, explanation given than that those human beings were to blame. As the Swedish author, essayist and politician Rolf Edberg (1976) says: 'our life becomes poorer, emptier, when one life form is lost and diversity reduced. Life forms lost can never be replaced. A lost instrument is a loss for the whole orchestra, and reduces the power of the symphony.'

A hundred species per million are currently estimated to be lost per year (Rockström *et al.* 2009). But what does this really say? There is no clue even about the number of species currently existing: 'Right now we can only guess that the correct answer for the total number of species worldwide lies between 2 and 100 million,' says the ecologists Michael Rosenzweig (Society For Conservation Biology 2003). According to the IUCN Red List of 2009, 17,291 species out of 47,677 assessed are under threat: 21% of all known mammals, 30% of all known amphibians, 12% of all known birds, 28% of reptiles, 37% of freshwater fishes, 70% of plants and 35% of invertebrates (UNEP 2010b). Coral species are declining most rapidly. Nearly a quarter of plant species is estimated to be threatened with extinction. The abundance of vertebrate species, based on assessed populations, fell by nearly a third on average between 1970 and 2006, and continues to fall globally, with especially severe declines in the tropics and among freshwater species (CBD 2010).

Meanwhile, one should have a reasonable perspective. Mass extinctions have taken place earlier. But the belief is that in earlier mass extinctions the rate was some 15–30 species per year, whereas now this is about thousands of species per year. According to the Millennium Ecosystem Assessment, the rate is some 50–1000 times higher than earlier. It estimates that the rate will increase further (MEA 2005). Some claim that 99% of all species have been extinct already over billions of years.[112] And most people are likely to be happy that the *Tyrannosaurus rex* is not on the loose; if it did I am certain that there would be calls for their culling. One often hears that there are chain reactions: if one species vanishes, other species also disappear, or the whole balance in the ecosystem is disturbed with unpredictable

---

to survive the predatory pressure from human beings. That was not the case for their relatives in the Americas (Lithander 2003).

[112]   Obviously, this figure is just a very rough estimate and there is no possibility to really know this. But all seem to agree that a vast majority of all species that ever existed have already been extinct.

consequences. And, for sure, it can be like that as there are such keystone species. The scenario varies, but the central idea remains that a single species has a fundamental impact on ecosystem functions. On the other hand, nature is very multifaceted, and at least in the longer view a new 'balance'[113] will arise. Also, species that vanish are, mostly, specialists with narrow niches, rather than generalists. If they are gone, a generalist will often take their place, at least for a while.[114] Human beings, mice and house sparrows are generalists that adapt themselves to many different environments (the mice and the sparrow seem to follow in human footsteps). Human beings have, accordingly, taken over the role of many of the predators they have exterminated.

So many species disappearing at such a rapid pace is worrisome, not only because they are beautiful, interesting or have 'a right to exist', but also because this can lead to a loss of potential raw materials, such as for food, medicines or pest-control agents. More than 50% of all medicines are produced by substances first detected in nature. More than 50,000 species are used for medical purposes, of which 3000 are traded internationally. Some 1000 of them are grown commercially (IUCN 2009). But by and large, it is not that one species is lost that a problem arises. If that were a disaster, the world would already be gone as most species have been extinct. The problem is the large-scale transformation of the whole planet to serve human beings, a process where not only a few species but also landscapes, whole ecosystems and their essential life-upholding services are exterminated.

Are we equal to the elk, the carrot or the stone?

There are many ideas about the relationship between man and nature. They are not only a matter of theoretical interest as those views colour society's standpoint in matters of great importance for the future; so we will take a look, admittedly a brief and somewhat simplistic look,

---

[113]   I put 'balance' in quotation marks because I don't think it is so apparent that there is anything such as balance in nature. Balance assumes stagnation, but nature and life are incredibly dynamic seen in a longer view. Living in a land that was covered with ice only some 10,000 years ago, a land that is under constant change, it is very hard to recognize what is balanced in this place (Sweden). Even most of our revered 100,000 lakes are slowly disappearing from natural processes.

[114]   The long-term tendency is undoubtedly towards more specialization; that is the whole essence of evolution.

at this relationship. The Norwegian philosopher Arne Naess (1974) and others have developed a worldview coined 'deep ecology'. In short it says that the human being is a part of nature and that all parts of nature have equal rights to develop and flourish, 'A classless society within the whole biosphere, a democracy where we speak about rights not only for humans, but also for animals, plants and minerals' (Naess 1974). The ideas also have some similarity with the rather famous *Gaia* theory of James Lovelock, where Earth is seen as one huge organism where different processes (consciously or not) contribute to the creation and maintenance of good life conditions.[115] Other, older, similar ideas were expressed, for instance, by Albert Schweitzer and the Jainites. One can have sympathy with the thoughts but, in my view, they are not really helpful for forming a view of nature that can help deal with current challenges of society. There is always the risk of drawing the conclusion that nature is 'good' or always 'right' and what is good for this abstract nature is always good for people. But one gets no help in how to manage conflicting interests by that view. It is hard enough to manage conflicts among human beings, and it is simply very difficult to envision a society where the interest of an individual human shall be balanced against that of a single tree, not to speak of a stone.

For the Russian-American author–philosopher and neoliberal icon Ayn Rand, ecology as social principle is dangerous: 'Ecology as a social principle [. . .] condemns cities, culture, industry, technology, the intellect, and advocates men's return to "nature," to the state of grunting subanimals digging the soil with their bare hands' (Rand 1976). And obviously, ecology shouldn't be exalted to a social principle; ecology is no principle[116] at all and ecology was never the teaching of how society shall be organized, even less than economy. But ecology teaches some rules or limitations and provides an understanding of a human being's role in ecosystems, which we, whether we like it or not, have to live by. These rules are, in the end, more determining than the economic 'laws' that many economists refer to.

---

[115] A very different, and thought-provoking, view of self-destructive life is proposed by the biologist Peter Ward in his book *The Medea Hypothesis: Is Life on Earth Ultimately Self-Destructive?* (2009).

[116] This observation is no critique of ideas such as ecological farming, agro-ecology or social ecology where the word ecology is used to indicate a certain perspective.

In my view, to equate us to an animal is to take away our responsibility, for ourselves and for other life forms. If we were like any other animal we wouldn't have to care about greenhouse effect or the extinction of other species. Do you think the lion consciously saves gazelles to ensure a sustainable kill? Do you think the cuckoo limits her number of eggs so that she will not out-compete the involuntary hosts of her genes? When we left Eden, that is when we left the hunter and gatherer society, innocence was left behind and the possibility to be equated to the mosquito, the elk, the snake and the apple was forever lost. It is equally delusional to project typical human properties onto animals and plants, which is to say that they are human beings in the same way as we are. In the same way as one hardly condemns the hawk for taking a field mouse should we condemn the human being who hunts a deer, or the evolutionary extension of it, breeds a cow for slaughter.[117]

Human beings are part of nature, have always been and will always be. The British social scientist Gregory Bateson points out that it is human habit to change the environment instead of changing oneself (which would represent a genetic adaptation): 'Man the outstanding modifier of environment, similarly achieves single-species ecosystems in his cities, but he goes one step further, establishing special environments for his symbionts [i.e. plants and animals]. These, likewise, become single-species ecosystems: fields of corn, cultures of bacteria, batteries of fowl, colonies of laboratory rats, and the like' (2000: 451). While this is what we try to do, luckily we fail, would be my reflection. It is not uncommon that species change their life environment, and to some extent, all life forms do that; the environment would be different without them. Mostly the change is limited, but sometimes it can have big consequences. Geoff Cunfer (2005) says that the bison probably changed the ecology of the North American prairies more than the cattle ranchers. The dams of beavers are well-known interesting pieces of engineering. Some eucalyptus species can spontaneously ignite. As they can stand the heat of the fire themselves, it is warfare against other species with enormous effects. Leaf-cutter ants manage advanced farms by the growth of fungi on leaves they collect.

---

[117]   This doesn't mean that human beings have to eat meat. One has a choice to change one's ecological role as well, for instance by not eating meat, if one sees that this is beneficial for humanity. This ability to redesign one's ecological role is part of the speciality of being human.

Still, the scale of human intervention is unprecedented or at least unique in the context of my discourse.[118] Plants and animals are enslaved to serve human beings, who make new ecosystems. Thousands of years ago, already, people had invaded almost all inhabitable spaces on the earth. Not even the vast Amazon rainforest is as pristine as many believe. People have lived there for thousands of years, they have farmed, they have hunted, and they have cut trees and collected plants. Many of the nature types we sees as 'natural' and want to protect are human creations.

Large-scale irrigation schemes are early examples of people engineering their environment. Venice has for a thousand years been engaged in engineering water, in particular the lagoon, half sea, half lake. Venice owed its rise to a trading power largely as a result of its salt works. Those salt works were threatened by the inflow of sweet water. To avoid this and to avoid siltification as well as to fight malaria, the Venetians rerouted several rivers. They also regulated the fisheries in the lagoon to preserve the stock, with similar measures as those used by 'sustainable fisheries' today, for example, mesh size of fishing nets to allow young fish to pass. Environmental policy and environmental engineering took a central place for the Venetians, as it does even today (Radkau 2008). The sinking Venice is a useful reminder for how difficult it is to fight the powers of nature.

To respect nature is to respect ourselves

Nothing human beings do can compete with a piece of nature in complexity. A square metre of land contains so many organisms with so complicated and complex relationships that one can't understand them. A gramme of forest soil may contain over a million bacteria colonies. It is estimated that several million individual animals and over a thousand different species may reside beneath a square metre of soil surface (Perry 1994); all of them are constituted by many, many cells and there is interaction between them, an exchange of energy and nutrients. Perhaps, a square metre of nature is more complex and contains more information than the whole Internet.

---

[118]   Again, Richard Ward (2009) gives interesting perspectives, of a longer view, of how, for example, microbes producing hydrogen sulphide inhibited life for long periods and how climate change and mass exterminations in many earlier periods were a result of life itself.

Apart from the direct and indirect services obtained from nature, nature also has great aesthetical and cultural value. It gives another perspective, many reference points; it is an unprecedented piece of art. Natural environments have, for centuries, been known for their healing properties for people in spiritual need or mental stress. It is good to have respect for nature. The respect should be based on what it is; for its beauty and for its value, both apparent values and those not discovered. This respect should of course also include ourselves. A level of development has been reached where we are already managing most of the terrestrial ecosystems. This gives human beings unique powers among the creatures on the planet and unique responsibility too. We can't pretend that we don't have that power or that responsibility; both have to be exercised with due diligence, humbly realizing that there are so many things we don't know.

It should be kept in mind that exploitation of nature is also a question of *whose* nature. Different groups are affected by destruction of nature in different ways, and normally the nature of the weaker group is destroyed while the nature of the rich is preserved. This leads some to believe that the rich 'care more' for the environment or that only when there is a certain wealth that human beings 'can afford' to care about nature, but this mixes up cause and effect to a large extent. The rich can exploit the nature of the poor and even export their waste to the nature of the poor, whereas the poor has no other option than the exploitation of nature, be it in the form or natural resources or in the form of their own labour. To some extent, the whole trick of modern civilization is about exploitation of nature; exploitation of historical production, of which fossil fuel is the primary example; exploitation of the future nature, by dumping waste. But it is also the history of a few people exploiting many other people. At the centre, apart from nature and energy, are technology and society. In Part 3 we discuss them and how they are interlinked.

Society has four big challenges regarding the use of natural resources; well perhaps using the term 'challenges' is an understatement, 'system errors' is perhaps a more suitable term. First, humankind uses already too much of the planet's resources; the capital is being consumed and it is not a question of *whether* society will go bankrupt[119] but *when* it will do so. Second, the distribution of these

---

[119]   The term is a bit unfortunate, the repercussions of an exhaustion of the natural capital is a lot more dramatic than a financial bankruptcy, which in essence only means a redistribution of money. The terms for bankruptcy with

resources is extremely inequitable, between countries and within countries; rich countries and rich people use a totally disproportionate share of the resources (and rich people in rich countries are apparently the worst). Third, economic growth is continuing and seems to be hard-wired into society, which means that resource use will also continue to grow.[120] Fourth, the world's population is primed to grow for at least 40 more years, perhaps longer. We can't negotiate with nature and no less can we expect that the poor will, or should, accept that common assets are wasted.

nature can't be negotiated. There is no bailout.

[120]   We discussed and showed earlier that even if the additional resource use per additional GDP unit is going down, the total resource use is still growing.

PART III
OUR SOCIETY

We have everything, but that is also all we have.

(Ole Paus)

In the preceding chapters, we discussed the relationship between man and nature, or what we often carelessly call the environment. I demonstrated what the extraordinary strain of the economic system, constant expansion of the population and human consumption patterns entail for nature and ultimately for society. I also showed how the capitalist market works, or rather how it does not work, as an institution for stewardship of nature, of ecosystems and mineral resources. This is hardly surprising because markets were never developed with the purpose of managing complex biological or ecological systems.

Now we move to how human beings relate to each other and to society at large. Also, here I discuss how society is impacted by the economic system. The economic system is a part of society, although our previous discussion showed that one problem is that the economy has gradually taken supremacy over most of society instead of the other way round. It is hard to understand why many people who are most against the individual being subordinated by society seem to find it natural that individuals should subordinate themselves to economy, be a slave under economic 'laws', laws that are just human rules in the same way as the laws or norms of society.

– 18 –
What Drives Human Enterprise?

We are people through other people.

(Bantu proverb)

Homo economicus and Homo geneticus

Two widespread perceptions or theories interpret or explain all human efforts and motivation from one single factor. One is socio-biology, which sees human agency as just a reflection of the selfish interest of genes to multiply, and the other is economicism and the invisible hand, which interprets all human actions as motivated by economic stimuli; that is, all traits that are 'profitable' will be chosen and will ultimately dominate. These two theories are not contradictory. Socio-biology is perhaps more applicable at the private level and economicism at the societal level. The more extreme proponents of both theories try to explain *all* human behaviours with their respective theory, for instance that we chose partners on the basis of economic considerations alone.

Socio-biologists, with some support from the theory of natural selection, consider egoism of genes to be the engine of all developments and actions. They don't deny that people can act altruistic but explain altruism as something that is driven by egoism in the first place. The fact that young men waste their genes as canon feed in a meaningless war — or in a dignified battle for freedom for that matter — is explained by the fact that they (or rather their DNA) can 'count' on the genes of their sisters being spread and, in such way, their own genes can live along. Or some other far-fetched explanation. That the world is not filled by rapists cannot be well explained by this theory but can well be understood with a societal perspective of human beings. If the desire for rape were genetic, only through a systematic killing of the infants born out of rape could it be contained, especially if there were no societal mechanisms against it (and such an infanticide would again be a societal response). Socio-biologists also fail to explain why women in high-income countries don't have so many children.

They could manage to raise four, and well up to ten, children and all would survive, spreading their genes much more effectively than with the current two children. Despite all these objections, it is quite clear that many human actions can be explained on rather simple evolutionary grounds — anything else would be sensational.

Simpleton economicists and their popularizers want to explain all human actions by economic drivers. Also, many Marxists tend to see the acts of different groups to be fully dependent on their economic standing in relation to production factors — here the concept of 'class' is critical. The reasoning goes from self-evident things to very cumbersome explanatory models. History shows that economic motives and personal gain were in no way important drivers before capitalism. Material wealth has, for sure, always been attractive, but social status and power have always been more important. And economic wealth was not the route to these; many cultures despised trade and entrepreneurship. The basic error is the perception that economic drivers are superior to all other drivers and that all other drivers can be 'translated' into money. The strength in the reasoning is that the economic driver is a forceful paradigm with many self-reinforcing loops. And as a society is built along this paradigm, human beings consciously and subconsciously create a world that rewards exactly this; in such a way it becomes self-fulfilling. But this is not truer than the statement that the most important driver for human development is the ability of an individual to throw a spear farther than anyone else. This ability was most likely very important in a hunter society or a war society, much more important than the ability to amass wealth.

Even the liberal economist and philosopher Friedrich von Hayek (2001) notes in his famous *The Road to Serfdom*, an attack on both fascism and socialism, that the ultimate objectives of a rational person are never economic; that there is no 'economic motive', but only economic factors that influence our ability to achieve other goals. Another version of economicism is the idea of 'rational choice'[121], that is, the idea that one can find a rational explanation to all human deeds. Also, this idea is to some extent totally self-evident and totally

---

[121]   It is ironic that in many cases the same people promote these ideas and the idea of the virtues of a free market, based on the individual's *free* choices, because as Nassim Nicholas Taleb sharply notes: 'If you believe in free will you can't truly believe in social science and economic projection. You cannot predict how people will act' (2007: 184).

meaningless in the sense that irrational choices most certainly will lead to failure in the long run as this is more or less the definition of 'irrational'; if it is successful it isn't irrational . . .

The arguments of the economicists and socio-biologists can complement each other and don't necessarily contradict each other. Also, the economicists and socio-biologists are often the same people, or the same people use their arguments alike. If economic wealth also leads to higher levels of reproduction, they most certainly complement each other. While there are some indications that success is attractive, it doesn't seem to translate into rich people or rich societies having more children than poor people or poor societies. Modern population patterns seem to support a totally opposite view. The poor have more children and today they also have more surviving children. Both theories forget the importance of the social and cultural context.

We are tamed by culture

Many look towards animals for guidance on which behaviours are 'natural', with the understanding that what is 'natural' is good or right or, at least, inevitable. And, for sure, many interesting parallels exist between human and animal behaviour. But the search in nature for examples of certain behaviours is often coloured by what people want to see. For some reason, many tend to look a lot more towards chimpanzees than bonobos. Both species are equally close to human beings, so why are comparisons drawn mainly with the aggressive chimpanzees? Chimps live in aggressive patriarchates whereas bonobos live in matriarchates, using sex for social interaction and conflict resolution. While social hierarchies do exist among bonobos, rank plays a less prominent role than in other primate societies. Isn't it so that human beings project what they want to see in the examples they seek? Space doesn't allow any major digression into gender, but if there is any area where the most meaningless comparisons with nature are made, it is here. Even human reproduction is described in terms of male activity and female passivity, while reality might be quite different.[122] Nature has such diversity that one can find almost any behaviour there.

---

[122]   In the article 'The egg and the sperm: How science has constructed a romance based on stereotypical male–female roles', Emily Martin (1991) shows that it is the egg that captures the sperm and not the sperm that penetrates the membrane of the egg.

The long adolescence of a human being is long and slow for a (evolutionary) reason. While one can envision human beings growing up in a few years, like horses or cows, or in half a year, like dogs or pigs, in reality human beings need a lot more time. A newborn chimp's brain weighs around 60% of a grown-up chimp's brain, whereas a newborn human's brain weighs just 24% of an adult human's brain (Gärdenfors 2003). It takes human beings some 12–18 years to grow up, and this time frame seems to get longer as society becomes more and more complex. Humankind has to learn so many things and, more importantly, human beings need to be 'socialized'. Some even qualify the human brain as a bio-cultural organ (Welzer 2011), with layers of accumulated cultural learning. Through culture, society harnesses and adapts some genetic heritage. For instance, let us look at reproduction. In almost all cultures, human beings wait to reproduce long after they reach sexual maturity; in many societies, parents try—admittedly, sometimes unsuccessfully—to control their offspring's choice of mate.

Most rules and customs in society are mainly about the preservation of society and not so much about biology. Ultimately, behaviours that are taught must, in the end, be considered as 'natural' as other behaviours that are more clearly genetically conditioned, and they are as much or as little 'good' as behaviours that are genetically conditioned. Society and culture are a part of the human species. Without society, without culture, without the mastering of fire or without language (which is a very good example of a social innovation that is essentially human), we are simply not human beings. Therefore, the argument that genetically conditioned behaviours are more important or more 'real' than culturally conditioned behaviours is a denial of humanity. This is not in contradiction to the survival of the genes because human genes are most common with those of the family, then with those of society and, finally, with those of humanity at large. It is also not in contradiction to the human quest for utility as individuals, which the economicists claim is the predominant force. Human society and cultures recognize both behaviours but have always reined in their forces.

Mutual aid

The Russian geographer and anarchist Peter Kropotkin wrote *Mutual Aid: A Factor of Evolution* (1902) in response to 'social Darwinism' and, in particular, to Darwin's supporter Thomas H. Huxley's essay 'The struggle for existence in human society' (1891), where Huxley applied

Darwin's observations to human society, citing 'survival of the fittest' as the predominant force in shaping human society and development. Kropotkin was strongly against Huxley's perspective. He supported Darwin's theories about natural selection, but believed that, at least in Huxley's version, it overemphasized the struggle between individuals of the same species. Kropotkin looked at other animals, but in particular at human society, and claimed that mutual aid is an important principle both in nature and in human development. He cited the cooperation in mediaeval cities, in village communities and workers' cooperatives as successful examples of mutual aid. He showed how such self-organized, voluntary systems operated the fisheries in the Caspian Sea, the Volga and the Urals.

Hundred years later in Japan, people created communal institutions to organize insurance, credit and health care, since the state prioritized the military–industrial complex (Honda 1997). And today, open-source software development is a good example of the successful application of mutual aid. 'Perhaps in the end the open-source culture will triumph not because cooperation is morally right or software "hoarding" is morally wrong . . . , but simply because the closed-source world cannot win an evolutionary arms race with open-source communities that can put orders of magnitude more skilled time into a problem' (Raymond 2000).

Recently, Sloan Wilson (2003) showed the importance of human cooperation (among others in the form of religion) in his book *Darwin's Cathedral: Evolution, Religion, and the Nature of Society*. He gives an example to better understand how 'groups' or 'societies' can be 'adaptive units', that is, subject to 'laws' of evolution at the level of the group and not only at the level of the individual. Males of some species can kill infants in order to mate with the mothers. This would be, and is, a successful reproduction strategy for them, but not for the mothers, the infants, the group or the species. Human beings (human society) have a moral revulsion against such practice and the person trying it would be excluded from the group. Such a group would be better adapted and more successful than a group that did not have such moral codes. In this way, moral systems have evolved by group selection to suppress self-serving (but group-damaging) behaviours. Sloan Wilson notes, however, that 'those groups of males who do not kill each other's offspring might well kill the offspring and appropriate the females from other groups' (2003: 38). This puts our solidarity towards our own group (people, tribes whatever category) and our hostility towards 'the other' in an understandable evolutionary context.

One can lift the discussion one level higher though, where the survival of the entire species depends on the human ability to adapt to new situations (such as climate change). Such a situation would then assign adaptive group properties to the whole species and be a ground for universalism. All these are just examples of what was concluded earlier, that societies also are evolutionary, adaptive organisms, where the cooperation of individuals are almost as important as the cooperation of cells within the body.

The unit of survival is not the individual

We have discussed at length in Part 2 how the human being changes her environment. Perhaps it would have been wiser or better to have grown a thick fur than to have built houses, and to have developed more body fat and flippers to swim better, but that was not the adaptive response of humankind. I venture that the speed of human development and the complexity of human society don't allow for a sufficiently rapid genetic adaptation of the individual. Gregory Bateson says: 'the unit of survival is a flexible organism-in-its-environment' (2000: 457). I believe that 'environment' should here be understood to encompass ecosystems, but also culture and society; all of these work together as adaptive responses, and it is only by taking care of all of them that we can meet the challenges of the future.

Human beings are governed and driven by DNA, by the brain and by human culture. It is all about information. DNA is coded information that governs some, but not all, of human actions. The brain receives impulses from the inside or the outside and processes information and takes action to improve a human being's chances of survival. It does that in sync with one's genes, but not as a slave under one's genes. Human culture is the collective memory of who we are, where we came from and where we are heading. This memory, this information, also guides a lot of human actions. Each individual is formed by culture during upbringing and, in this way, each generation is adapted to the prevailing culture. Culture and society are rather extensions of human bodies; we don't really exist without them.

The lone able savage (as depicted, for instance, by Rousseau or his later followers) is a construction—the human being was never alone. Society is the organizational form so to say and culture is the information in its wider sense (including religion, economy, etc.). Therefore, one can't pitch the individual against society. Sure there will be, and has always been, a tension between the independence of the individual and the demands of submission of society (I am an example of an individual who most reluctantly submits oneself to the demands and expectations of society), but this tension is part of the project of being human. It can also be seen as similar to genetic mutations. Individuals revolting against society are mutations ensuring variation and development; like most mutations, most of these revolts will be unsuccessful, but a few will succeed. To see the individual and society as integral parts of being human is a first step in thinking about how to build a better world. And this transcends both the socio-biological idea that an individual only acts in the self-interest of his or her genes and the capitalist ideology that each person acts as an individual maximizing his or her economic gains.

19

Globalization

Globalization today is not working for many of the world's poor. It is not working for much of the environment. It is not working for the stability of the global economy.

(Joseph Stiglitz,
Globalization and Its Discontents,
2002)

Globalization is worn out as a word. It is spoken about as if it is something new, but it isn't. Take a look at the quote below from *The Manifesto of the Communist Party* by Karl Marx and Friedrich Engels; it could as well have been from Naomi Klein, Walden Bello or Vandana Shiva.

> All old-established national industries have been destroyed or are daily being destroyed. They are dislodged by new industries, whose introduction becomes a life and death question for all civilised nations, by industries that no longer work up indigenous raw material, but raw material drawn from the remotest zones; industries whose products are consumed, not only at home, but in every quarter of the globe. In place of the old wants, satisfied by the production of the country, we find new wants, requiring for their satisfaction the products of distant lands and climes. In place of the old local and national seclusion and self-sufficiency, we have intercourse in every direction, universal inter-dependence of nations. And as in material, so also in intellectual production. (Marx and Engels 1848)

Some components in what is called globalization include an increased global trade leading to a world market; that is, a price for a good becomes more or less the same, corrected for transport costs (which are ever decreasing) and tariffs, ownership across nation borders and international capital flow. Other components include an intensive exchange of ideas and a harmonization of consumption pattern, values, culture and entertainment. Finally, there is direct communication between people in different countries, by means of migration travel and virtual contacts.

Some find that there has been a qualitative jump in globalization as a result of the Internet. Be that as it may (and we come back to the issue later), some areas of the world today are less globalized than they were some 100 years ago. Large parts of the world simply had no border controls for people; Europe was, in principle, free from passports from the mid-1800s to the First World War, when passports were introduced to keep spies out and useful labour in. We have described earlier the long history of international trade flows. Also in the Americas before 1492, mother-of-pearl from the Gulf of Mexico ended up in Manitoba and Lake Superior copper in Louisiana (Mann 2005). Later, triangular trade tied together North America, Africa and England. Foreign investments and know-how played a big role in the development of Sweden in the seventeenth and eighteenth centuries. 'The economic development of Sweden in the seventeenth century was a Dutch by-product' (Danielsson 1977).

Even food self-sufficiency was left as a goal in the commercialization of Dutch farming. Holland was dependent on grain imports from the Baltic areas in the sixteenth century, while Dutch farmers grew flax or made cheese (Bath 1976). Ideas, innovations and germs were early globalizers. As political scientist Joseph S. Nye points out: 'The oldest form of globalization is environmental. For example, the first smallpox epidemic is recorded in Egypt in 1350 BC. It reached China in 49 AD; Europe after 700; the Americas in 1520 and Australia in 1789' (2008). The speed of spread is what is unique today. While it took three millennia for smallpox to spread to all continents, it took less than three decades for AIDS and half a year or less for the flu to go round the globe.

However, notwithstanding how impressive early trading empires were, despite all the might and splendour of Venice, Bruges, Genoa, Seville and Amsterdam, 'life and survival of most people was grounded, far into the nineteenth century, on local and regional self–sufficiency. [. . .] Before the age of the train, the steamship, and the truck, it was difficult to transport bulk commodities over long distances: at that time free trade did not yet mean what it means today' (Radkau 2008: 41). Use of energy is really what has speeded this up. The cost of transporting one T-shirt from China to the Netherlands in 2011 was 2.5 cents (*The Economist* 2011c).

The first wave of globalization

The period between 1850 and 1914 showed many of the signs of globalization. Foreign capital corresponded to 30% of the GDP in developing countries compared to 20% today. Migration flows were proportionally higher than they have been after the Second World

War. During a long period, there was an international currency, the gold standard. Improvements in transportation technology, in particular steamships and railroads, were both a driver and a prerequisite for the increased trade. A bushel of wheat cost 60 cents in Chicago in 1870 and double in London. By the end of the century, transport costs, and thereby the price difference, had shrunk to 10 cents (FAO 2003). Not only transport but also new, or more efficient, methods of preservation, such as freezing and heat preservation, developed. Australia exported 7200 kilogrammes of preserved meat in 1866 and 9900 tons five years later. It was assessed that England, at the onset of the First World War, was just 25% self-sufficient in food, lower than today (Tansey and Worsley 1995). International capital markets also developed. Karl Polanyi writes:

> [By the fourth quarter of the nineteenth century] the repercussions of the London money market were daily noted by businessmen all over the world; and governments discussed plans for the future in the light of the situation on the world capital markets. Only a madman would have doubted that the international economic system was the axis of the material existence of the race. (1944: 18)

In the period from 1850 to 1914, global economic growth was bigger than ever. A few countries in Latin America, Canada and Australia caught up with the wealthy European states (de Larosière 2007). There were also tensions similar to current times. European farms suffered competition from its former colonies, in particular the United States, Canada and Australia. Halfway into the period, protectionist measures were taken in many countries. Should one conclude that the wave of globalization was therefore over? No, certainly not. Those measures were taken because of the storm-like strength of globalization, to allow it to continue, but to limit the damages. Regulations thus helped to overcome its internal destructivity, similarly as social welfare legislation was needed to mitigate the social effects of labour markets. The system could survive only by regulating and limiting total liberalization.

The earth is flat

In 2010, I visited Berthe Kvarn, a family operated mill in the south of Sweden. Per Stenström, a member of the family, told us how this mill now was one of a dozen mills in Sweden. Until the railroad was built, mills mainly bought grain from the surrounding farms and sold it to local clients and the closest towns, and there were thousands of mills in Sweden. The railroad changed it all. With the railroad, the mill could buy grain from far away, even from other continents, and, as importantly, they could compete in a national, and later global, market. In a few decades, most mills were closed and the remaining ones grew in size.

The most important function of economic globalization is to create a world market with 'free' (that is unlimited) competition. As such, globalization has the same effect as other measures that strive to open up new markets, for example, building of roads. Roads and railroads are essential for development; they certainly speed up the exploitation of nature's resources (which sometimes is their stated objectives in the first place), for example, the Eire Canal and the Trans-Amazon highway. The welfare gains are often very uneven, and it is not sure that there are any welfare gains at all. The World Bank (2009) notes that, in almost all cases, the wealthy will draw more benefits from improved communications than the poor and the differences in income tend to increase rather than decrease. The same holds true for globalization in general.

The debate about globalization is polarized and lacks nuances. The proponents of globalization don't want to admit that not everybody is a winner, that the environment is harmed by the ever-increasing transportation of goods, and that globalization so far has been driven primarily by big companies and has also benefited them. And ultimately, they forget that the empirical evidence in favour of globalization is scant, if any; no evidence suggests that globalization delivers great welfare gains. In their discourse, all progress is assigned to globalization and all problems have other reasons. It is indeed possible, even likely, that an economic analysis shows that globalization, through increased trade, delivers a higher GDP. But to some extent this is only the trick of an illusionist. Increased trade means more shipping and airfreight, and more people travelling. Of course, this will increase GDP and more people will be employed. But does this mean that human welfare or the standard of living rises? That seems hard to prove.

Critics accuse the high-income countries of dumping agriculture products, undercutting farmers in low-income countries that have abolished their protection. This is a fair criticism, but it is rather an effect of too much protectionism and market meddling in the high-income countries (of which agriculture is the worst example); so it is more a result of too little free trade than of too much. The transnational companies are criticized for social and environmental dumping when they work in developing countries; they don't adhere to the same conditions as those in their target markets or where their head office is located. Asocial actions by companies like United Fruit[123] in Latin America are extrapolated to be the normal behaviour of transnational companies, but that is a far too simplistic picture. To begin with, as long as nation–states exist, it makes a lot more sense to follow the laws of the country where one lives or operates than the laws of another country; anything else would be another form of imperialism. But, more importantly, social and environmental standards are mostly remarkably higher in transnational companies than in local companies that are not at all subject to the same scrutiny by public opinion. Transnational companies often impose stricter standards on their suppliers as well. Foreign direct investments are important for economic development and transnational corporations bring with them new technology and know-how, often superior to what was used before.

Even with protectionist measures, like tariffs, the lower costs of transport and the increase in communication lead to stiffer global competition than before. Most industrial processes are standardized and very similar in all countries. As a result of lower transport costs and better information flow there is little difference between the processes in different countries, and key staff are easily moved. There is not much space for dramatic change in most classic consumer goods, whereas there is still a lot of innovation in modern goods such as cell phones and other electronic gadgets or in biotechnology. A factory in Thailand can be owned by an Indian company, have an American boss, and use raw materials from Zambia and technologies from China.

It is thus no longer the in-house capacity of the factory that is a determining factor in competition. Actually no factory is needed at all; a lot of hassle can be avoided by simply buying production capacity. On the country level, it is therefore not the entrepreneurship or even the capital that is the most interesting strategic advantage for industrial production, it is simply to offer the best institutional framework and the provision of labour (and the latter is only for labour-intensive production; for other types of production, the whole labour

---

[123]   Nowadays Chiquita. The company was known to meddle in the politics of the countries in which they operated and to treat workers badly. In Colombia, 47 banana plantation workers were killed in 1928 when they protested against low salaries. The expression 'banana republic' has its origin in those conditions.

force can also be imported).[124] In manufacturing, it is not companies that compete in an international market, but rather countries and political systems that compete. Countries, as well as companies, most easily compete by 'distorting' the competition, for example, by having no company tax, tax breaks, low salaries, differing standards, good educational systems and other services that make the country cover costs for the companies, that is externalizing costs. From this perspective, industrial production is even more exposed to commodification than agriculture, where it is basically only the products that are reduced to commodities; the production itself is still site-specific as long as agriculture is land-based.

What is striking is that the same groups that emphasize that there are only winners and no losers in globalization, that it is beneficial for everybody, constantly remind us of how important it is to 'keep our edge', or 'improve our competitiveness', to upgrade society so that 'we' are in the forefront of research and technical development. The columnist Thomas L. Friedman writes in *The World Is Flat: A Brief History of the Twenty-First Century*:

> So if the flattening[125] of the world is largely (but no entirely) unstoppable, and holds out the potential to be as beneficial to American society as a whole as the past market evolutions have been. How does an individual get the best out of it? What do we tell our kids? There is only one message: You have to constantly upgrade your skills. There will be plenty of good jobs out there in the flat world for people with the knowledge and ideas to seize them. I am not suggesting that this will be simple. It will not be. (2005: 235)

This is echoed by politicians from all kinds of camps and by institutions such as in the European Union's vision of 2020:

> Our economies are increasingly interlinked. Europe will continue to benefit from being one of the most open economies in the world but competition from developed and emerging economies is intensifying. Countries such as China or India are investing heavily in research and technology in order to move their industries up the value chain and

---

[124]   The discussion is, admittedly, a bit simplistic to make the point clearer. In reality, there are many nuances, but the underlying tendency is clear. This is also why 'China' is the emerging global power and not a range of companies from China to Bolivia.

[125]   The 'flattening' is how Thomas L. Friedman describes globalization enhanced by the Internet and other modern technologies. He also refers to it as Globalization 3.0.

> 'leapfrog' into the global economy. This puts pressure on some sectors
> of our economy to remain competitive, but every threat is also an
> opportunity. As these countries develop, new markets will open up
> for many European companies. (European Commission 2010b: 5–6)

Clearly, if it is so important for the United States and Europe to be ahead of the pack, it must be because the ones falling behind are 'losers', that is, globalization is actually more of a threat for them than an opportunity. Why can't people in the rich countries just relax a bit, spend energy on cleaning up the environment and reducing their ecological footprint, and enjoy a good life instead of being coerced into working harder and harder to keep the edge?

Inherent inequalities drive global trade. The global 'division of labour' that is the basis for globalization is as much a global 'division of power' as the difference between the capitalist industrialist and his workers. Alf Hornborg (2006) shows how, in 1850, when England swapped cotton fabric for £1000 with cotton lint for the same sum, it corresponded to an exchange of 4000 hours of work in a textile mill with more than 32,000 hours in tropical cotton production, as well as the resource of 58 hectares of land. All factors indicate that the patterns of exploitation inherent in globalization are the same today.[126] It is mainly because of the big difference in the purchasing power of the 'rich' as opposed to the poor that the rich can benefit from the cheap coffee from Nicaragua, the cell phone from China or the service of the call centre in the Philippines. And it is the knowledge of this that makes it so important 'to keep the edge'.

We protest against burqas and watch Big Brother

One of the most striking features of this wave of globalization, and one that makes it different from the earlier version, is the unregulated flow of information. Internet certainly has helped this, even if the process started already with the daily newspapers, the radio, the telephone and telegraph. This flow of information has not only aided the development of trade, science and technology, it has also led to universal values, such as respect for basic human rights. It is harder and harder for dictators to limit the flow of information to their subordinates; it is harder and harder for them to commit atrocities or

---

126   It is very difficult to make these kinds of calculations in the incredibly complex and interwoven value chains of today.

violate human rights without the whole world knowing. The same harmonization of values, which is seen as something good when speaking about human rights, is perhaps less appealing when it comes to culture. It seems like a sign of poverty that half of the world listens to American hits, another half follows the English league in soccer and a third watches the reality TV series Big Brother. We risk losing a lot of cultural diversity in this process.

Another 'civilizing' side of globalization is interdependence. One of the central tenets for the foundation of the European Union was that through the creation of a common market and business integrated over the borders one would create a common interest in peace and this would put an end to the nationalist wars that haunted Europe, and spilled over to the rest of the world for centuries. So far it has worked, even if it is perhaps unlikely that there would have been any war without the European Union in the period 1945–2011. The same logic can also apply globally. When individuals are all part of an international web, which is essential for livelihoods, one is hopefully less inclined towards conflict. On the other hand, the dismantling of both Yugoslavia and the Soviet Union shows that these things can be quite easily reversed. Globalization is also the awareness that human beings need each other and that some problems require global solutions; in that sense there are parallels between the apprehension of the interdependency of each other and the interdependency of man and nature. Both are essential for sustainability.

Further, there are moral reasons in favour of a free exchange between people. Trade is an exchange between people and should as such be seen as a basic human right. Why would I be denied the right to change products with someone in another country? Similarly, the free movement of people is to a large extent a moral issue. In both instances, it is important first to set the principles and thereafter discuss the limitations that might apply. This is not much different from the requirements for other human rights. Many of them have limitations too.

Looking into the crystal ball, it seems as if globalization as a political economic 'project' has lost its speed and credibility as a universal truth. This is because of a combination of factors, such as fear of global terrorism (e.g. 9/11), fear of global epidemics (e.g. AIDS, repeated flu scares, SARS), fear of a country's competitiveness, mistrust of unregulated markets in general and financial markets in particular (e.g. the Asian crisis at the end of the 1990s, the global crisis in 2008/2009) and an increased interest in the local. Countries take countermeasures, like they did at the end of the nineteenth century.

But as it did then, the undercurrent of globalization of information, trade and people will continue. Only a really big crisis, a global ecological crisis, such as a global food shortage, shortage of fossil fuel or of a strategic mineral such as copper or phosphorus or a third world war, will stop it. None of these are totally unlikely, and a world war is perhaps most likely to occur as a result of one of the others.

I am not *against* globalization. I believe that free movement of goods — and of people — are human rights. I believe that, in total, globalization has more benefits than drawbacks, but then I speak about globalization as more than a narrow economic thing. I think of the globalization of human rights, of the Internet, of the fact that dictators all over the world can't get away so easily any more. I think of globalization as a force undermining the authority of the nation-state and nationalism. So while there are, in general, drawbacks and benefits of globalization, depending on how the rules are bent globalization can be good for one and bad for another. Globalization at present has been driven or, rather, hijacked as a capitalist project, opening up all aspects of human life to exploitation. As such, it deserves the protests.

20

Development Issues

Consider the old adage, 'Give a man a fish and he is fed for a day, but teach a man to fish and he can feed himself forever.' Is this adage really the right approach? If you teach a man to fish, does he have a line and net to be able to catch the fish? Does he have access to water? Can he get his fish to market to earn income? If the man fishes, does any of the fish get to other members of the family? And does the man even like fish at all?

(John Clark,
Third CIVICUS World Assembly, 1999)

In conjunction with the independence from colonialism, there was optimism, and a forceful economic growth was predicted for most developing countries. But that didn't happen. Many parts of Africa are today poorer than they were in 1960, and the gap between them and the high-income countries has increased. At the time of the First World War, colonies represented one-third of the global trade. At the end of the 1970s, the total share of developing countries (including also some countries that were not colonies, such as China and Brazil) in international trade was 10% (Hart 2000). Most African countries are very dependent on exports to high-income countries, but these high-income countries are dependent on Africa to a very little extent; the value of African exports is negligible in the global economy (Bigsten and Durevall 2007).

In absolute terms, some 900 million live on a daily income of below US$ 1.08 (the World Bank's lowest poverty threshold) and some 2.1 billion below US$ 2.15 per day, that is, almost one-third live on 'almost nothing'; they earn *per year* less than I earn *per day* and what some charge *per hour* of their work, and yet very few earn per minute. The proportion of poor people shrank from 37% to 30% in 2002; most of the improvement was in China. In Africa south of the Sahara, 51% are poor (World Bank 2007). Landes (1998) estimates that today the ratio of income per head between the richest nations, such as Switzerland, and the poorest nations, such as Mozambique, is about 400 to 1; 250 years ago, this was perhaps 5 to 1. The ratio between Europe and India and China was perhaps less than 2 to 1 (Radkau 2008).

Significant differences in income exist within countries too. In the United Kingdom, GDP per person in the richest region is almost 10 times as high as in the poorest. In the United States, the difference in income between the District of Columbia and Mississippi is fivefold, and increasing. Shanghai in China is set to overtake Mississippi in 2015 and people there already earn more than people in a quarter of the regions of the United Kingdom and Italy (*The Economist* 2011b). Against this backdrop, it is hardly surprising that global development issues are quite high on the international agenda and that disparity in income and wealth has become a main topic of global concern.

Free trade—a development opportunity?

Some claim that the main reason for why some countries are lagging behind is that there is not enough capitalism and globalization. Some support is found for this opinion. Most countries with slowly growing economies are authoritarian with excessive governmental regulations. The costs and procedures of starting a new business are excessive in Africa and Latin America compared to high-income countries. It takes 6 days to register a new company in the United States whereas it takes 149 days in Congo and 195 days in Haiti. The cost of registration is above the average annual income in many developing countries whereas it is 0.7% of the average income in the United Kingdom and the United States and 0% in Denmark (World Bank 2010a).

But when it comes to the miraculous growth effects that *trade* liberalization is supposed to deliver, the arguments stumble. It is true that most of the high-income countries have chosen to liberalize their economies and be open for international trade. But the cause and the effects are not so clear here; most of them were not particularly liberalized in the early stages of their development, but they became more positive towards liberalization when they had become strong actors (Vylder 2002). The Victorian Great Britain and the United States were strongly protectionist during their early periods of growth. Trade liberalization followed rather than led the rapid economic growth in the 1950s and 1960s (Arrighi 2010). Further, the gains from free trade are not as important as many believe. Even the free-trade-oriented World Bank states that the costs for the current situation compared to those for full liberalization correspond to 0.7% of the GDP (World Bank 2007), hardly a make or break situation.

The advantages of trade liberalization are perhaps more apparent when all play in the same league. To match an ice-hockey team that can hardly afford to buy skates, not to speak about the necessary protection, against an NHL team is not fair play, is it? There are, for example, no indications that Africa has been able to draw many benefits from the increasing world trade. Its share of world export has gone down from 7.3% in 1948 to 2.4%[127] in 2000. In absolute numbers though, there was an increase from US$ 4.2 billion to US$ 142 billion in the same period (Gibbons and Ponte 2005).

That capitalist economy and capitalist values are slow in penetrating Africa can, to a very large extent, be explained by the fact that societies are not adapted to the ideals and organizational forms that are part of capitalism. It is quite likely that the introduction of capitalism, at least initially, could lead to worsened conditions. And even if it were to create economic growth, its social costs can be high. Polanyi explains:

> 'These institutions [the social institutions in the colonies] are disrupted by the very fact that a market economy is forced upon an entirely differently organized community; labor and land are made into commodities, which, again, is only a short formulation for the liquidation of every and any cultural institution in an organic society' (1944: 167).

Development starts in the field

By looking at the share of agriculture in the GDP of high-income countries, one might draw the conclusion that agriculture is not so important for development. On the contrary, a closer look reveals how important agriculture is for the early stages of development, and also for industrialization. The elites in developing countries show little interest in the agriculture sector of their countries; farm projects often fail; social and environmental effects of large-scale farm projects have often been dramatic and subject to criticism; there is political resistance to farm projects from the farm lobby in donor countries. All these contribute to the neglect of farming. Still, growth in farming benefits the poor more than other growth according to both the World Bank (2007) and FAO (2003). The role of agriculture in development, over and above the production of food and maintaining the landscape, is

---

[127]   Arne Bigsten and Dick Durevall (2007) say it went down to 1.4%.

manifold. Agriculture should also produce raw materials for industry (wool, dyes, etc.) and biomass for energy. The agriculture sector can also be the birthplace for new industries[128] and can accumulate surplus to be invested in other sectors. Finally, it is a market for industrial products (hoes, ploughs, etc.) and a source of labour for industry.

The view of agriculture and its role in development has been subject to many fashions and fads. It is important to realize that conditions are quite site-specific and what works well in one country doesn't necessarily work well in another. Many development projects focus on technology and agriculture methods. Of course that matters, but mostly not so much as what agronomists believe. They are merely parts of a much more complex process leading to increased production and productivity.

In early days in Europe, like in most other places, farmers totally lacked incentives for improving production and all added surplus was taken by landlords or the state. In addition, lack of demand mostly meant that any increased production would only lead to falling prices (Gadd 2000). In England, Denmark and Sweden many farms were left idle in the seventeenth century just because the rent was too high to enable any profitable farming (Bath 1976; Myrdal 1999). A similar situation prevailed in India during the nineteenth century where land rent often amounted to more than half the production value. In north Awadh in India, farmers were left with as little as one-sixth at certain times (Clingingsmith and Williamson 2005). The same happens today in many countries, and the lack of incentives is a core cause for little development in the agriculture sector. To change this requires a change in ownership structures, control of land and taxation systems. Without this change, introduction of new technologies or farming methods are almost doomed to fail.

Almost all farms are today linked to the world market, in one way or another. Even if they don't produce for the world market, the prices for the staple foods they produce are affected by world market prices. If local prices drop, exports commence and prices move towards world market prices; if local prices rise, imports press prices down.[129] In this

---

[128]   While the entrepreneurs of the late twentieth century sat in a garage working with computers or music, many of the earlier industrial entrepreneurs started their business in a barn or a stable.

[129]   Transport costs and delays in price information to trickle down or filter up are, in addition to any tariffs or quotas, the main factors that cushion this.

way, the global market can impact prices all over the world even if rather small quantities are actually traded. Still, many smallholders have a substantial part of their business outside of the market economy; they produce for self-consumption, have no hired labour and use no credits, all of which leads to their integration in the world economy being weak.

To this can be added the institutions they live in. The extended family is normally their only safety net and the group to which they pay allegiances; the learning needed for the job is mainly done on the job, and if they ever get a substantial surplus the pressure from the extended family is normally to share the wealth rather than to invest it in new production. They are also quite independent from the state, and the state normally doesn't care much about them in the first place. The smallholders abstain from possible future fortunes rather than exposing themselves or the family to risk. They value the traditional society they live in, and if they have to leave home to seek a job in town, they come back to their village to participate in ceremonies marking the rites of passage (such as funerals, weddings and circumcision), to work on a joint project or to help a sickly relative.

This is not unique for smallholders in today's developing countries; it was the same for farms in England or Russia during the transition to a market-based agriculture (Hylland Eriksen 2008). Many believe that the only way to get smallholders into the modern economy, and through that into 'development', is by coercion and force, even though not all express it as straightforwardly as the historian Ronald E. Seavoy: 'Commercializing agriculture means ending subsistence agriculture. How can subsistence agriculture be ended? The simple answer is by huge amounts of compulsion administered by central governments. Only compulsion will induce most peasants to increase per capita production for sale on anonymous markets' (2000: 93). This compulsion can be administered in many different ways. Taxes and rents are the more obvious ones, but there are also good-hearted versions such as the introduction of mandatory schooling (see the text box 'Schooling and modernization of farms'), the offer of generous micro-credits, etc.

> **Schooling and modernization of farms**
> Compulsory schooling often plays a central role *to stimulate and force* farming communities out of their subsistence farming. This works in many different ways. It costs money for parents to send their children to school, to buy books and clothes, to pay school fees or tip the teacher now and then; all of this makes it necessary to get cash income. When

kids go to school they can't work on the land, so farmers face a shortage of labour and incentive to either employ or mechanize, both cost money. This means that farmers have to increase income or take loans. Education itself influences the young, who will also become farmers, to 'modernize' their farms and the prevailing ideas on what is a good farm is promoted in schools, including the use of fertilizers, improved seeds, etc.; all of these are based on a further integration in the market economy. Some children continue to higher grades in district towns and even country capitals. Caring for them costs even more money. Children also change their values and attitudes in a more modern way, start to be embarrassed over how primitive their parents are and seek another lifestyle for themselves.

Still, even if all the forces of coercion are brought to play and farmers try to commercialize production, it is not clear whether they will actually be successful. Most farmers will simply not survive in this struggle for modernization. If they do, there will be enormous over-production. In addition, the competition and prices of world markets make it very hard to accumulate any surplus to invest. Mazoyer and Roudart (2006) explain the case like this. At the rate of US\$ 20 per 100 kilogrammes of grain, it would require one life of labour (33 years) for a manual farmer in a developing country to acquire a pair of oxen and small-animal-drawn equipment supposing that this farmer can devote all of the monetary income to this purchase.[130] It would require 100 years to acquire sophisticated animal-drawn cultivation equipment. It would require 300 years of labour to buy a small tractor and it would require 3000 years to buy the complete set of motomechanized equipment comparable to that of a European or American farmer (Mazoyer and Roudart 2006).

In the high-income countries of today, farmers who could not climb the ladder of development had somewhere to go, to the rapidly growing industries, which at that time also were labour-intensive. Farmers were not subject to competition from other farmers who were 100 times more productive. European farms had difficulties coping with competition from North America, especially after the introduction of steamship transport, and still the pressure of competition was a lot lower for them than it is for poor farmers in developing countries today. The response was to introduce protectionist measures. In addition, because of the productivity gains in developed countries, agriculture prices dropped by some 60% in the period 1960–2000. As the productivity of the poorest

---

[130]   Which I believe is completely unrealistic.

farmers remained much the same, it is obvious that they lost out. To believe that these farmers would be able to compete on staple food in free world markets is simply very far from reality. The inconvenient truth is also that the billions of people who would be made redundant in a large-scale modernization of the farm sector in developing countries simply are not needed for *any* production. And that free markets don't offer them any solution.

Equality is good for growth, but growth is not necessarily good for equality

An increased income for farmers and farm workers stimulates demand for goods and services by local artisans (blacksmiths, construction workers, seamstresses and brewers among others) and can in this way induce a virtuous cycle. A dollar in increased income can in this way easily become two. Local wages also increase. A big difference in this regard arises between the situation when growth is by hundreds of smallholders or when it is in a big plantation. When a plantation increases its income, most of the money is spent on imported inputs and machinery as well as on luxury products for private consumption, with little positive impact on local trades. There is thus a strong link between equality and local economic development (FAO 2003).

Economic growth has not helped the poor in countries with big inequalities. Despite that Brazil had a GDP per person that was almost three times as high as the one of Vietnam in the early 2000s; the poor of Brazil were as poor as the poor of Vietnam[131]. Compared to other high-income countries, United States is doing badly: a child born in a family of the richest 5% of the population will on average live *15–20 years* longer than a child from the poorest 5% (UNDP 2005). Sen (1999) compares the states of Punjab and Kerala in India and finds that Punjab has reduced poverty by means of economic growth, whereas Kerala, despite lower growth, has reduced poverty more than any other state in India, mainly through education, health care and land reform. This data doesn't mean that growth is *bad* for the poor, but it does show that growth itself doesn't eradicating poverty and that poverty can be eradicated even with no or moderate growth given the right policies. These policies don't include maintaining income disparities.

---

[131]   It appears that Brazil finally implemented some successful poverty policies with the Bolsa Familia during the Lula presidency.

There are common interests among the local elites and foreign actors. Large sections of the elite in developing countries have strong links to high-income countries. Some of them date back to colonial times. Today these links are maintained by local elites studying at universities in the high-income countries, and their children going abroad to study or to 'international schools' in their home countries; they also go shopping in high-income countries and are part of an international jet set. This leads to the local elites in developing countries (in the same way as the elites in most countries) sharing joint interests and values with the elites in those high-income countries rather than with their own population. That democracy is badly developed doesn't make it better.

One can blame, sometimes justly and sometimes unjustly, high-income countries for some of the misery of the low-income countries, but that doesn't free governments of the latter from responsibilities. There is, especially among those who reject globalization, a romanticized view of the nation-states as guarantors for the interest of the people. If only that were the case! Many governments in developing countries are both corrupt and incompetent and most are at least one of them, and their politics are only rarely pro-poor.

Some claim that so-called Western democracy doesn't fit in many countries. Perhaps that is the case. One tends to believe that that kind of democracy is the only model, but that is hardly the case. The party model certainly is no prerequisite for democracy. More important is perhaps liberties of various kinds, and systems that give people the possibilities, in daily life, to realize their dreams, or at least to try to realize their dreams. Freedom of speech and a free media are certainly important ingredients; so are the freedom of seeking income in the trade of choice, the right to live where one wants and female emancipation. In *Freedom as Development*, Sen says: 'freedom is viewed as both (1) the *primary end* and (2) the *principal means* of development' (1999: 36).

Collateralize the slum

Some admit that capitalism doesn't work well for the poorest and promote new ways to make it work. The micro-credits of the Grameen Bank in Bangladesh are world famous and the initiator, Muhammad Yunus, jointly won the Nobel Peace Prize in 2006. The system had more than 7.4 million members in over 80,000 villages in 2009, with a credit volume of US$ 529 million. A large share of the borrowers comprises women (Grameen Bank 2009). With this and other examples as role models, micro-credits have expanded rapidly, and there are few other things that are acclaimed by both the left and the right.[132] It is still very hard to assess the real impact of micro-credits. There are fears that they are used for consumption rather than investment[133] or that they are used to pay off other credits. Most micro-credits are driven by development aid, and, like all other such projects, they can become self-motivating and stimulate credits even if there are no needs or real business ideas.[134] Debt is also one of the strongest powers to force people into the market (money) economy as we discussed before.

According to Peruvian economist Hernando de Soto, in *The Mystery of Capital: Why Capitalism Triumphs in the West and Fails Everywhere Else* (2000), the poor have all the material resources they need to secure their livelihoods. The key is that developing countries and countries of the former Soviet Union have insufficient arrangements for property and ownership. What they miss is formal, officially acknowledged ownership to their assets, primarily land, shops and slum dwellings. This means that they can't be the basis for capital accumulation, in particular they are not accepted as collateral for loans; they are 'dead capital', and the informal sector is thus under-capitalized and therefore less productive (Soto 2000).

Both Soto's ideas and micro-credits build on the perception that lack of capital is an important factor limiting peoples' ability to rise out of poverty on their own. There is certainly no doubt that lack of capital—the lack of money—is part of the burden of poverty, yes perhaps even the proper definition of being poor. But it is not equally clear, even if it is plausible, whether limited access to credit keeps people in poverty. Lack of capital and lack of credits are part of a larger poverty complex that also has a lot of other components, such as illiteracy, political isolation, personal stigma and low-nutrition food leading to low energy. One should not forget that if I can use my illegal slum shack, or my land, as collateral for a loan, I can also *lose* it when I can't pay back. Ultimately, it is only by making the property formal that capitalists can take it over; without title they certainly will not buy

---

[132]   Between writing the sentence the first time and the final editing, some criticism against Muhammad Yunus has been aired.

[133]   Sometimes this can be merited and sometimes it is even hard to make a distinction. If money is used for better nutrition for children or for medical care or for providing clean water, it can actually be seen as an investment in human capital, which also is useful for 'production'.

[134]   I have encountered a number of NGOs engaged in micro-credits that are desperately seeking ideas so that the money can be spent, which for me has always been a signal of something flawed.

it. Once it is sold, the individual loses the moral right to the property; if it can't be sold, the individual's right to the place can't be lost. Allocating land or property to the poor might in this way ultimately only feed the rich with more property. There are clearly similar risks with payments for ecosystem services, as discussed earlier.

Failed development—failed states

Urbanization in the now high-income countries was part of their economic and social development; by and large, it was driven by the opportunities in the cities, primarily in industries and services. In most developing countries, the rate of urbanization is very high. But when there is no associated industrialization or modernization of the agriculture sector, this is simply not sustainable. In the world today, there is no need or place for the kind of very labour-intensive industries that were the basis for the industrialization of the high-income countries of today. Many industries today have been automated and no matter how cheap labour is, one can't compete. This means that a lot of people and countries are not *behind* but *beside* the development path. Another obstacle is that natural resources are simply not enough to allow all states to develop, and those natural resources that were easy to exploit have already been exploited by the rich countries. Finally, there is no market left for simple industrial products; China and India are already flooding global markets with goods for mass consumption.

One can compare the leap from the hunter and gatherer societies to the agrarian societies; not all societies managed to do that, the laggards more often were simply forced away by the expanding agrarian cultures. There are still, however, groups that never transcended to the agrarian culture. Almost exclusively they live in areas where conditions for farming are too harsh, such as deserts, tundra and rainforests. The question is whether there are, analogously, a number of societies that will never be able to take the leap into industrial society, because *they lack both comparative* and *absolute* advantages. This can explain why some of the poorer countries, which, according to classic economic theory, should export food to pay for imports of other more sophisticated products, end up also importing food. Automation and ever-increasing productivity in high-income countries mean that there is no need for so many factories. Low salaries are not important enough as a competitive factor when oil is still so much cheaper than human labour. Tourism and culture can of course still be a niche, and they are for a few countries. But also in tourism, middle-income countries fare a lot better than the least-developed countries. It seems that the only option left for these countries is to slowly build up a local market. But how is this achieved when people are so poor that they have no purchasing power and when cheap imports flood the domestic market? The trials with closed borders, import substitution, farm regulation and overvalued currencies of the 1960s and 1970s failed across the board.

Everybody is perhaps not invited

Integration to the global economy has meant a radical increase in wealth for many people, most clearly seen in the case of China. But it has come with a high social, cultural and environmental price tag. Some seem to be able to benefit from globalization, others not. The gaps between rich and poor people, rich and poor regions, and rich and poor countries clearly have increased from a long-term perspective, simultaneous with the increase in capitalist penetration. Thus, the explanation that countries or people are poor because they are not capitalist enough is flawed. Notwithstanding, several countries could do a lot better by improving the conditions for the private sector, especially by reducing corruption and Byzantine administration. While development in general is not a zero-sum game where the success of one must mean the exploitation of the other, there are certainly many instances where this is the case.

– 21 –
Growth and Well-Being

No one has more missed opportunities than the one with the most choices.

(Hylland Eriksen,
Jakten på lycka i överflödssamhället,
2008)

In this chapter, we look at what makes us happy or feel well and how that relates to the main driver in this society, economic growth. Would it not be a good idea to look into what makes us feel good and then optimize resource use for that, instead of using money and consumption as proxies for well-being?

What do we need?

The hierarchy of needs is a theory proposed by Abraham Maslow in his 1943 paper 'A theory of human motivation'. Its essence is that needs at a lower level have to be satisfied before needs at a higher level become important. The needs from the bottom of the hierarchy, sometimes depicted as a pyramid, are:

Physiological needs, such as food, air, shelter and sex.

Safety needs, such as personal security, health and well-being.

Love needs, such as love, affection, belongingness, friendship and intimacy.

Esteem needs, such as respect and self-esteem.

The need for self-actualization, that is, 'What a man *can* be, he *must* be,' such as longer-term aims and morals.

A lot of the literature still refers to this model. Intuitively, it might seem very reasonable; who can think about friendship if he can't get food? But if we look around, we see many instances where the theory doesn't hold. The need for spirituality or religion seems to be a very distinct need among many, even among those who miss most things. It is exactly those who are the poorest, without food and without safety, who tend to be most religious[135] (Norris and Inglehart 2004). That love can come before almost all other needs is not a phenomenon only in novels. Or as Rousseau says, '. . . the less natural and urgent the needs, the more the passions augment . . .' (1964: 195).

There are other models for analyzing human needs. The Chilean economist Manfred Max-Neef makes a distinction between needs and satisfiers (see Max-Neef *et al.* 1989). Human needs are seen as few, finite and classifiable as distinct from 'wants' that are infinite and insatiable. Needs are constant through all human cultures and across historical time periods. What changes over time and between cultures is the way these needs are satisfied. Human needs are understood as a system—that is, they are interrelated and interactive. There is no hierarchy of needs apart from the basic need for subsistence or survival. The same satisfier can relate to many needs; for example, work can give meaning and security, satisfy subsistence needs and be a place for creativity and identity. To have friends is not only about affection but also about identity, leisure, participation and protection. Breast-feeding not only fulfils the child's need for food but also contributes to safety, identity and hopefully relaxation (Max-Need 1987). Most of the satisfiers are of a social nature, such as sharing, friendship, games and sense of belonging. It is no coincidence that social rejection, expulsion or ostracism was seen as a horrendous punishment throughout history. Without being flawless or above challenges, Max-Neef's model is at least a lot more useful than Maslow's hierarchy, and as such it gives better directions for how human beings can improve the feeling of satisfaction in life.

Wealth or well-being?

There is no direct correlation between an individual's or a country's economic (material) wealth and their sense of happiness, satisfaction or well-being.[136] This is an old observation. The more balanced defen-

---

[135]   Americans are an exception to this.

[136]   Notably this is a linguistic minefield with, often badly defined, terms such as happiness, satisfaction and well-being (measured subjectively or objectively), where the results for what makes people happy are not necessarily the same as what satisfies them or makes them experience well-being. Constraints of space don't allow a deeper exploration of this here. In referring to research, I have used the terms used by the actual researchers.

dants of capitalism also agree; for example, the economist Joseph A. Schumpeter (1942) says that people in industrial societies don't have to be happier or experience more well-being than people in the Middle Ages. Today, only the very rich in the high-income countries have access to — can afford — what is natural for most poor people, to have someone at one's deathbed. An American study from 2004 states that while economic wealth has increased threefold in 50 years (see Figure 21.1), people's feeling of well-being has been constant; in fact, mental health has deteriorated and social networks are weaker (Diener and Seligman 2005). Between 1958 and 1991, the average income of the Japanese increased sixfold; still, in 1991 they were as satisfied (or as little satisfied) as their parents were in 1958 (Hylland Eriksen 2008). Another study compared the 400 richest Americans with East African Masaai and found both groups to be approximately equally satisfied with their lives (Diener and Seligman 2005).

Figure 21.1 Well-being and gross domestic product in the United States (1957–2008)

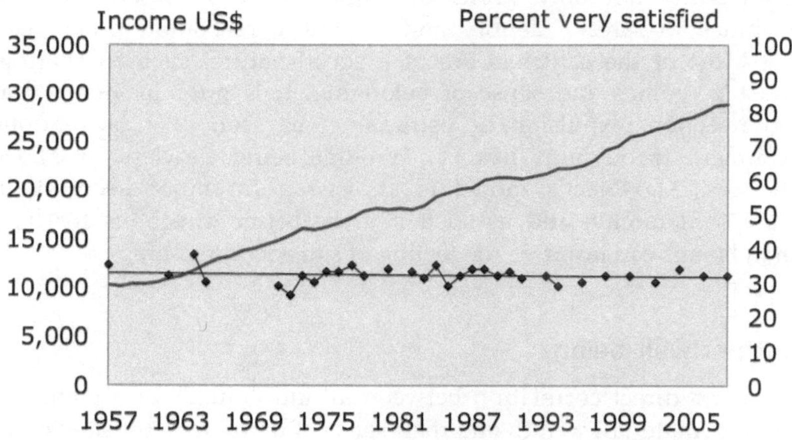

Source: Diener and Seligman (2005).

Among low-income countries, there are countries where people are very satisfied and those where they are not as satisfied. This is not surprising and supports the thesis that there is no direct correlation between wealth and well-being. Low-income countries are often haunted by war and natural disasters and have oppressive political systems, all of which obviously affects well-being negatively. What seems to be rather clear is that there is a reduced marginal benefit of higher incomes, which appears around US$ 10,000 per year. Above this level, there is hardly any increase of well-being. And if one considers the effects of health, human rights, democracy, gender equality, equality in general[137] and governance, one sees almost no influence of the GDP as such. The GDP is, however, to some extent positively correlated with those parameters (Diener and Seligman 2005; Inglehart *et al.* 2008; Wilkinson and Picket 2009).

Most factors that are important for the well-being of society are about human, and in particular, social capital. Most likely, there is causality both ways. More social capital leads to more well-being and better governance and vice versa.

Figure 21.2 Factors influencing subjective well-being in Great Britain.

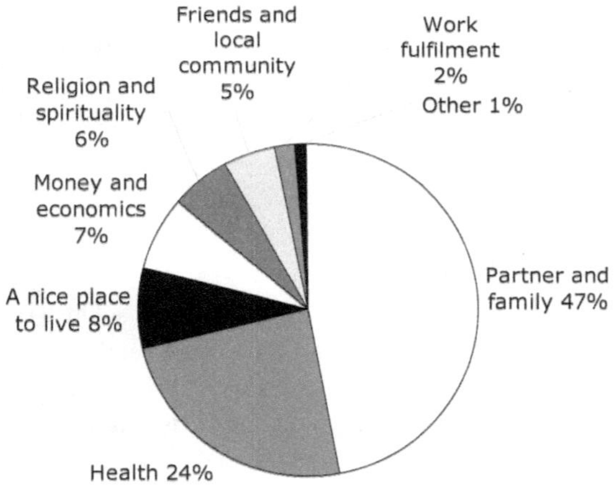

Source: Jackson (2009).

From the analysis of Max-Neef *et al.* (1989) of basic human needs and their satisfiers, we can see that most of the needs have social or relational satisfiers. And one doesn't have to be a psychologist or behavioural scientist to see the accuracy of this in daily life, as is shown by Figure 21.2. Even at work — something by definition done to get a salary — the most important factor for most people is not the level

---

[137]   This is in particular stressed in the study by Richard Wilkinson and Kate Pickett (2009); we will return to their results as presented in *The Spirit Level*.

of pay,[138] but appreciation in their work, supportive supervisors, respect and status relations with colleagues, etc. Thus, it is strange that society—and the individual—continues to act as though material wealth is most important. Perhaps it is because non-material satisfiers are considered more 'private' and therefore have been excluded from public discourse, which is totally dominated by economist, or I would rather say economistic, perspectives.

Comparison and thereby associated status drives a lot of material consumption. It is first when one's neighbour has a swimming pool that the need emerges for oneself, or perhaps one builds the first pool in the neighbourhood. And it is the same with choice. To have no choice is limiting, but endless choices are therefore not a recipe for satisfaction, or as psychologist Brian Schwarz says: 'There is no question that some choice is better than none, but it doesn't follow that more choice is better than some choice. [. . .] I am pretty confident that we have long since passed the point where options improve our welfare' (2005). When choices are unlimited, one will always fear having made the wrong choice.

Unequal societies are not happy

In *The Spirit Level: Why More Equal Societies Almost Always Do Better*, Wilkinson and Pickett (2009) show for area after area that unequal societies perform worse than more equal societies (their research is limited to the high-income countries): life expectancy is three to four years shorter in less equal societies; crime rates and the number of persons in prison are considerably higher in unequal societies (highly unequal United States has 14 times as many people in prison per 100,000 inhabitants as equal Japan); child mortality is higher in unequal societies; and education gives lower results in unequal societies, just to mention a few of the observations. Yes, even obesity is affected. Obesity among men and women as well as calorie intake and deaths from diabetes are related to income inequality in high-income countries. In addition, obesity in adults is also related to inequality in the 50 US states; and the percentage of children who are overweight is related to inequality both internationally and in the United States.[139]

---

[138]   Admittedly this can also be because the level is quite satisfactory while other aspects of work have been set aside.

[139]   Some critics have pointed out that the plotting of data that Wilkinson

Some claim that income disparities (or low taxes or low marginal taxes or any other proxy for the same thing, that some get more money than others) are good for economic growth, mostly motivated in a way where entrepreneurs need to be rewarded for risk and that with increased division of labour it is just natural that differences will increase. On the other hand, inequality also affects trust between people and the social capital, so people in unequal societies are less likely to trust each other. This in turn will badly affect economic performance. If inequality is caused (or causes, the direction of causality makes no difference here) by high unemployment or bad health, people will work less, which should be negative for growth.

My hypothesis is that inequality is slightly positive for growth, but not for the reasons normally stated. The growth-driving effect of inequality has to do with status and human desire for ever more things to consume. There is quite general agreement that status is a relative thing and what people think is a reasonable standard of living depends on what they expect and what other people have. Therefore, it is only natural that inequality spurs more desire for consumption and in this way also growth. *It is because we are less happy and more frustrated in an unequal world that inequality drives consumption and, hence, growth.*

'High salaries are a threat to living standard' or 'High salaries are a threat to growth' is commonly heard from company bosses. But what does this really mean?[140] Isn't it very hard to comprehend how one could get a higher standard of living by lower salaries? The argument undermines the credibility of exactly what capitalism claims to deliver — an increase in wealth for the masses.

Mass production made products cheap. Social activism and trade unions opposed the appalling conditions and low salaries in the industries and forced an increase in salaries. In that way, workers could to a larger extent harvest the fruits of their own work, and a self-

---

and Pickett (2009) use as support for their statements does not necessarily *prove* that inequality causes or augments all those negative things; that is, causality can also be the other way round, say that health is good for equality as healthy people can work, or that there are other factors that affect both equality and the factors in question. Nevertheless, the totality of correlations is quite convincing and they serve a lethal blow to any claims of the opposite, say that increased income disparities have any benefits.

[140]   This discussion may build on the (false) premise that growth is inherently good, something that is disputed in other parts of this book. This discussion is about the hypocrisy of those who defend high-income inequalities with arguments that it is good for the economy, and not about whether growth is good as such.

reinforcing process started. Increased salaries ▶ increased purchasing power ▶ increased consumption ▶ increased productivity ▶ lower prices ▶ increased salaries. A few times this cycle has stalled, for example, during the Great Depression in the 1930s. At other times the process has gone wild, mainly because some external factor has disturbed the balance. For instance, if households experience that their standard of living and possibilities are substantially higher than justified by productivity increases they might over-consume. This can be caused by radical tax breaks, a credit boom or a housing market bubble. This was very much the case in the run up to the 2008/2009 financial meltdown.

Economic growth leads to a scarcity of time

Philosophers, politicians and economists have all given voice to the idea that once we have been freed from destitution and poverty, through the blessings of industrialization, we would get time to spend for art, spiritual development, games and exploring the deeper meaning of the concept of freedom. But this didn't materialize. Already in 1969, economist, and later trade minister of Sweden, Staffan Burenstam Linder (1970) described in *The Harried Leisure Class* how consumers in high-income countries have become more stressed and restless in their efforts to increase productivity of their leisure time. The productivity of work has increased tremendously through mechanization and use of external energy. This creates a corresponding pressure on one's time off to also be more 'productive', influencing both household work and pure leisure. That there is more money to dispose of doesn't alleviate the situation, rather the contrary, as all that money needs to find its use.

There are only so many hours 'to spend'. Through technical development and efficient management, an awful lot of things are produced today, a lot more than before. But human beings work more or less the same. This means that there are more and more things to consume and services to buy in the same limited time. No wonder we get stressed. In the same way as productivity in industries and offices has increased tremendously by more tools and machines, one's time off is also increasingly dominated by them. More and more tools and gadgets are used and activities that have no corresponding gadget lose their appeal; for cooking, a food processor or an ice cream maker is preferred over a straightforward slowly simmering stew. Sports, even the 'simple outdoor life', is dominated by designer brands and special technical aids. One helmet for ice hockey, another one for a bicycle, a third one for driving a motorcycle and a fourth one for the piste are needed, just to mention the most common ones in a cold climate.

In a similar way, we are supposed to constantly upgrade our competence. We are said to be moving towards a knowledge society. Earlier, some five to seven years of formal education was enough for many trades, followed by apprenticeship on the job; today, schools reach some twelve grades in most countries and a bulk of the people are supposed to continue into higher education. People need to be better and better and not fall off the merry-go-round that is spinning quicker and quicker. This competition is another side of the growth paradigm and a cause of insecurity and loss of well-being. Not only do people want more things all the time, they should also know more and be attractive in the market (be it the labour market or the dating market). This contributes to stress and to some actually falling off altogether. Our ability to integrate those who are a bit different, the odd ones, is diminishing by the year, both at work and socially. The time for taking care of an old relative or for just sitting and listening is constantly shrinking.

I consume therefore I exist

> If I were a rich man
> If I were a rich man,
> I see my wife, my Golde, looking like a rich man's wife
> With a proper double-chin.
> Supervising meals to her heart's delight.
> I see her putting on airs and strutting like a peacock.
> Oy, what a happy mood she's in.
> Screaming at the servants, day and night.

Sings Tevye in the *Fiddler on the Roof* (1971). Today, Tevye's thoughts are believed to be universal throughout the history of humanity, but there is little support for that assumption. However, current society behaves like it. The thought of what one could or should do if one had more money keeps children and adults alike wake at night; it crashes marriages and forms the foundation for an enormous growth in lotteries, betting and gambling. The global gambling market was worth some US$ 335 billion in 2009 with casinos, lotteries and gaming machines representing the three biggest shares (*The Economist* 2010b). In Great Britain, people spent £28 billion in gambling in 1999 and £53 billion just five years later. The Brits are spending more per person than the Americans but less than the Australians or the Swedes (Gambling Commission 2006). Well, at least it contributes to a substantial increase in the GDP.

According to the political scientist Farhang Rajaee, 'People have their radios, stereos, televisions, air conditions, fans and similar appliances on from the minute they come into the house, not because they need them or want to use them, but because they constitute their very existence (2000: 67–68). In the 1990s, the Americans spent 40% of their leisure time in front of the television and with that rate the average American would be exposed to 1.75 million shots of commercials in a lifetime (Carley and Spapens 1998).[141] In 2008, almost US$ 130 billion was spent on commercial ads in the United States (Nielsenwire 2010), ads spreading the view that people would be happier if more is consumed. Is this quest for more things hardwired in humankind, or is it something that has been created — or at least augmented considerably — through a constant simulation? I believe more in the latter.

---

[141]   With the Internet, these figures are certainly changing, but overall the exposure to commercial messages is not decreasing, rather the opposite.

– 22 –
Common Resources

Land belongs to a vast family of which many are dead, few are living and countless members are still unborn.
(Nigerian proverb)

Even in today's world of privatization there are many common resources. Most, or perhaps all, of nature is a common resource. Human society is a common resource. The accumulated knowledge of humankind is a common resource. The question of how these common resources are managed — and indeed where the line between common and private, individual resources lies — is of fundamental importance. Commons have been eroded all over the planet, mainly as a result of encroachment by private ownership. Commons are also monopolized by dictatorial governments and polluted by human consumption. The future of the commons will play a big role in human society.

The tragedy of the tragedy of the commons

The expression 'tragedy of the commons' was coined by the ecologist Garrett Hardin (1968), in an article of the same name, to describe the need to protect and monitor common resources from over-exploitation. Many have used his article, or at least the catchy phrase, to argue that common resources should be privatized because that would mean that they would be taken care of; sustainability would be guaranteed through the profit interest of the individual. It has also been used against common management or public management of resources. Hardin (2003) himself clarified in a later article that what is needed is that the commons are managed: 'A "managed commons" describes either socialism or the privatism of free enterprise. Either one may work; either one may fail: "The devil is in the details." But with an unmanaged commons, you can forget about the devil: As overuse of resources reduces carrying capacity, ruin is inevitable.' He also stated that:

> The more the population exceeds the carrying capacity of the environ-
> ment, the more freedoms must be given up. As cities grow, the
> freedom to park is restricted by the number of parking meters or fee-
> charging garages. Traffic is rigidly controlled. On the global scale,
> nations are abandoning not only the freedom of the seas, but the
> freedom of the atmosphere, which acts as a common sink for aerial
> garbage. Yet to come are many other restrictions as the world's
> population continues to grow. (Hardin 2003)[142]

Very seldom does one hear those who talk about the tragedy of the commons quote these statements. We have discussed earlier that privatization of the commons played a big role in the development of capitalism in England. It is worth noting that there is no indication that those commons, before they were privatized, represented any tragedy, that is, that they were mismanaged in any way. They were common in order to protect them from overuse rather than the opposite (Montgomery 2007). On the Solomon Islands, the white beech is a rare tree essential for the construction of canoes. When someone knows they will need a new canoe in the future, they mark a young tree to inform the others of its future use, and still ask the chief for permission before finally cutting it (Satoyama 2010). The same is reported from many other places:

> Well-organized village communities were frequently in a better posi-
> tion to care for the common forest in keeping with their needs than
> money-hungry lords who wished to fill their coffers with the forest
> and who often did not even know the woods they were supposedly
> protecting . . . It was precisely the division of the commons that was
> often followed by the cutting of the forest. (Radkau 2008: 73)

Institutions in society are also a kind of commons created by human beings. The last decades have experienced a widespread battle over how to manage resources such as schools, hospitals, utilities, postal service and water works just to mention a few.

Experiences of how to manage common resources point in differ-ent directions and we should not jump to conclusions about how they should be managed in order not to be destroyed. There are options other than privatization and state control; many common resources, probably most, have been managed by local communities, and it is rather recently that they have been either privatized or taken over by

---

[142]   My references to Hardin here should not be seen as an endorsement of his ethics regarding how to behave in an overstretched world, some of which I find questionable to say the least.

the state. The economist Elinor Ostrom (2009) has shown how cooperation in the management of common resources is the rule rather than the exception. What is of critical importance is that the users have real say in the arrangements. This is hardly anything new; Kropotkin (1902) said the same almost 150 years earlier (see Chapter 18). For sure there will still be no guarantee that the management will be perfect, people can act egoistically or foolishly both as individual owners and as members of a community. The difference is that in a community, the majority must act egoistically and foolishly for it to dominate.

The dictatorship of intellectual property rights

> Shortly after a large-scale clinical trial in 1955, the first inactivated polio vaccine was being injected into tens of millions of people around the world—possibly the most successful pharmaceutical product launch in history. Asked why he had not obtained a patent on the phenomenally successful vaccine, Jonas Salk reportedly replied, 'That would be like patenting the sun.' A few decades later, this view seemed absurdly quaint. (Alan Dove, quoted in Science Commons 2010)

Some problems are associated with the fact that nobody owns a certain resource, but certainly greater problems are associated with privatization of common resources such as knowledge, natural resources, innovations and technology. While there might have been secrets of the trade or knowledge that was monopolized also earlier, in no society this was done from the perspective of ownership. At a certain stage society started to protect intellectual property in order to motivate investments by assigning monopoly to certain people who, falsely or rightly, were seen as innovators or originators. In England, patents in the modern sense originated in section 6 of the 1623 Statute on Monopolies, which described patents as 'monopolies' and exempted them from the general ban on royal grants of such rights.

Mostly intellectual property rights have been established not on the basis of any idea of the 'rights' of the originator but rather on the basis of a utilitarian perspective that it is beneficial for society to assign such rights (compare with the discussion on privatization of nature's resources).[143] It is by protection of the interest of the originator that they will be stimulated to be more creative, innovate and bring to the

---

[143]   Notably, in continental Europe, the perspective of a 'natural right' to the originator was more prevalent.

market new products. Gradually, over the course of history, this perspective was replaced by the notion that rights to control the use and dissemination of information are forms of 'property' rights[144] (Fisher 1999).

The 1790 Copyright Act of the United States established a copyright term of 14 years. Copyrights acquired today will last for the life of the author plus 50 years. A less straightforward but equally important issue is the definition of a copyrighted 'work'. Until the middle of the nineteenth century, a copyright owner enjoyed little more than protection against verbatim copying of his or her language. So, for example, in 1853 a federal Circuit Court rejected the claim of Harriet Beecher Stowe that a German translation of *Uncle Tom's Cabin* infringed upon her copyright.

Today the story is quite different and the kinds of works to which copyright laws may apply have also grown enormously: in 1884, the Supreme Court concluded that photographs could be copyrighted; in 1971, Congress decided that musical recordings should be shielded from copying. In 1979, computer software was added to the list of protectable works. Like copyright, patent laws were gradually extended. In 1842, hoping to provide 'encouragement to the decorative arts', Congress extended the reach of the patent statute to cover 'new and original designs for articles of manufacture'. Until the early twentieth century, plants were considered products of nature and hence unpatentable. The Plant Patent Act of 1930 overrode this principle, extending a modified form of patent protection to new varieties of asexually reproducing plants. In 1970, Congress went further, reaching new and 'distinct' sexually reproducing plant varieties (Fisher 1999).

As the world is getting more and more complex, the tangling web of patents and copyrights is getting more and more impenetrable and it is reasonable to ask which of the original motives for these rights are still valid, if any. Innovation, art and culture all existed before intellectual property rights, and the empirical evidence for the value of them for society is largely missing. If the manufacturing of a product needs a number of components or technologies patented by others, it is a complicated process just to negotiate with all patent holders. Addition-

---

[144]   In a way, the same process has taken place with ownership of land, which was also a common resource that was gradually privatized.

ally, all the licensing fees have to be paid. One single microchip can 'contain' more than 5000 patents; for a cheap DVD player patent costs are as high as manufacturing costs (Wikipedia 2009). In practice, this could mean that good products never reach the market despite both demands from consumers and interest from manufacturers. In the words of economists, a clear 'market failure'.

The use of the neem tree as a fungicide was patented by an American company, despite that neem has been used as a pesticide for 2000 years. A coalition of the Greens in the European Parliament, the Research Foundation for Science, Technology and Ecology and the International Federation of Organic Agriculture Movements sued the company to the European Patent Office. After a 10-year-long process, the patent was declared void (IFOAM 2005). At a certain time there were concerns over this kind of bio-piracy, that is, that companies would capitalize on indigenous knowledge. This led to the possibility of patenting that kind of knowledge. Even if the intention was good, it seems as if, ironically, these measures paved the way for exactly what should have been prevented, a privatization of these resources.

Patent rights can mean that fewer medicines are produced and that they are produced for a higher price, because patent rights limit competition (remember the origin in England in the Statute on Monopolies). A special case is the antiretroviral drugs used for HIV/AIDS. For long the 'giant pharma' refused to let generic copies be produced and sold cheaply in low-income countries, which meant that millions didn't get access to them, again a typical 'market failure'. In this case, public relations of the pharmaceutical companies ultimately became nightmarish and they backed considerably.

It is indisputably the case that a lot of pharmaceutical research would not happen unless there was some protection for the innovation. On the other hand, the research is not necessarily geared towards what is best for the patient or society. There is, for instance, much less interest in drugs that, once and for all, cure an ailment than in those that need to be administered for the rest of your life. As can be seen with malaria treatments, there is little private investment in it because the clientele is so poor — which is also the case for why so much public and charity money is spent on this. As well as being an argument for patent rights, the situation with medicinal drugs seems to provide at least equally strong arguments for publicly funded research and common access to the result. For example, the effects of antibiotics are threatened by resistant bacteria. To be able to still treat the growing number of infections with such bacteria, we need to develop antibiotics that are only used as a last resort (so that no bacteria develop resistance against them too). But there is clearly little commercial interest to develop a new antibiotic and then not use it actively. Only governments or other public interest organizations will do that.

Who is served by intellectual property rights?

Intellectual property rights clearly serve the interest of the holder/owner; that is what they are there to do. But the question is: do they harm others; are there losers? The economist Helpman (1993) concludes that developing countries have nothing to gain by stronger intellectual property rights. The perspective that net producers of patents have more to gain from property laws than those using the patents is supported by the history of property laws in the United States, according to Professor of Law William W. Fisher III:

> Until approximately the middle of the nineteenth century, more Americans had an interest in 'pirating' copyrighted or patented materials produced by foreigners than had [sic] an interest in protecting copyrights or patents against 'piracy' by foreigners. The shift in the 'balance of trade' had a predictable effect on the stance taken by the United States in international affairs. In the early nineteenth century—as Charles Dickens learned to his dismay—the American government was deaf to the pleas of foreign authors that American publishers were reprinting their works without permission. In the late twentieth century, by contrast, the United States has become the world's most vigorous and effective champion of strengthened intellectual-property rights. (1999: 11)

The United States took thus a very hard line in the negotiation of the TRIPS agreement, insisting that other nations give in their generous version of patent and copyright laws. One could quite easily assume that the same relations that exist in the international arena also apply within a country and that there are reasons to doubt that non-patent holders, or society at large, actually benefit from intellectual property rights.

The movement for open source in software shows that technological development can be driven by forces other than rent-seeking patenting. Famous examples are the operating system Linux and the Open Office software package. In 2011, two-thirds of websites ran on the free Apache HTTP Server program (Wikipedia 2011). One very interesting example is the online encyclopaedia Wikipedia. Since its creation in 2001, Wikipedia has grown rapidly into one of the largest reference websites, attracting nearly 470 million visitors monthly as of February 2012. More than 77,000 active contributors are working on more than 22,000,000 articles in more than 285 languages. As of December 27 2012, there are 4,129,594 articles in English. Every day, hundreds of thousands of visitors from around the world collectively make tens of thousands of edits and create thousands of new articles to augment the knowledge held by the Wikipedia encyclopaedia (Wikipedia 2012). Other examples include the Science Commons, a project for designing strategies and tools for faster, more efficient web-enabled scientific research (Science Commons 2010). On a lighter note, there is even an Australian 'open-source beer', Brewtopia, allowing individuals to design their own beer, on the basis of a publicly available basic brew recipe (Brewtopia 2010).

Some believe that the open-source movement represents fundamentally new ways of production, which challenges or transcends the capitalist model. I am less convinced about that. It is more like people working in a new organizational form to create something that they need and which they think should be publicly available. From this perspective, it might not be so radically different from a communally managed beach, a garden or sports field? Both Linus Thorvalds, developer of the Linux operating system, and the founders of Wikipedia later worked for pure commercial operations, a support to the notion that phenomena such as Linux and Wikipedia are not a radical departure from capitalist commercialism.

In 2004, Jack Messman, president of the software company Novell, said that commercial software companies must work higher up the value ladder, the ones working with open source work with infrastructure (Friedman 2005). It is just another twist of the eternal debate of whether the private, the public or civil society shall manage something. What is interesting is not so much that this is done outside the market, and outside the state, but that it shows a new range of work for 'civil society', and a new creation of 'community' and it is there that its potential for change lies. The other conclusion from open source is that honour, a sense of meaning and other peoples' appreciation seem to be sufficient motivation for driving development and continued innovation. Linus Thorvalds is perhaps as satisfied with his life as Bill Gates even if Gates has become one of the richest persons in the world and Thorvalds is 'merely' an icon for open source.

Our biggest asset is our society

The concept of 'social capital' is about the institutions, norms, laws and networks that help to develop both society as well as the human, manufactured and financial capital. A very important aspect is that social capital is about trust between individuals. If trust is very low, it is hard to do even simple transactions. Do I dare to give you the money before I have the goods in my hand? Do you dare to give me the goods before you have the money in your hand? It is almost impossible to manage businesses or any other cooperation without trust. Political institutions in society and trust in them is also an important side of social capital. Another side is the many civil society organizations established to pursue joint agendas.

"Back of the envelope" calculations show that in 192 countries (except for a few raw materials exporters), human and social capital equals or exceeds natural capital and produced assets combined (Serageldin and Grootaert 1999: 42).

The social justification for intellectual property rights is weak. The idea that people need a profit motive for wanting to change their lives or innovate is absurd and contradicted all the time in real life. A lot of innovation is driven by curiosity, by engaged individuals, by public institution or by civil society (like in the case of open source). The importance of innovations driven by capitalist markets is overrated. In already developed technologies and institutions innovation driven by profit motives can be very big, but this is rarely the case for the big breakthroughs (see e.g. Hourihan and Atkinsson 2011). Investors and entrepreneurs normally come in later and try to profit from it one way or the other. Two of the most groundbreaking modern innovations are the results of public research and funds – the Internet and penicillin. The justification for assigning intellectual property rights to nature and to life forms seems even weaker, apart from the ethical dilemmas that patents to life entail. Ultimately, we are coming closer to a point where knowledge as such will be privatized.

One of the central tenets of capitalism and modern society is that private ownership is essential for the individual's and therefore humankind's progress. Things that are privately owned, or at least controlled in a way where the benefit accrues to an individual or a private unit, are often managed better. This forms the basis for a lot of entrepreneurship that has made the world a better place. The discussion in this chapter and the earlier discussion show that there are many instances where private ownership is in direct conflict with the interest of the general public and that there are many examples of alternative ways of managing resources than private ownership.

Private ownership certainly implies no guarantee for sustainable management. This is particularly the case for management of resources with a long-term perspective, such as mining, forestry or land care. And private ownership is no guarantee that the benefit for society of a resource is optimal. Maathai talks about common resources in today's Kenya: 'As we were to learn, if you can sell it, you can forget about protecting it' (2006: 173).

Conversely, private ownership of things that are of immediate interest for oneself and of little interest for others, for example, clothes, a house and tools for production, makes a lot of sense and has proven to be successful throughout a long period of history. The more the ownership of a resource influences others, the more questionable private ownership is from both a utilitarian and a moral perspective. Continuing the expansion of private ownership into more and more areas and in particular into common social or natural resources is not a project to enhance the well-being of all, but a project to increase the wealth of a few. To realize the limited marginal utility of extension of private ownership and the inherent weaknesses of it would be a good start to think on new tracks.

The future of civilization *lies in the expansion of the commons*, bringing back power to communities and developing new ways for management and governance of commons that before were unmanaged. It is power interests and lack of imagination that make people believe that a resource can only be managed by governments or by private ownership (markets). Both are rather recent phenomena and are certainly not the only ways, even if many political parties want people to believe that. In the last part of the book, we will come back to this discussion.

Distribution of Resources

When I give food to the poor, they call me a saint. When I ask why the poor have no food, they call me a communist.
(Dom Helder Camara)

Earlier we discussed the distribution and use of natural resources. Largely, an individual's use of natural resources reflects the economic resources of the individual. Here, we will discuss the distribution of economic resources, equality or inequality if you so prefer. Almost all countries have some systems to redistribute resources or at least a certain protection for the poor. It seems that in some countries, notably the United States, relative poverty is not so much on the political agenda, whereas the discourse in other countries, such as in northern Europe, considers inequality and income disparity as problematic.

There are also differences in how income inequalities are tackled. Some countries and cultures seem to put more effort into avoiding inequalities in the first place; for example, in some cases the salary gaps between bosses and workers is relatively small (which is the case in Japan), while others try to 'correct' disparities through taxes and support (in particular Scandinavia). The middle class and upper class have had more motivations than altruism to be sensitive to the fate of the poor as they might (and have) rebel against too miserable conditions.

Silk stockings to factory girls

As long as there were pure national economies, one could possibly apply a national perspective on income distribution. In feudal times, it was quite reasonable to compare the living conditions of the feudal lord with his underlings. For sure, they were not isolated from the rest of the world, but a very small part of their 'economy' was open and thereby influenced by the rest of the world. Once the nation-states became the rule, it was natural to expand the point of reference to the whole nation and compare the income of a farm worker in the fields of

California with that of not only the farmer who employs him but also the finance magnates in New York, who were all part of the same economy.

Today, in the integrated global economy, where individuals are dependent on each other, where my call to a customer service centre is responded to by a call centre in India, where my green beans are grown in Kenya, where my clothes are sown in Bangladesh and half of all the stuff I have is *made in China*, where television shows are mostly from Hollywood and where my fancy bicycle is from Italy, it is actually quite valid to make global comparisons. Today, one cares about Chinese girls who are locked up in their textile mills overnight and, apart from lousy salaries, who also risk getting caught in fire with no escape, or about children sowing footballs for Nike. And not only do the rich see the poor, but also the poor all over the world see the rich. Through television, magazines and the Internet, the image of how the rich live is spread to all corners of the world. It must be much more provocative to see this daily than to be blissfully unaware of the extent of the gap.

Poverty can shrink while the gaps increase and the gaps can narrow while poverty increases (all become equally poor). But one has to be more nuanced when discussing poverty. Relative poverty is real poverty, and gaps matter. Difference in income has a moral side, a social side and an economic side. Smith noted that: 'No society can surely be flourishing and happy, of which the far greater part of the members are poor and miserable' (1776: 88), and gave the example of a linen shirt:

> A linen shirt, for example, is, strictly speaking, not a necessary of life. The Greeks and Romans lived, I suppose, very comfortably, though they had no linen. But in the present times, through the greater part of Europe, a creditable day-labourer would be ashamed to appear in public without a linen shirt, the want of which would be supposed to denote that disgraceful degree of poverty, which, it is presumed, nobody can well fall into without extreme bad conduct. (1776: 399)

It can be very limiting to be poor in a high-income country even if an individual's absolute income is high in an international comparison. In a high-income country, one needs a substantially higher income to buy the goods and services[145] that are linked to a certain standing in society; therefore, one should not ignore poverty in high-income

---

[145]   E.g., what one is expected to provide one's children when they go to school.

countries by comparing it one-to-one with poverty in low-income countries.

Even if there is economic growth, there are no guarantees that this will help the poor. The appalling inequality can nullify all the possible wealth for the poor. Well, now some may object and say that there is a clear correlation between, for instance, GDP and life expectancy, and there is one. But there is a much stronger correlation between income of the poor and public expenditure and life expectancy. A society with slow or no economic growth but with equality and a good public health care system will have a higher life expectancy than a society with high growth rates but with no public health care and continued poverty among large groups. Studies from Great Britain show that during the two great wars, life expectancy *increased* markedly. Despite a limited supply of food, undernourishment decreased. The reason for this surprising pattern is likely that solidarity, sense of community and social responsibility increased by the external pressure of the war. Public health care and support to the poor increased remarkably during the same periods (Sen 1999).

> **Ten million own one-third of the wealth of the planet**
> Global household wealth amounted to $125 trillion in 2000. The wealth share estimates reveal that the richest 2% of adult individuals own more than half of all global wealth, with the richest 1% alone accounting for 40% of global assets, according to a study of the World Institute for Development Economics Research of the United Nations University. Adults with wealth of US$ 2138 are among the wealthiest half of the world. More than US$ 61,000 was needed to belong to the top 10% and more than US$ 510,000 per adult was required for membership of the top 1% (Davies *et al.* 2008).

The figures above are about wealth. One can present other data reflecting income instead; the data show slightly less differences but the disparities are still huge. Some 2 billion people live on less than US$ 2 per day globally. It is hard not to see this in contrast to the absurd income that some others benefit from. Some 145,000 American earned more than US$ 1.6 million in 2002. Their average income was US$ 3 million (*International Herald Tribune* 2005) and thus the total income was US$ 432 billion, which was more than the total income of the poorest billion people.

While there were many positive sides of the dismantling of the Soviet rule, the gap between the poorest and wealthiest groups of society is now significantly higher than it was pre-transition. For example, in Russia in 1991 the poorest 20% received 12% of total national income, whereas the richest 20% received 31%. By 2003 the income gap had widened significantly, with the poorest 20% receiving only 6% and the richest 20% receiving 47% (UNEP-EEA 2007).

Despite what many critics say, there should be no doubt that industrialization has 'given' the majority of people improved living conditions measured as material consumption, that is, how many things one can consume. Schumpeter (1942) noted that in the 1940s the relative share of income between workers and industrialists (capitalists) was more or less the same as 100 years earlier. But he also notes that industrialism has made many products a lot cheaper. While the upper classes could already in the old society live a comfortable life, mass production made things that were luxuries earlier into everyone's property. The increased salary levels made it harder for the upper classes to buy the services and the luxury craft they wanted; so relatively, even with the exploitation of labour, the labourers benefited more. Schumpeter writes:

> Electric lighting is no great boon to anyone who has money enough to buy a sufficient number of candles and to pay servants to attend to them. [. . .] Queen Elizabeth owned silk stockings. The capitalist achievement does not typically consist in providing more silk stockings for queens but in bringing them within the reach of factory girls in return for steadily decreasing amounts of effort. (1942: 67)

I believe that Schumpeter is essentially right, but note that he seems to confuse nylon with silk. Nylon is a typical industrial material suited for mass production (and emblematically made from oil), while silk is a natural product—and therefore more limited in access—and considerably more expensive.

Do gaps increase or decrease?

For most periods in world history, it is hard to estimate whether people got it better or worse within the frame of a generation or two. Certainly, there were events, such as wars, natural disasters or epidemics that can be noted and that made mortality rise or poverty increase, but longer-term trends are hard to determine. We have earlier discussed how the transition from hunting and gathering to farming hardly meant an improved living standard, most likely the opposite for the masses. In agrarian societies, there was political oppression, inequality and exploitation in a way previously unknown. Productivity increased gradually, but the ruling class and its army of support staff normally appropriated all the increased wealth and the majority — the peasants, serfs or slaves — continued to be very poor.

With industrialization, productivity increased dramatically and bigger groups could get a better living. Initially, the fate of the new working class, the labourers, was bad, even if it varied in different countries. The situation in England was particularly bad from 1820 to 1850, which also was the period that influenced Dickens and Marx, an influence that in its turn influenced millions. A good proxy for living standards from data-deficient countries and periods is body height as it reflects nutritional status and health (with some delay in time). Body heights shrank considerably in England from 1820 to 1850, in the United States from 1830 to 1890 and in Australia from 1867 to 1893.

Studies from France, the Netherlands and Sweden do not show the same picture. Countries industrializing later have normally shown better conditions in the transition (Steckel and Floud 1997), perhaps more because of improvement in health and sanitation than because of the benevolence of industrial tycoons. Slowly, the share of the added value in industry that goes to the workers increased, but, more importantly, the overall production and productivity increased manifold, in some sectors several hundred times, thus naturally benefitting the workers even if they get a minor share of the gains. The improved conditions and salaries in the industry influenced other sectors as well, so that those supplying services to households, barbers, bicycle repairmen and knife-grinders could get higher incomes. Also, farmers and farm workers slowly benefitted from increased income.

If we look at the United Kingdom over a longer period of time we see that income share of the richest 1% shrank from almost 20% in the mid-1920s to around 8% at end of the 1970s. But after that, the gap increased again and reached almost 14% at the end of the 1990s. As can be seen from Figure 23.1, there was a rather similar development in the United States, Canada, Australia and New Zealand (Atkinson and Leigh 2010).

Even the supposedly egalitarian Sweden showed a similar picture. In 1950, the elite of the country had an average annual income corresponding to 11 industrial workers. This shrank to 5 by 1980 but income differences then increased rapidly so that the elite earned as much as 18 industrial workers by 2007 (LO 2009). An interesting question is why this pattern is like that.

The two factors that come to mind are the intensified global competition that followed in the collapse of the Bretton Woods system[146] and the political neo-liberal 'revolution' under the leadership of Margaret Thatcher and then Ronald Reagan. Those two factors are also linked to each other and mutually supportive.

Figure 23.1 Income share of richest 10% in Anglo-Saxon countries (1921–2002).

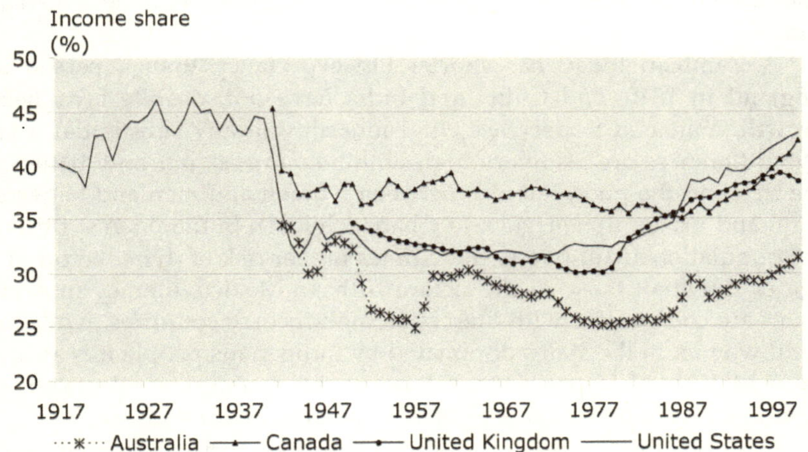

Source: Atkinson and Leigh (2010).

We can also look at other indicators than money and see whether or not one is making progress. It is a mixed bag. The United National Development Programme (UNDP) follows, measures and documents the development in a number of areas since 1990. UNDP (2005) described the changes since 1990 as follows:

Much has been achieved since the first *Human Development Report*. On average, people in developing countries are healthier, better educated and less impoverished — and they are more likely to live in a multiparty democracy. Since 1990 life expectancy in developing countries has increased by 2 years. There are 3 million fewer child

---

[146]   The international monetary system established after the Second World War. The chief features of the Bretton Woods system were an obligation for each country to adopt a monetary policy that maintained the exchange rate of its currency within a fixed value in terms of gold. The United States left the system in 1971, which signalled the end of it.

deaths annually and 30 million fewer children out of school. More than 130 million people have escaped extreme poverty. These human development gains should not be underestimated.

Nor should they be exaggerated. In 2003, 18 countries with a combined population of 460 million people registered lower scores on the human development index (HDI) than in 1990 — an unprecedented reversal. In the midst of an increasingly prosperous global economy, 10.7 million children every year do not live to see their fifth birthday, and more than 1 billion people survive in abject poverty on less than $1 a day.

A Zambian today has shorter life expectancy than a person in England in 1840, and China and India have not, despite breakneck growth, managed to decrease child mortality in any substantial way. The differences are also very high within countries, not only between the rich and the poor, but also between women and men and between rural and urban. In Senegal and Ghana, children in the poorest fifth of the population are at two to three times higher risk of dying before the age of five than those in the richest fifth. In Mexico, literacy in some states are comparable with literacy in high-income countries, while for rural women in the states dominated by indigenous people it is as low as in Mali. In Pakistan, 64% of boys in cities complete all grades at school, whereas only 17% of girls in rural areas do the same.

Undoubtedly, there is a certain connection between female empowerment and economic growth, but it is certainly no strong correlation. Only two countries have gender balance in their parliaments, Rwanda and Sweden, one in the high-income bracket and the other one at the lowest end. UNDP has developed a gender empowerment measure, in which Costa Rica and Argentina do better than Japan and Italy and Muslim Malaysia scores as high as Christian Greece, despite much lower GDP (UNDP 2005).

Lost opportunities

A widespread myth is that United States is the country of opportunity; the success of the individual is in his or her own hands. This was in particular repeated in conjunction with Barack Obama being elected as President, and by Obama too in his speeches. One is told that American society is a model for how people can be successful if they just work hard. Perhaps this was true — for whites not subject to indentured labour; certainly not for the Native Americans or the African slaves — in nineteenth–century United States when conditions were very different and where there were almost inexhaustible natural resources waiting to be exploited, especially compared to Europe, which was still half feudal. But the growth of the welfare state in Europe and the closing of the frontier in the United States changed this radically. Studies comparing the United States and Great Britain, on the one hand, with Canada, Germany and Scandinavian countries, on the other hand, show that social mobility is considerably higher in the Scandinavian countries and Canada and to a lesser extent in Germany than in Great Britain and the United States (Blanden *et al.* 2005).

This pattern coincides with the level of inequality so that the countries with the lower equality show less social mobility. Unfortunately, many Americans still seem to believe the opposite and nurture the myth that the only thing that needed to get rich is hard work and dedication. Consequently, even if rarely spelled out as clearly, if one is poor one is to blame oneself. Those who 'win' are in some way seen as 'better' and therefore they have the moral right to their wealth. This also seems to be the message from Christian fundamentalists who have adopted the virtues of wealth and success as expressed in Proverbs 10:22 'The blessing of the Lord brings wealth, and he adds no trouble to it,' instead of the dull moral of Matthew 19:24 'Again I say to you, it is easier for a camel to go through the eye of a needle than for a rich man to enter the kingdom of God.'

An American's footprint is 10 times as big as an Indian's

Instead of looking into money, we can also look into the actual use of resources. Many economists and apologetics of capitalism don't think one can speak of 'development space'. They frown at the idea of zero-sum games when applied to economics. They believe growth is good and that one can't assume or infer that someone's wealth is built on the exploitation of others. On the contrary, growth and profit will lead to all getting a better life: 'a rising tide lifts all boats'[147] or 'trickle down'[148] are popular slogans. This perspective might have some justification for a shorter period in a certain context, but as we noted above, economic

---

[147] An expression attributed to John F. Kennedy. However, it appears that the context where this was used by Kennedy actually was very different from how it has been used by neo-liberals. Kennedy used these phrases in the context of the government investing in certain infrastructure or programmes as pointed out by Donald Lazere (2009).

[148] The idea that if the rich can increase their income, for example by tax breaks, a substantial proportion of that will trickle down to the poor.

growth with low income-equality is less beneficial than growth with a good distribution of income.

We have already in Part 2 looked into the big difference in resource consumption and emissions between the rich and the poor. People in high-income countries consume up to 10 times more natural resources than those in the poorest countries. On average, an inhabitant of North America consumes around 90 kilogrammes of resources each day. In Europe people consume around 45 kilogrammes per day, whereas in Africa people consume only around 10 kilogrammes per day (SERI 2009). Global estimates suggest that 884 million people lack access to clean drinking water and 2.6 billion people do not have access to improved sanitation facilities such as running water and toilets, etc. (UNEP 2011). At the end of the last century, the 1.1 billion 'consumers' in the rich countries represented 19% of the world's population. They had 83% of the world's GDP, 83% of world trade, 95% of commercial lending, 92% of all cars, 80% of all iron and steel and 81% of all paper (Carley and Spapens 1998). As we can see, there is an incredible difference in the access people have to the resources of the world.

Another way to express human use of natural resource that has gained considerable currency in recent years is the 'ecological footprint'.[149] The ecological footprint of humanity is assessed to be at least *30% more* than the total capacity of the earth[150] and if one continues on the same track humankind will use two planets by 2030 (WWF 2008). The average inhabitant on the planet has some 2.6 global hectares (the measurement used in the footprint system) as a footprint (mind you, this is already too much), which corresponds to the footprints of Costa Rica, Turkey, Iran and Albania. Americans use 9 global hectares, the British 6.1, the Chinese 1.8 and Indians 0.8, respectively.

Many, for instance Julian L. Simons, Johan Norberg and others, try to show that discussions about scarcity of resources or environmental space are flawed. Julian L. Simons says: 'Raw materials — all of them — have become less scarce rather than more. The air in the US and in other rich countries is irrefutably safer to breathe. Water cleanliness

---

[149]   The ecological footprint is tracking flows of resources and carbon emissions through production, consumption and trade to show where ecological assets are available and where they are being used.

[150]   This is possible as footprint accounts take into consideration use of stocks (fossil fuel and minerals) or unsustainable emissions (e.g. carbon dioxide above certain level).

has improved. The environment is increasingly healthy, with every prospect that this trend will continue' (Wired 1997). Norberg (2001) shows that the price of electricity in Sweden has dropped to a fifth of what it was in 1900. Since 1920, the time a worker in Sweden had to work for 1 kWh of energy in gasoline has fallen 10-fold (Günther 2011). Since wages have increased faster than the price of gasoline, by 2010 an unskilled worker in the United States spends less than two-thirds as much, as a percent of wage, for a gallon of gasoline than a worker in 1949 (Williamson 2011).

One can easily be lulled into the perspective that there is no worry. The average income of a US worker was some US\$ 400–500 at the turn of the last century compared to some US\$ 40,000 a hundred years later. Using the perspective of farmers from developing countries, whose income has perhaps not even doubled in 100 years, one gets a very different perspective, and it is easy to understand why they still can't afford electric light or fossil fuel. That there are real conflicts in interest of the rich and the poor in the discussion of resource use, and that the discussion about environmental space is 'real' was shown clearly in the failure of the Copenhagen Climate Summit 2009 to reach an agreement about which countries should shoulder the 'burden' of reducing its emissions.

Hundred years ago, a lot of the minerals, coal and the oil came from the industrial countries (mind you, England and the United States were the two biggest fossil fuel producers 100 years ago), and the favourable fall in relative prices of those resources is not only a result of higher labour productivity, but also probably as much a result of terms of trade where resource extraction has moved to poorer countries with less bargaining power. A similar discussion also applies for future generations. If human beings continue large-scale mining of some minerals that will run out, future generations will be deprived of the same mineral.

The problem doesn't fix itself

'It is all wrong to have millionaires before you have ceased to have slums,' said British entrepreneur John Spedan Lewis (1957). Capitalism has failed to deliver the wealth to the many that it promised. True, in a few countries in North America, big parts of Europe and Japan there is a remarkable wealth and most parts of the populations have a good standard of living. Capitalism, however, has not spread the wealth. On the contrary, the majority of the population in the world is still very poor. The wealth that has been created is built on a non-sustainable exploitation of natural resources, in particular fossil energy. It is also based on a bipolar world where expensive stuff is sold to the masses in low-income countries that supply cheap goods, masses that have no real prospects of radically improving their situation. If these people had the same standard of living as those in the high-income countries, those in the latter would not at all get the same value for money either at home or abroad (when they go for holidays).

In this way, not only capitalists but also workers in high-income countries 'profit' from the lousy conditions of other workers. I don't say this to blame the workers in high-income countries, and we should also remember that the worker in the export industry of China probably is still more happy with this job in that export zone than the work she or he had in the countryside for half the pay. But this doesn't take away the fact that the wealth of high-income countries is built on the backs of low-income countries. Unequal access to resources is a real problem and there are no inherent mechanisms in the market economy and capitalism to correct this. Rather the opposite, an unfettered capitalism will (and has) increase the differences rather than decrease them. That the poor are not even poorer is a result of social activism, that is, actions in opposition to the system rather than a result of the system itself.

– 24 –
Man and Technology

I have no doubt that it is possible to give a new direction to technological development, a direction that shall lead it back to the real needs of man, and that also means: to the actual size of man. Man is small, and, therefore, small is beautiful.
(Eli F. Schumacher,
Small Is Beautiful, 1973)

For many, technology is definitively a good thing; for others, technology is neutral; for yet others, technology is a strong factor that determines society and at least some technology is inherently harmful for man or nature, often for both. Technology, usually combined with new energy resources (compare discussion at the end of Part 1), has led to an enormous increase in productivity. But it has also led to the breaking down of natural resources and societies. Some technologies and technical complexes have clearly a very limited value or are negative. The main effect, to increase productivity and thereby allow for higher production or more wealth or more profit is nothing one needs to question in general, the proof is all around. All in all, when looking into and assessing technology one needs to look at the whole package, the technological complex (which was discussed in Part 1), the effects on work organization, the interaction between man and technology and division of labour, the energy sources needed and, ultimately, the whole interaction between technology, energy, nature, man and society.

We have seen earlier how technology has increased labour productivity in agriculture so that a modern farmer or farm worker produces 1000 times more food than a manually working farmer. A similar development was seen in the early industrialization of textile manufacturing, where productivity in spinning increased 1000-fold in two generations from the end of the eighteenth century in England (Ayres 1989). Other sectors have been slower to transform. Productivity gains in Swedish forestry, has been around 5% per year throughout the period 1950–2000, and thus increased more than 10-fold. At the same time, capital intensity has also increased tremendously. Earlier, a saw, an axe, a horse and some simple barracks were all that was needed to keep a worker going. Today, workers drive expensive harvesters or other sophisticated machines that the laymen don't even know exist. The people left in the forest are few. Productivity in manufacturing has clearly increased tremendously. Mature manufacturing industries, say of cars, show productivity gains of a few percent per year. According to Worldwatch Institute (2004), 12 hours of work in the United States in 2000 produced as much as 40 hours of work in 1950, and (obviously depending on the industry) some 200 times as much per hour as in the eighteenth century.

Technology in the hands of the power

Technology is rarely neutral by itself. When the effects of building the highway system in Mexico between 1945 and 1970 were investigated, the conclusion was that less than a percent of the population travelled more than 24 kilometres in an hour in 1970. For the overwhelming majority, it would be much more beneficial with a larger net of good, but narrow roads and better bicycles than with multilane highways (Illich 1973). If you travel in an African country, you can see the same. Roads are filled with bicyclists and pedestrians who rapidly have to throw themselves aside to avoid being hit by the limited numbers of 4-by-4 cars driven by the country's elite or by development agencies. The car and its system is in general an interesting reflection of society. When few have cars, and they convince the state to build a road network for their use, it is quick and convenient to drive a car. As many more get a car, roads get crowded and the advantage of driving goes down. The whole society adapts itself to car use as the norm while the marginal utility decreases. The elite then fly helicopters and private jets or build a parallel toll-based network in order to keep its advantage.

There is a frequent saying that governments should not choose technologies, let the market do that. This perspective nullifies most of the development of human society, where technology development, to a very big extent, has been commanded by governments or public institutions. The dominating client for early technology was governments, the church and the ruling classes (who again formed the government), in particular the military played a big role.

Not only has labour productivity increased so that each worker produces a lot more, but also units of production, factory sizes, have increased as an effect of the advantage of the scale of production. This phenomenon can be observed in most industries and sectors, even in professional services such as auditing, where a handful of firms control most of the markets. The British economist E. F. Schumacher (1979) tells how the normal brick mill in the nineteenth century produces 10,000 bricks per week. At the end of the century, it had reached 100,000 per week. In the 1970s, the standard size was between 1 million and 2 million per week. The biggest brick factory in 2009 produced 16 million per week (Bigsiteofamazingfacts 2009).

With an increased scale follows a more developed division of labour and a bigger distance between workers and leaders. Often, but not always, there is also a division between the managers and the owners. Big units often constitute monopolies. Even if they are not de facto monopolies[151] in the consumer market, because of international competition, they have a monopoly role towards their suppliers (who have no one else to sell to) and even more towards their employees (whose skills have no other value than for this company in the vicinity) and towards the local communities where they operate (which are utterly dependent on the factory/company for survival, taxes and others). This is an often-overlooked perspective of monopoly, while its effects are dramatic.

The business plan of the factory is to produce externalities

To a very large extent, many industrial technologies are about producing 'externalities'. The industrial capitalist model extracts resources in distant places (destroying other peoples' nature and possibly livelihoods), has workers do something with the resources and pays them only part of the added value; the cost of pollution is carried by the local communities, the cost of health care is footed by society, the cost of schooling the workers is covered by society, etc. The products are sold at a profit and the waste is someone else's problem. Natural and social 'commons' are resource pools and dumping grounds for the factory.

Those that are most successful in externalizing their costs are the most successful. 'The rationale of machine technology is to (locally) save or liberate time and space, but (crucially) at the expense of time and space consumed elsewhere in the social system,' as summarized by Hornborg (2006: 80). When one realizes this, one sees that what economists call 'externalities', as if they were some kind of mistake in

---

[151]   This kind of monopoly is mostly called monopsony.

the process, are part and parcel of the business plans that are behind factories. Of course, there are tools and machines that produce very few externalities or where those externalities are very small. By and large, however, competition drives industries into increasing external-ities. The most apparent example is relocation of factories to emerging economies, where salaries are lower and regulations lax. Many more trends in modern manufacturing can be viewed from that perspective. However, there is always a counter-movement from society to pressur-ize industries to take care of those externalities, as they are indeed damaging.

The unequal exchange underlying machine technology can be ex-posed by measuring the net flows of biophysical resources such as energy, matter, embodied land (ecological footprints) or embodied labour. The mechanical 'power' of the machine is to a large extent also an expression of the economic and ideological 'power' through which it is sustained. Ultimately, machines are kept running by global terms of trade, where the poor get a raw deal. Like with so many other discus-sions it is hard to discern cause and effect. Is it technology, the machine, as such that creates this imbalance in power, or is it the imbalance in power that skews the application of technology into favouring the powerful? I think that they are mutually reinforcing, like so many other issues. In the marketplace, it is clearly technology — and organizational form — that tilts the terms of trade in one's favour that will survive.

Technology has created the illusion that economic processes can 'produce' resources, that technological and economic transformation of raw materials into consumer goods can make more resources available; for the price of a particular consumer good, one can buy more materials than are used for the production of the good. This is also reflected in the extraction of added value from labour. The value (price) of a good is higher than the work embedded in it, after deduction of other costs, which means that the capitalist can buy more work. These are also the mechanisms by which capitalism, technology and economic growth are intrinsically intertwined, and one more reason for why 'sustainable capitalism' is an oxymoron.

Technology as a driver of growth

The application of technology in production has largely the objective to increase efficiency, which in our modern society translates into productivity per person-hour. The increase in productivity means that more goods are produced. Earlier, this was often accomplished by simply increasing the pressure on the worker, using the whip so to say, and this is still not uncommon in parts of the world. In the rich economies, the same is accomplished by rationalization, which boils down to the application of energy and technology. This increased productivity could lead to three different scenarios. First, people can work shorter hours. Second, people can produce more. Third, people can be laid off. In reality, all three results occur, but the prevailing one is to produce more because that makes best use of the resources available and maximizes the profit of the owners.

In addition, because of competition, prices of comparable industrial goods are constantly slumping, so the value added by the workers for the same quantity will drop. This also means that the worker has to produce more *just to maintain the same economic productivity*. In this way, competition and technological progress are both powerful engines for growth, which mutually reinforce each other. As Meiksins Wood (1999) says, 'Only in capitalism is it necessary to grow just to stay in the same place.' Expectations that technology will solve the problems caused by growth are therefore wishful thinking, at least as long as the development and application of technology are under capitalist imperatives.

The car drives in the speed of a pedestrian

The Austrian philosopher and priest Ivan Illich showed in several of his books how modern technology and its institutions can reach a critical point where they become counterproductive. He showed how schooling created obstacles for learning, how the medical industry produced more illness than it cured and how the motorist in his car actually didn't travel at a quicker pace than a pedestrian (see below). The explanation is that advances made in the beginning make people believe that more of the same will give better results, while in reality most technologies and techniques show a decreasing marginal utility. Special interest groups emerge around new technology (professionals, inspectors, consultants, suppliers) for which continued investment and expansion are aims in themselves, while returns steadily drop (Illich 1973).

People confuse the aims and the means: more doctors and expensive technology are demanded instead of health; better cars and highway grids are demanded instead of easy access; more teachers and computer-aided education are demanded instead of more knowledge or insights; sophisticated kitchen appliances, prefab and support services (recipes on the Internet, television chefs) are demanded instead of profound knowledge and understanding of the nature of raw materials and human nutrition.

In a 40-year-old example, Illich (1978) showed how inefficient the automobile society is.[152] A typical American uses 1600 hours per year for his car; that is more or less the same as the hours he works. He (it is mostly a 'he', and it certainly was 40 years ago) sits in it while driving; he parks and cares for it. Of course, he also has to work to earn the money not only to buy it (or make the monthly payments, because he bought a car he couldn't afford) but also to pay for petrol, road taxes, insurance, etc. In total, this becomes 1600 hours. And in a year, he drives some 12,000 kilometres. This corresponds to an average speed of just above 7 kilometres per hour (12,000 kilometres divided by 1600 hours). This is more or less the speed of a brisk walk (Illich 1978) and considerably slower than going by bicycle. Ultimately, the performance of the car ride is even worse than that estimated by Illich's analysis. A lot of the work in the production chain of a car, for example, the plantation worker tapping rubber, the sweatshop worker sowing car upholstery in a distant country, etc., is most likely done by people with lower salaries than the car buyer. Because of unequal terms of trade, the consumer in a rich country can buy other peoples' work for a much lower price, so the embedded work in the car is much more than what he pays for.

Societies have persisted in their one-sided investments in car traffic,[153] which de facto amounts to a discrimination of those who don't drive, largely the poor. Their more limited mobility results in fewer opportunities. Studies in Boston, Los Angeles and Tokyo show that those without a car have more limited professional choices, more so in the United States. The billions spent on roads and car infrastructure are often motivated by the fact that it is profitable for society at large, as people spend less time commuting. There is no support for that notion, however: for each dollar spent on roads in the United States, the users save 11 cents (UNEP 2009). The car doesn't deliver what it is supposed to, efficient transport. In addition, it does deliver a lot of so-called externalities such as pollution, noise and accidents. By 1990, traffic death was the ninth most common cause of death in the world. Each year, 1.3 million are killed in traffic;

---

[152]   The fundaments of the discussion seem to hold also today, even if there might be an error margin of say 20%.

[153]   At the time of writing, there actually seemed to be a bit of a paradigm shift on the way, triggered by the combined challenges of the financial crisis of 2008/2009, the climate change and peak oil. Let's just hope that this can prevail.

by 2020, this figure is estimated to be 1.9 million, an increase of 65%. Death is certainly not the only human suffering; some 50 million are injured annually, many with lifelong impairments. In Canada during 1999–2000, motor vehicle collisions represented 48% of all severe injuries (CIHI 2002; Ny Teknik 2008; SvD 2009a). If accident tolls don't go down,[154] one-third of all people living will have been injured in a traffic accident in 50 years.

Medical science truly has done wonders. Immeasurable human suffering has been cured or avoided; quality of life has improved and many lives have been prolonged by medical science and health care. This doesn't mean that all is well, however. The modern health industry is to some extent counterproductive. It has created a lot of new diseases, caused for example by antibiotic-resistant bacteria or by medicines themselves. By means of more and more sophisticated analyses and tests more and more abnormal conditions are discovered,[155] which then have to either be cured by medication or surgery or be further investigated.

Up to one-fifth of all elderly hospitalized in Sweden are there, at least partially, because of side-effects of medicines (Sjukvårdsrådgivningen 2007). In the United States, more than a fifth of the elderly gets 'potentially inappropriate' medicine (Chunliu *et al.* 2001). An average of 195,000 people in the United States died as a result of potentially preventable, in-hospital medical errors in each of the years 2000, 2001 and 2002 according to HealthGrades (2004). Research in 20% of American hospitals in the early 2000s showed that '. . . medical injuries may add to a total of 2.4 million extra days of hospitalization, $9.3 billion excess charges, and 32 591 attributable deaths in the United States annually'[156] (Chunliu and Miller 2003: 1872). The Centers for

---

[154]   They do go down in the more advanced economies, for example, fatalities went down from 50,000 in the late 1970s to below 40,000 in the late 2000s in the United States, even more in Japan. Road fatalities per 100,000 inhabitants per year are at around 4–5 in countries such as Switzerland, the Netherlands and Japan, 12 in the United States, above 30 in many developing countries and an appalling 48 in Eritrea (Wikipedia 2010a).

[155]   The cost for a 'full medical screening' was £2750 (some US$ 4000), as reported in an article in the *Times* (2006).

[156]   There are apparent discrepancies between various studies here. One cause for this is how one defines deaths caused by medical injuries or errors, and what one includes, for example, whether one includes 'neglect' and 'wrong diagnosis'. In addition, and perhaps most importantly, whether one compares with 'do nothing' or with the 'best available'.

Disease Control and Prevention estimates that roughly 1.7 million hospital-associated infections, from all types of bacteria combined, cause or contribute to 99,000 deaths each year (*New York Times* 2010a).

Medical science and the health industry have developed a creed that all deficiencies can be fixed, and 'deficient' today means everything that isn't perfect. Plastic surgery and the whole battery of medication for mild depressions and conditions like ADHD (or other letter combinations in fashion) are examples of this. Treatments often lead to new needs, complications or need for continued, lifelong medication, which is of course the real boon for the pharmaceutical industry. That hospitals and clinics are increasingly privatized doesn't make things better. As expressed by Swedish author Göran Rosenberg (2002): 'the business idea of the medical and healthcare industry is based on the fact that there is unlimited demand for health and, therefore, the market has no limit.'

The development of antibiotics revolutionized the treatment of infections; thanks to antibiotics, disease caused by bacteria could, to a very large extent, be treated and cured, cheaply and effectively. This opportunity is now threatened by the increased resistance of bacteria to antibiotics. Every year 25,000 people in the European Union are hit by multi-resistant[157] bacteria. Only the health-care costs for them are in the range of €1.5 billion (SvD 2009b). Already some bacteria are resistant to all known antibiotics. The issue has now reached the echelons of politics and was one of the topics in the EU–US Summit (2009), which agreed 'To establish a transatlantic task force on urgent antimicrobial resistance issues focused on appropriate therapeutic use of antimicrobial drugs in the medical and veterinary communities, prevention of both healthcare- and community-associated drug-resistant infections .'.

Radical monopolies

With *radical monopolies*, Illich meant a technology or organizational form that has been given or has acquired such an importance or status that it is virtually impossible to do without it and where other forms of supplying the same results are not recognized, legal or even possible.[158] Police and the military are two very obvious such

---

[157]   Bacteria that is not only resistant to one kind of antibiotic, but to many, most or all of them.

[158]   The main difference to a normal monopoly is that when a company

examples. Human beings are prevented from executing legal rights themselves (say by lynching the thief) but have to hand that over to the police. So far, so good, perhaps. After the introduction of books and widespread literacy, one can say that the written word becomes a radical monopoly; it is almost impossible to live in a modern society without knowing how to read or write. Oral tradition has no value and is not recognized. In modern society, radical monopolies have expanded into areas such as learning; one has to use the school method and institutions to acquire certain knowledge or skills in order to be recognized and schools are compulsory. The impact of compulsory schooling as a disciplinary tool and a tool for impressing values as well as competitiveness should not be underestimated. In a short time, the Internet too has become a radical monopoly; it is increasingly difficult to get access to information or contact with essential services if one doesn't have Internet access. Money is a special kind of radical monopoly. Political parties have gained a radical monopoly over politics in parliamentary democracies; other forms of exercising democracy are illegal or not recognized.

Standardization and certification are growth businesses in the radical monopoly sphere and, as such, they drive concentration. One example is food hygiene standards.[159] Both society and supermarkets have increased the demands and the number of standards a food producer has to fulfil. Apart from, hopefully, providing consumers with safer food, this has also lead to many small-scale producers quitting as they can't invest in upgrading their technology to live up to the ever-increasing demands. In the same vein, traditional food processing methods are indirectly becoming illegal or impossible in the marketplace. In response to an oil contamination scandal in 1998, the Indian government imposed rules that every oil mill must have its own laboratory and chemist, which led to a million small mills closing down (Shiva 2005). Control over farmers' seeds is gradually lost to huge corporations as a result of increasing regulation of seed markets and through standardization and certification, implemented under the pretext of helping farmers.

---

has a monopoly on a certain product or a service we can often do without the product or service all together; for example, if there is a monopoly in train traffic between two cities you can go by bus instead, and even if there is a monopoly in cotton shirts you can buy a nylon or linen shirt instead. It is first when you are virtually forced to use the technology or organizational form that it constitutes a radical monopoly, even if there are many suppliers.

[159]   Such as the HACCP (hazard analysis and critical control points) system, meat inspection, BSE (bovine spongiform encephalopathy, commonly known as mad cow disease) testing, public and private systems, etc.

By and large, even the market itself is a form of radical monopoly. Rather than being an opportunity, markets are compulsory for the most basic forms of survival; there is hardly any option for anybody to stand outside the market anymore. Exchanging goods and services outside the normal market is 'black market' or 'tax evasion'. The result of these radical monopolies is a limitation of the liberty of people and sometimes they entail real compulsion, such as compulsory inoculations, compulsory pension savings or insurances, or prohibition for letting chicken go out.[160] Many, perhaps most, of these rules are benevolent in purpose and are fairly reasonable, but they can de facto still be oppressive and, more importantly, force people into production and consumption patterns mandated by society. They have also been used for forcing people away from their traditional occupations and lifestyles.

People living on hunting, gathering, fishing and pastoralism have since long seen their customary rights being gradually undermined, their land taken and their kids forced into schools. North American Indians were enclosed in reservations but were also expected to apply the 'normal' standards; in most countries in Europe gypsies and travellers, apart from general discrimination, have been forced to sedentary lives to allow their children to go to school. Traditional whaling or hunting methods have been banned, and collectors of wild resources deprived of their access to land. All these examples show that the right of the individual, the freedom, the liberty, all of which are said to be essential for and hallmarks of market societies, is a rather thin veneer. Society uses a battery of compulsion and discrimination to force human beings to behave in the desired way.

Three very different conclusions are possible from this discussion. First, that free markets and freedom of enterprise are illusions. Someone always regulates the benefits and the base rules of the game. That is how it always was and most likely will continue for eternity. Instead of maintaining the illusion of liberty, we should discuss the conditions and the limits for it. Second, as so many limitations in individual liberty have already been accepted for the benefit of the economy, of society and of human health, we should be more than willing to also do that for the environment and other aspects of human well-being. Third, in opposition to the two others, we should vigilantly

---

[160]   Imposed by a number of governments during an avian flu scare in the mid-2000s.

oppose the expansion of these radical monopolies and rather limit them as much as possible to defend the free individual. I believe there is a grain of truth in all three perspectives.

Technology is not neutral

Technology is not neutral. Certain technologies fit in certain societies; they even contribute to the formation of societies and certain societies support, encourage or enforce certain technologies. When new technologies are introduced, we should be more observant to the possible side-effects on the environment, on society and on human relations. Some technologies, such as nuclear power, can't be assessed in isolation with a simple checklist because they pre-suppose so many other institutions and technologies. Nuclear power also pre-supposes eternal stability of society which is utterly ahistorical. Introduction of new chemicals and biotechnology should be subject to prior approval procedures before release, while it is not realistic, nor desirable, that all kinds of innovations should need some explicit approval before use. Still, technology should be subject to a lot more critical analysis. Governments, as they already do today, will continue to play a role, but perhaps the criteria used for why some technology is promoted and another is not should be made clearer. In particular, vigilance is needed against technologies becoming counter-productive and technical complexes and radical monopolies forcing us to use a certain technology.

Ultimately, we should also realize the exploitation inherent in a lot of the mechanization and that 'rational' is a deceitful concept, which always has to be judged by asking 'beneficial for whom?', 'exploiting whom?' In a world where technical development aimed at the creation of most well-being and the preservation of nature, industries and machinery would follow a very different logic than they do in a context set by competition and profit-seeking.

– 25 –
Work and Leisure

Work is the refuge of people who have nothing better to do.
(Oscar Wilde,
The Remarkable Rocket, 1888)

To work, to exercise a profession was degradation according to Plato. What made life worth living was, in his view, things that we associate with leisure—sports, debating, art and philosophy. Remember that his society was one largely built on slavery; slaves had all kinds of functions, not only heavy manual work but also administration and commerce was to a large extent done by them. In this way, there was an upper class of people who didn't *have to* work, an ideal that later on was called 'economic independence'. Society has gone quite far from that view today. Increasingly work, even in the limited meaning of salaried labour, is portrayed as a kind of human right. To be 'un-employed' is one of the most deprived positions, almost on par with being a slave. On the one hand, work is seen as something individuals have to do in order to earn a living; on the other hand, it is something that is important for the self-realization of the individual. But only up to a limit; one is suddenly expected not to work when one reaches the age of retirement.

Who is the producer?

Female entry to the labour market is seen as an important indicator of gender equality, or perhaps it is a driver rather than an indicator. In the United States, the percentage of the total female population active in the money economy rose from 9.7% in 1870 to 44.7% in 1990 (Giampietro and Pimentel 1993). To be subject to exploitation is, in this paradigm, a human right. Part of the reason for all this is, of course, the mixing up of productive occupation and labour, a mix up that to a large extent is a construction and a late construction. The ruling class now wants to be seen as productive as opposed to their predecessors who resented 'production'. A 'producer' today is a capitalist, whereas a producer 100 years ago was a worker or a farmer.

What is really work and labour? One is easily fooled to say that it is production of goods and services. But can we even distinguish between production and consumption? What is one individual's consumption is another individual's production. Animals inhale oxygen and exhale carbon dioxide; plants do the opposite. One activity is called work and the other consumption. Well, we should perhaps not become too philosophical here; it is just enough to see that the perspective of production and consumption is not very straightforward. This also applies to the division between 'work' and 'leisure', which is very much a modern idea.

Up to the breakthrough of industrialism, there was little strict division between them in most daily-life situations, especially for the farming population, which constituted the majority.[161] When things were less hectic, people were occupied by maintenance of clothes, houses or tools. The clothes of the worker were as important as the hammer or the scythe. Not even holidays were separated from production. Religious rites in conjunction with sowing or harvesting were perceived to be as essential for the production of food as farming. In churches, temples or mosques important messages were announced; after service people exchanged important information. This information and these contacts were perhaps as important for the livelihoods of people as shoeing the horse.

In a traditional society, the perception that cooking food is leisure while growing it is work would be strange; both are part of survival, and to prepare food is as essential as growing it. Even eating itself is as essential as producing food. Production, consumption and re-production, work and leisure they were — and still are — all part of the same web of life.

Division of labour

Division of work or labour can be applied at a lot of different levels: within a profession (different workers do different tasks in production, in the extreme form the assembly lines), between professions (some are carpenters others are blacksmiths), between members in the household (parents, children, grandparents and, of course, man–woman), between social classes (rulers and ruled, farmers and landlords,

---

[161]   This holds still today, which I can attest to after 30 years of farming.

capitalists and workers), between societies (some tribes or states specialize in certain produce and sell it to others who specialize in other produce) and, perhaps, also between species if one applies this far (between cows and human beings, between wheat and human beings).

There are very far-reaching repercussions of the division of labour. It is most likely the most defining social process or institution. Far beyond its influence on productivity and efficiency, it has been the cause of, or basis for, gender roles, exploitation and oppression. It seems likely that some division of labour was introduced for practical reasons. Most indications point to women, because of childbearing, child carrying and breast-feeding, participating less in dangerous tasks or tasks that demanded speed or force, in particular hunting. Most animals, especially primates, too have some division of labour. It is therefore likely that human beings had some kind of division of labour between males and females of the species, even before they became human[162] so to say. It is known that collection societies were rather egalitarian and that division of labour was little developed. Most certainly, however, there were people who were experts, that is, better than others, in arts such as finding prey, making arrows, detecting water underground (even through vegetation over ground) or sniffing up edible mushrooms. Small farms rarely developed division of labour; even between sexes it was, and still is in most places, little developed compared to many other industries. Women have often done heavy digging; men have participated in wood-carving, spinning and weaving. In large farms, with slaves, serfs or employees, division of labour has often been common.

Some kinds of division of labour are clearly driven by practical needs. Even in a primitive mine, it is mostly more efficient for some workers to dig the ore, some to transport it out of the shaft and some to work on the final extraction. I would be inclined to see war as one of the most defining drivers for division of labour. Even egalitarian hunters and gatherers would normally choose some kind of com-mander when faced with a war; in the heat of battle there is simply not much space for palaver; the enemy certainly will not stand idle while one tries to reach an agreement on tactics or strategy. In more advanced wars, there are also differences between the skills and arms most useful for offensives and for defence. Different kinds of soldiers

---

[162]   And because we are human, we can chose to reconstruct it as well.

emerged such as archers, lancers, knights and operators of catapults; others supplies the combatants with munitions and food. All this led to success in the battlefield and the strength and symbolism in war as a phenomenon made it easy to adapt the rest of society to the logic of war — in a similar way as society today adapts all aspects of life to the logic of the market.

Religion is another cause for division of labour. Some individuals were assigned (or took the right to?) to devote their time and energy to communicate with the spirits, interpret their messages and apply different rituals, for the benefit for society, or at least that is how it was seen. All in all, a social/political/administrative division of labour surely was more important and relevant than an economically motivated division. The early agrarian states instituted a number of professions; different kinds of warriors, tax collectors, scribes, envoys, messengers, priests. It was rather social division that inspired the economic division that came later.

Industrialism and capitalism gave division of labour new meaning and expanded it to formerly unknown heights.[163] In the first sentence (after the introduction) of *An Inquiry into the Nature and Cause of the Wealth of Nations* (1776), Smith shows how important division of labour is with an example from the making of pins. The link between the rise of industrial capitalism and division of labour is strong. So is the link between division of labour and inequality and oppression. Smith argued that the difference between a street porter and a philosopher was as much *a consequence* of the division of labour as its cause, unlike Plato who saw their different roles as granted, given by nature, which well suited Plato's elitist views.

The introduction of machines was a major blow to the proud and well-organized craftsmen who could be replaced by cheaper and

---

[163]   Sir William Petty (1623–1687) was an English scientist who took note of the division of labour, showing its existence and usefulness in Dutch shipyards. Classically, the workers in a shipyard would build ships as units, finishing one before starting another. But the Dutch had it organized with several teams each doing the same tasks for successive ships. There were other examples of 'factories' organized also before the Industrial Revolution, such as the Venice Arsenal for shipbuilding. At the peak of its efficiency in the early sixteenth century, the Venice Arsenal employed some 16,000 people who apparently were able to produce nearly one ship each day, and could fit out, arm and provision a newly built galley with standardized parts on a production-line basis not seen again until the Industrial Revolution (Wikipedia 2010c).

unorganized labourers. Control of work is one of the main driving forces for mechanization and division of labour; it is by cutting up work into small tasks that one can control speed and quality. Notably, it is also a lot harder for workers who don't understand the whole production process to grasp the degree of exploitation they are subject to or to get ideas that they could perhaps manage the factory themselves. In this way, division of labour certainly plays in the interest of the rulers and those who exploit other people's labour.

Division of labour leads to an interdependency between people. This can, on the one hand, lead to less freedom and independence and a feeling of dependency. On the other hand, exactly this dependency is perhaps the basis for a society where an individual cares for the well-being of others, because people know they need each other; in a way a confirmation of the discussion that mutual aid and cooperation is an essential part of human progress.

Is scale a problem?

Smith (1776) criticizes the division of labour saying it leads to 'mental mutilation' in workers; they become ignorant and insular as their working lives are confined to a single repetitive task. This criticism was later repeated by many; for example, Marx (1844) wrote that 'with this division of labour', the worker is 'depressed spiritually and physically to the condition of a machine'. This observation has been made repeatedly, and particularly in conjunction with the introduction of assembly lines. Today, even if almost everybody seems to recognize the problems associated with, at least an exaggerated, division of labour, it seems to be almost a non-issue, presumably because people are so dependent on it; it seems almost inevitable and therefore there is little point in discussing it. Still, again and again, people express awe for situations that are defined by much less division of labour, such as handicrafts, managing a household or small-scale farming, which are the main pockets of resistance. The connection between division of labour and scale is strong. The bigger the company, the longer the supply lines and more extensive the division of labour. It was against this backdrop that Schumacher (1973) famously formulated 'small is beautiful'.

One can ask whether the modern information and communication society causes more or less alienation than the earlier smokestack industry society. On the one hand, manufacturing seems to move towards less assembly lines and more automation and qualified work, which should make work more interesting and rewarding. On the

other hand, complexity is increasing all the time and even big producers are rather assembly operations for components produced from
a large range of suppliers. Even information technology is fragmented
into thousands of new professions; gone are the days when geeks like
my oldest brother sat in a garage and built personal computers,
including painting the box, writing the operating system and making
interesting applications. Today, no one masters 'the system'. That we
understand less and less of the technosphere we have created is
worrying. The work of the steam engine was still quite comprehensible
by relating to everyday life experiences, notwithstanding how dreadful
the machine was. The computer terminals operating today are nothing
that can be 'understood' in a similar way.

There is no end

Ancient farmers, and still many farmers today, work fully manually,
with a hoe and a scythe as their main tools. Modern farmers drive an
air-conditioned tractor, run on diesel from one country, while listening
to music from another continent. They don't hear the birds (if there are
any left in the industrial desert) or barely step out of the tractor to feel
the tilth or smell the soil; data stream from a display, telling the farmer
to adjust the speed or increase the doses of fertilizers. Most people
would not choose the life of the ancient farmer; but, also, most people
would not choose the life of the most mechanized farmers either.
While people do make a lot of small choices, the logic of the market
and the economy often drives them into places where they didn't want
to be in the first place. But in a world where everybody competes with
the other one, there is no end in sight, there is no 'enough' or 'good'
level of mechanization and division of labour — it just goes on and on.

Labourers have not become the automata that Marx envisioned,[164]
but nevertheless the side-effects of the division of labour are problematic. While it is clear that there are economic gains from division of
labour, there are also losses, but the losses don't affect the bottom line
of the companies very much. At the same time, it is a completely
essential element of the Industrial Revolution. It is only in an economy
and production modes that are built on a totally different scale that
division of labour can be radically reduced.

---

[164]   Ironically, the Soviet Union brought the division of labour to new
heights and also enforced it in the agriculture sector in a much more radical
way than in the capitalist economies.

Probably, division of labour, like so many other things, has diminishing marginal utility. The negative effects increase with ever-increasing divisions, but the gains are biggest with the first steps. In the existing system, it is only if the *economic* gains decline that there are incentives to reduce division of labour. There are strong indications of social, cultural or human 'losses' or 'externalities' caused by division of labour. In my view, the technology used, the scale, the degree and way of division of labour are intricately linked to each other and to how society is built and how nature is exploited. Today, division of labour is almost uniquely explained in terms of efficiency and economy and rarely as a means for creating or upholding inequality.

The strong link between division of labour and the emergence of the factory makes it rather clear that exploitation of other peoples' labour is a main driver for division of labour in modern times (compare with the analysis of how one of the main drivers for a factory is the externalization of costs, as explained in Chapter 19). The discussion about division of labour, technology or scale is not an either–or discussion. Interesting questions are how labour ought to be divided if there was a free choice and what scale of organization and which technologies would one feel comfortable with. It is about the societal, economic and cultural context under which these develop and the self-reinforcing processes linked to this. They are cause and effect at the same time. Ultimately, the main point of intervention to change the logic is by changing the drivers, which include inequality, exploitation, profit-making and competition. If they were different, the logic of division of labour would also be different.

Up till this point, we have discussed the capitalist market economy mainly on the basis of its effects, in particular its side-effects. It is now time to take a closer look at this creature and how it lives up to its own stated logic.

26

Capitalism and Market Economy Today

Markets may lead to the underproduction of some things— like basic research—and the overproduction of other—like pollution.

(Joseph Stiglitz,
Globalization and Its Discontents,
2002)

In society today, capitalism, market economy, human rights and democracy are portrayed as being part of the same modern projects. Many use the terms casually and sloppily, perhaps intentionally, and mix up capitalism and market economy and democracy and liberty or human rights. Democracy can exist without liberty (e.g. in Russia in the period after the collapse of the Soviet Union, or in other half-authoritarian regimes), and human rights can exist without democracy (e.g. freedom of speech without universal suffrage). Also, market economy can exist without capitalism and vice versa, depending on the definition. China has a market economy despite being rather socialist (e.g. one can't buy agriculture land). Apart from capitalist companies, other models such as cooperatives can be found for actors in the market. Many seem to see capitalism as a perfection of the market, but perhaps it is the other way round, that capitalism distorts markets so they are no longer a place for exchange between equal parties, but an instrument for exploitation, locking people in unequal relationships. Or perhaps the market paradigm, the market society, the competition and the money institution have nurtured capitalism.

The market as the normality

Market terminology is strong and the paradigm of the market economy is such that the market is seen as the normality. When pollution of the environment is discussed, we talk about 'externalities' and 'market failures', words that are apparently used because the

market is the norm. The question is whether one can even envision a
society without markets. Probably not, but a society with markets is
not necessarily a market economy. Markets have existed for a long
while, but it is only recently that something like a 'market economy' can
be discerned. That there were *some* markets that existed also during the
feudal period doesn't mean that there was a market economy, because
most products were exchanged or distributed outside that kind of
market. We discussed the emergence of the markets for goods, labour
and land in Part 1. Here, we continue the discussion, but look more
closely into the nature of these markets today.

In theory, a market economy doesn't have to be a capitalist
economy. One can envision a system where there is no accumulation of
capital and where market exchange is just a way of distributing goods,
or at least that is assumed by many. For now, let us agree that there are
very, very strong links between capitalism and market economy. A
market economy, as I use the term here, means that consumption of at
least goods and possibly services are regulated by prices that are
determined in the marketplace, at least in theory, by supply and
demand. Even when there is no labour market, the actual pay for work
is indirectly valued by the market after the value of the products, in a
way that, for example, a farmer, who is not employed, will de facto get
a certain hourly pay for his work; so even a market economy without a
labour market has a reach into work.

We have seen that in the old empires and in most feudal societies,
markets were not the main method for distribution of goods. And
there are few indications that those systems were 'inefficient' or
wasteful. In the modern complex world, however, planned economies,
or, rather, centrally planned economies, have failed. There are many
arguments against central planning, over and above the arguments
about freedom. Perfect planning supposes perfect information. But
there is no such thing as perfect information. Central planning is slow
to react to changes in demand and in the environment. Centrally
planned economies have failed, in particular, in agriculture. Important
factors such as weather can't be controlled, and therefore there are no
good conditions for planning. Finally, central planning is more
exposed to corruption than a non-planned economy.

Having acknowledged that, one also needs to accept that *almost all
economic activities are subject to planning*. Companies certainly make
plans for production and produce according to those plans. (How else
would they organize production?) And the whole process that started
modern industrialism, the establishment of manufacturers, was a

process to enable more exploitation—through planning and control—than was possible when weavers sat scattered in homesteads. The 'economies of scale' that are referred to in most mergers are an effect of having identical functions for marketing, administration, etc., instead of having two competitors; that is, one company can do the same with less effort—clearly the result of a planning process and not a market process. Half a billion farmers who plant their crops each season certainly plan their production. Most transactions between companies are also planned, that is, are based on long-term agreements of supply, for example, as between the supplier of car parts and the brands that assemble cars.

The biggest client of them all, the public sector, offers contracts of a duration of several years, even up to decades, of specified goods or services (such as supply of food to hospitals or guns to the army), and effectively suspends the market as an institution for distribution. Within the family, one distributes food according to needs and justice and not according to a 'market', and there are many other instances of non-market distribution. All this seems to work quite well, so perhaps this discussion for and against a 'free' market economy versus a 'planned economy' is somewhat of a smokescreen, hiding the real characteristics of the modern market economy.

Everything for sale

The history of capitalism is a history of expansion. It has to expand and find new areas for profit or new markets. And if it can't find it, it will invent it. There has been a geographic expansion, so capitalism now controls most parts of the globe; there has been an economic expansion, so more and more aspects of economy are subjugated capitalism; and there has been a social and political expansion, so bigger and bigger parts of society are enclosed in capitalism. Thus, gradually, capitalism finds new ways of making profit where there was no profit before. It creates markets, needs and consumption as much as it invests in production.

The first and easiest way to accumulate capital is by trade, that is, by buying and selling a product for profit. This was the origin of commercial capitalism. Then there is the high finance capitalism bankrolling governments or other capitalists. The next step is the organization of production based on wage labour—industrial capitalism. But there are many more ways to generate a profit and they are increasing by the day. Financial capital is nowadays not only about bankrolling governments and huge trade operations. Today, small firms and private persons borrow money on a large scale in particular to finance their housing. In the United Kingdom, lending to the housing market in 2008, just at the start of the crash the same year, corresponded to 85% of the GDP. The United States has the world's largest mortgage market, one of the few countries with a mortgage-to-GDP ratio over 100%. Mortgage debt rose from 61% of GDP in 1994–1997 to 103% of GDP in 2007, before falling slightly to 101% of GDP in 2008. Only 4% of new houses are bought for cash (Global Property Guide 2010).

There are also ways of getting rent from patents, that is, investments in so-called intellectual property, which we discussed quite extensively in Chapter 22. Different forms of franchises and licensing (e.g. McDonald's) are yet another version. Carbon trading and other forms of natural resources trading (e.g. trade in environmental services) are on the increase.

As we discussed in Part 1, there was wage labour before capitalism, but it was mostly of limited importance, and there were certainly nothing like a 'labour market'. With capitalism, this changed. Marx and others thought that capitalism would lead to the emergence of two distinct classes, capitalists and labourers, where labourers sold their work in an open labour market. Well, certainly, these classes emerged, but the labour market itself never became that totally free and many people work for the public sector, where salaries are as much set by politics as by the market. By and large, the majority of people are still outside those two classes. In the middle of the 1990s, some 2.5 billion people were working in the economic sectors producing goods or services, not including household work (mainly women). Of these 380 million worked in the industry, 800 million in services, including public services, and more than a billion in agriculture. In industries, the majority of workers are wage labourers; in services, the number of self-employed is high; and in agriculture, most work as (self-employed) peasant farmers. In total, the number of wage labourers was fewer than 900 million, somewhere around one-third of the work force (Filmer 1995). Adding household workers, less than a quarter of all work is done by wage earners and many of them are employed in the public sector, where the capitalist–worker dichotomy is less relevant.

Before wage labour, a majority of the population worked in some kind of forced work situation. Still there are places and sectors where forced labour is common, such as in the countryside in South Asia or in international trafficking. In Europe, women's work was not paid and went largely unrecognized. Still in many countries, women are more or less prevented from seeking work, sometimes by law, more often by tradition, religion and prejudices. To some extent, one could claim that wage labour liberated women. From a broader perspective, it is doubtful that selling one's body to 'the man' can be seen as true emancipation.

Through the alienation caused by division of labour and technification of work, wage labour also has the tendency of becoming a meaningless and endless toil; it is done primarily just to get a salary, not to fulfil dreams or any particular need. The organization and activism of unions mitigate some of the worst exploitation and, in countries with weak unions, governments mostly intervene by setting at least minimum wages. Many other work-related issues such as duration of workdays, holidays, retirement age, rules for hiring and firing are subject to government regulations in most countries. The reason is that a 'free' labour market doesn't work.

As noted above, the majority of working people are not wage earners but self-employed. In reality, while they are 'free' they are often more exploited than the workers, at least in most parts of the world. The main mechanism by which they are exploited is through competition. While governmental labour regulations are there to protect workers from the effect of unlimited competition between workers for jobs, the self-employed barber or smallholder has very little of a similar protection, even though the pressures are quite the same.

The big conquest of human capital by capitalism was wage labour linked to industrialism and use of fossil fuel and, thereby, the associated possibilities of profiting from the value addition. Thereafter, there has been an expansion in the various businesses that are associated with human capital, such as education and health care. Stealthily, there is an expansion of the reach of capitalism to the human body. It has many fronts, but most of them are associated with health care or medicine. Apart from normal medical care, which has been turned into a business far from the vocation of the original healers, there is an expansion of the market into reshaping human bodies to the prevailing ideal of beauty, whatever it is. Reproduction itself has become another 'market' with fertility treatment and surrogate mothers.

In the U.S., a full-service egg implantation—including a donated egg, the lab work, and the IVF procedure—costs upward of $40,000. In Cyprus, you can get the same service for $8,000. [. . .] An American woman gets an average of $8,000 per batch of eggs, but can ask upward of \$50,000 if she's an Ivy League grad (a 100-point increase in SAT score correlates with a \$2,350 rise in egg price); on the other hand, an uneducated Ukrainian flown to Cyprus for the extraction process will get a few hundred dollars—and a few days in the sun—for her eggs (Fast Company 2010).

Adoption is also a booming market. Organ markets are surfacing as a result of improved transplantation techniques. One can point out that there is nothing entirely new here, that wet nurses have sold their milk for millennia, that prostitutes have sold access to their bodies also for millennia. Ancient rulers, I guess, should not have hesitated to take one, or both of the kidneys of an underling to save themselves, their children or some other important person—if they just had had the capacity to do it. But there are both qualitative and quantitative differences here.

Every day on the news, 'the markets' are spoken about. For most people the term 'markets' leads thoughts to some colourful display of farm produce, a place where people buy and sell goods. But in our economistic world, 'markets' now means the trade in financial instruments. The stock market goes up or down with this or that; currency becomes cheaper or more expensive. In the course of a few hours, individuals can lose or earn thousands of dollars. But do we? Is there any difference in what I have or don't have just because my stocks or bonds (if I have any) plummet by 20%? Actually no; these papers have no real value. It is hard to understand how this market contributes to wealth and added value of anything. And it is quite easy to find examples of how financial markets themselves are to blame for great economic shocks and disturbances.

The claimed purpose of capital markets is to provide capital for entrepreneurs. However, today there is almost no link with this original purpose, if ever there was any.[165] Originally, financial capital was supposed to be a reflection of the underlying values of other capitals, mainly land and manufactured capital. Today, absolutely no such relationship remains. The value of swaps and derivatives at the end of 2007 was US\$ 454 trillion (ISDA 2009). This corresponded to some eight times the world GDP and around four times the global

---

[165]   We have discussed in Part 1 that the *haute finance* originally was oriented to bankrolling governments, financing wars, colonization or government investments rather than to production of goods and services.

household wealth (Davies *et al.* 2008). The value of stocks in the 54 biggest markets in 2007 was more or less on par with the world GDP and the value of currency trading amounted to more than US$ 2 trillion *per day* (Reuters 2007), that is, more than 10 times the world GDP in a year. In the mid-1970s, almost all currency exchanges were to finance international purchase of goods and services. In the late 1990s, less than 0.1% of the international money transactions had anything to do with that purpose (Hart 2000). The sociologist Richard Sennett writes in *The Culture of the New Capitalism*: 'An enormous surplus of capital for investment was unleashed on a global scale when the Bretton–Woods agreement broke down in the early 1970s. Wealth which had been confined to local or national enterprises or stored in national banks could more easily move around the globe. . . . The empowered investors wanted short-term rather than long-term results' (2006: 37).

In Part 2, we already discussed at length the perspective of ecosystem services and trade with emission rights. Other 'markets' for ecosystem services are rapidly forming and are worth tens of billion dollars (TEEB 2010); see the example of 'wetland banking' in the text box 'Paying for mud'.

> **Paying for mud**
> A mitigation bank is a wetland, stream, or other aquatic resource area that has been restored, established or preserved for the purpose of providing compensation for 'unavoidable impacts' to aquatic resources permitted under state or local wetland regulation. Mitigation banks are a form of 'third-party' compensatory mitigation, in which the responsibility for compensatory mitigation is assumed by a party other than the permittee. This transfer of liability has been a very attractive feature for exploiters, who would otherwise be responsible for the design, construction, monitoring, ecological success, and long-term protection of the site. In 2005 there were 450 approved mitigation banks and an additional 198 banks in the proposal stage. They represent an increasingly important economic component of the environmental consulting sector, showcasing the synergies that can arise between environmental protection and economic expansion according to the US Environmental Protection Agency. Sixty two percent of the banks were privately-owned. Mitigation banking is now an industry contributing hundreds of millions of dollars annually to the GDP. (EPA 2010)

There is still a battle between those who want to curb and regulate environment destruction with taxes and regulations such as outright bans or maximum permitted emissions and those who want to apply market-based solutions, even if there are few who actually propose only market-based solutions and rather few who totally oppose any market-based solutions. The logic for trade in emission rights, 'cap and trade' mechanism, is quite clear from a capitalist perspective. Ultimately, instead of being a liability, the right to pollute becomes an asset in the balance sheet of the companies. But don't they represent a cost? Well they do, in the same way as the registration of a patent does, but ultimately it becomes an asset that can be bought or sold and made profit from, its historical origin is of little relevance any more. The value of the permits adds to the GDP of the country, so it is a win–win situation. The actual effects on the environment and the long-term implications of capitalization of these kinds of rights are still unclear. All in all, the ironic thing is that what capitalism can be blamed for to a large extent, that is, destruction of the environment, becomes a new and lucrative field of expansion for capitalism; and the more the destruction, the bigger the business.

Give me freedom and regulate the others

'The only place you see a free market is in the speeches of politicians,' said Dwayne Andreas, former CEO of US agribusiness giant Archer Daniels Midland and a major political campaign donor (*Mother Jones Magazine* 1995). *All* societies have regulated markets, to correct one or the other deficiency or externalities. The reality is that the capitalist market economy doesn't work and can't work unless there is also a society that takes care of its externalities, but the market is also leading to the breakdown or weakening of society. That is one of the core contradictions in the market system, in capitalism.

Polanyi (1944) explains how the belief in unfettered capitalism could get its strongest hold in the United States. First, there was no feudal tradition or traditional nobility or its institutions. Second, there was 'unlimited land'[166] up to the end of the nineteenth century. Third, until the First World War there was a constant flow of immigrants which also meant that labour availability was good, and the newcomers filled up the labour market from the bottom, so that the 'older' workers could graduate and become farmers, small traders or even industrial capitalists. Through its constitution, the separation between

---

[166]   .That it was stolen from the indigenous population is another story.

economy and politics was more marked in United States than in any other place and private property got its strongest position so far. When all land was occupied and immigration ceased in conjunction with the War, the United States initiated a lot of regulations, culminating in the New Deal of the 1930s (Polanyi 1944).

At the turn of the twentieth century, American entrepreneurs had to abide by 130,000 pages of governmental regulations. Norberg (2001) in his *In Defense of Global Capitalism* sees all these regulations as unnecessary government meddling. I am inclined to agree with Norberg that many of them most likely are meaningless, some of them even counterproductive, both economically and socially, but Norberg fails to acknowledge that the system would not work without many of them.

L'État comprime et la loi triche, L'impôt saigne le malheureux (The State oppresses and the law cheats, Tax bleeds the unfortunate) starts the third stanza of the socialist anthem, *L'Internationale*. In the current Left–Right political discourse, taxes are linked to 'anti-capitalism', so those propagating for more taxes (i.e. big governments) do that to limit the reach of capitalism. But this is not the whole story. Historically, taxes played a big role in forcing people, especially farmers, into the capitalist system (see earlier discussion about smallholders in Chapter 20 and the line from *L'Internationale* above). By forcing people to pay taxes the government makes them work more than otherwise and perhaps brings the woman of the household into the labour market.

Denmark, United States and Japan are all capitalist countries, but their tax levels are different; Denmark has the highest level and Japan the lowest, but that hardly makes Japan more capitalist than Denmark. Most of the money the government takes in as taxes it spends again by buying services and goods. Capitalists normally benefit from large-scale government interventions and programmes. They either lend money to the government and get their rent or sell goods and services to the government and thus benefit; often they do both. Already Marx noted that the only share people have in public wealth is national debt and that public debt plays crucial role in capital accumulation. Capitalists certainly didn't cry when Kennedy declared that the United States should be the first country to place a man on the moon, despite the gargantuan expenditure of the government. And capitalists certainly don't cry over military aggression.

Capitalism and wage labour penetrate increasingly large parts of everyday life, often via the government. Care of the sick, the elderly and children was originally an activity dealt with within the family or sometimes by communities. This was not normally a wage-labour occupation. Gradually, the government got involved and built hospitals and nursing homes, using wage labourers and financed by taxes. Now, in many countries these hospitals and nursing homes are privatized. The same goes for schools and a host of other social services, which are now subject to capitalist exploitation, thanks to the 'socialization' of homework that preceded it. Like with the commons, capitalism needs the state to produce new markets and to transfer activities that are outside the reach of the market into something that can be exploited. This is now visible for ecosystem services, where governments establish new lucrative markets such as carbon payments or wetland banking.

To subjugate important society mechanisms under capitalism is, of course, the final resolution of any contradictions between capital and society, but it is also the final blow to any values and basically a negation of any politics, because if everything is managed by 'the market' there is no need for politics any more. Ultimately, this will never happen, simply because there are far too many 'market failures', and so many things that markets need to work out. As political scientist Charles E. Lindblom (2001) points out, the market system can't get people together to write a constitution, it can't allocate land for major infrastructure projects and it can't organize children's education for parents who can't afford to buy it.

It is not the role of the government as a tax-seeker or a buyer of services or a factory-owner that is a threat to capitalism. It is rather its political role as an interventionist that could pose a threat to capital-ism. The anti-state and anti-regulation rhetoric of capitalism is a delusion, a manipulation, a remnant from a time when capitalists were indeed in conflict with the state. Similarly, the idea of certain groups of the Left that big governments are the best counterforce to capitalism is a delusion. The state and capitalism are good bedfellows.

Towards a society of services or servants?

Marxists claim that capitalism itself will go under because of inherent conflicts in its system, and each time a global financial crisis emerges, some are quick to declare the end of capitalism, a death that again and again seems prematurely declared. Capitalism has developed over time and adapted itself to new environments, or perhaps rather created new environments. Arrighi (2010) identifies four stages or cycles of 'global' capitalism and suggests that it has gradually taken over a bigger share of human society, by gradually controlling more and more economic processes. These are the Genoese cycle (1350–

1640), the Dutch cycle (1560–1790), the British cycle (1740–1914) and the US cycle (1870–2010?). Arrighi makes two important observations about the four cycles. First, the period of supremacy of the leading power has shortened considerably with each cycle, so British supremacy lasted only half the time of the Genoese supremacy, and the hegemony of the United States now declines rapidly.

Second, at the end of each cycle, there has been a tremendous financial expansion. Once the leading power, or rather their capitalists, can't extract more wealth from trade or production, because the material expansion has reached its limits, they are left with financial intermediation and speculation. To earn money from money and not having to bother with the material base might appear to be the ultimate victory, but these periods of financial expansions characterize *the end* of each cycle. The expansion, and ultimate collapse, of financial capitalism is likely to be a much more rapid process than in the earlier stages. With information technology, the speed in speculation, the number and volumes of transactions that can be carried out, has exploded, which means that limits will be encountered rather rapidly. Perhaps we are there already. Arrighi finally notes that for each new cycle the leading power has to have more human and natural resources than its predecessor, and, in this way, it is rather logical that China emerge as a main global actor. Although I am not sure of every detail of Arrighi's view, and he himself wasn't sure either, its ability to explain what has passed and what is happening now is unmistakable.

Capitalism has many different faces and those different faces are referred to both by critics and by proponents in a confusing way. In a debate, one speaks about sweatshops in China and the other speaks about Internet start-ups in Silicon Valley; one speaks about farm workers in Kenya and yet another of Hungarian entrepreneurs. Early industrialization was very much about sweatshops and smokestacks. The idea was to squeeze in as many workers as possible in a crowded space and churn out as many things as possible, with crude exploitation of both workers and natural resources. The critics want people to believe that this is still the reality for almost all production.

Many industries went another way, however, by increasing mechanization, made possible by the use of fossil fuels; they increased the productivity per worker tremendously. Workers' skills were also upgraded, not to the level of pre-industrial craftsmen,[167] but still

---

[167] Certainly today's workers often know a lot more than the craftsmen

enough to ensure that they were not so easily replaced by new people. The capital investments per labourer are very high, up to millions of dollars per workplace. This also meant that the conditions for the workers improved radically. In this version, profit is not so much a result of exploitation of the workers, or at least it doesn't appear to be so in the same way.[168]

These models are still there in today's capitalism and they work in parallel in many places. But most profits today come from neither trade nor production. The enormous fortunes of Bill Gates (Microsoft), Ingvar Kamprad (IKEA) and Warren Buffett (Berkshire Hathaway) are not generated by *production* at all. Manufacturing goes towards the same fate as farming — ever lower margins and overproduction. In the case of Gates, it is innovation and the associated property rights, licensing, that generated (and generates) the income. Kamprad revolutionized furniture marketing, in particular by turning furniture into commodities and by a new way of distribution. IKEA has hardly any own production — they shop round the whole world to find the best (cheapest) producers. Buffett represents another form of capitalism, that of the pure investor. He has, according to media, received a 20% annual return on his investments for over 40 years. There are, of course, many other ways to earn money, in particular through pure speculation. This dissociation from production is widespread because there are no profits in manufacturing. The margins in industrial production itself are rather low and global competition keeps them that way.

> The problem for the platform manufacturer is how to make differentiation profitable . . . The Volkswagen corporation has to convince consumers that the difference between a modest Skoda and a top-end Audi — which share about 90 percent of their industrial DNA — justify selling the top model for more than twice the low-end model. How can a 10 percent difference in content be inflated into a 100 percent difference in price? . . . Imaging differences thus becomes all-important in producing profits. (Sennet 2006: 145)

Stagnating demand will affect large segments of the industry in a similar way as agriculture suffered from lack of demand and falling prices for 100 years, despite an enormous population growth. The

---

of earlier days. But they hardly master their trade in the same way — mainly as a result of the division of labour.

[168]   'Exploitation' in the sense of added value taken from the worker is much higher per worker. But as a result of the enormous work productivity the worker can still land a very decent salary.

proportion of (not the absolute volume though) industrial goods in human consumption is falling. The combination of increased competition, increased productivity and automation means constantly falling prices up to point where there is no profit to be made. Most production today is also a commodity; production capacity is hired wherever there is spare capacity to the cheapest price. As long as cheap oil subsidizes transport it doesn't matter much where that capacity is.

Reflecting this, most capital flows today have nothing to do with production, not even with the buying or selling of any products, as we saw before. Profits made in the industrial sector are more related to branding, selling services and support and those in the intellectual property rights sector are related to all the patents that are used in production. Those rights are also commodities and are sold or licensed globally.

It is said that we are moving towards a society of services. There is some truth in this, even if manufacturing still plays a bigger role than many believe. Globally, manufacturing is actually not shrinking, even if it goes down in high-income countries. On the global level, the striking change is from agriculture to services. The growing service sector has four distinct segments. The first segment comprises professional services oriented to business and industries; it includes auditing, engineering and back-office functions of many kinds. As per statistics, these same jobs would qualify as manufacturing jobs if they were performed in the factory. These kinds of services are generally well paid. The second segment comprises the whole apparatus of wholesale, transport and retail, together representing 17–18% of the work force in the United States. This sector includes a lot of low-paid and low-skilled work.

The third segment encompasses the police, health care and education, services that in most countries are performed by the public sector. Often these jobs are quite badly paid, but have better employment conditions than the retail sector or restaurants. The fourth segment is the pure personal service sector, that is, cleaning, massage, shoe-shining, etc., tasks that, historically, were done by servants. Here, conditions are often bad; in some countries, a lot of the work is done by migrant workers, legal or illegal. Those who predict that there will be a lot of growth in this sector also predict that big differences in income will remain because it is basically differences in income that drive and make possible most of these services. Most people would do their own cleaning if they had to pay the cleaning lady the same salary as their own — it is as simple as that. In manufacturing, labour productivity has increased 100-fold and more with industrialization, while 'productivity' in 'personal services' such as health care or cleaning clearly hasn't increased much. This means that personal services are relatively much more expensive now than earlier.

Borrowed feathers—capitalism and democracy

Despite unprecedented global wealth and production and the 'victory of capitalism', a vast share, perhaps a majority, of people are deprived of fundamental human freedoms. While freedom of speech and freedom of property are highlights of liberal freedom, freedom in the sense of the ability to pursue a dignified life is taken away from all those who go hungry to bed every night. To be poor is to be severely limited in freedom. In addition, in many parts of the world there is social discrimination (gender, caste, religion, ethnicity, etc.) and direct oppression of the citizens.

Democracy is part of the 'modern' project, but only partly in the way the proponents of capitalism want us to believe. There are no democratic nation-states that are not market economies, but there are market economies that are not democratic. The contribution of capitalism in democracy has mainly been through the crushing of old authoritarian rules and economic relations. We already discussed in Part 1 how the parliamentary state grew first not as a democratic institution but as a power-sharing arrangement between different elites, and there was nothing in capitalism that advocated universal suffrage, not for men and certainly not for women. Conveniently, the inventors of 'democracy', the Greek city-states, had perverted the concept already from the beginning by excluding both women and slaves. The pressure on the social structure of emerging capitalism was enormous. People organized themselves in trade unions to balance the power of the industrial capitalist. Capitalism reshaped all parts of society. People turned to the state for more influence over legislation and to develop new social institutions instead of the old ones that were collapsing under the pressures of the market.

Because of growing individualism, which was in a self-reinforcing relationship with capitalism, family institutions also weakened and society had to take over issues previously managed by the family, such as learning, nursing children and the elderly and health care. The emerging new institutions and the associated welfare state not only were a result of more real democracy, but also provided services that actually were needed for capitalism to work well, which the more foresighted liberals saw early.

Inside the companies, there is very seldom any kind of democracy.[169] The same system that claims to promote democracy for running society thinks it is a very bad idea for running a company. In addition, companies freely use their resources to manipulate democracy, either by direct campaign contributions and lobbying or by support to various think tanks or political advertisements. A city can debate whether they want to use their common resources to build a tramway, a new school or a baseball arena. But if a local company decides to sponsor one of them by 50%, it heavily manipulates the outcome of the decision—and the town ends up with a baseball arena carrying the name of the company. Corporate interests sometimes cooperate with democracy and sometimes undermine it.

Over the last years, the idea that societies and public services should be run as companies has gained ground, which means that the reach of democracy is reduced. In the liberal view, markets are an expression of freedom that should not be vulnerable to political decisions, and protecting this becomes a criterion of democracy. This also makes it possible to *invoke democracy against the people* in the economic sphere. By emphasizing liberal, individual rights as key for democracy, and the right of the economic sphere to act without public intervention, liberalism is at the same time working against democracy in the sense of power of the people. All in all, capitalism borrows the feathers of democracy; it can take little credit for the expansion of democracy.

---

[169]   A few countries, such as Germany and Sweden, have rules about workers representation on company boards; for example, in Sweden workers must be represented on boards of companies with more than 10 employees.

27

The Legitimacy of Capitalism

When we buy stock we are not contributing capital: we are buying the right to extract wealth.

(Marjorie Kelly,
The Divine Right of Capital, 2001)

The legitimacy of capitalism rests on two arguments, closely related to each other. The first is a system that in a rational way, through peoples' self-interest, efficiently organizes production and consumption, with little or no government control. This system creates increasing wealth. We have discussed this in quite some detail and the result is not very convincing. Although this system has created enormous wealth and a tremendous increase in useful as well as useless 'things', the wealth is very unevenly distributed, resulting in enormous gaps in society. It has also not made people more satisfied. Meanwhile, it has created havoc in nature, from pollution of a local water source to disruption in life-supporting cycles such as the carbon and nitrogen cycles. 'As efficient as markets may be, they do not ensure that individuals have enough food, clothes to wear, or shelter,' says Stiglitz (2002: 222). There is a need for constant government intervention to correct all so-called market failures. It is wise to exempt large parts of society as well as nature from the logic of the market. Most people who are essentially positive towards capitalism and market economy realize that.

The second argument is a moral defence both for the accumulation of capital and for making profit. Even if it can appear a bit unfair, these mechanisms are needed to stimulate innovation and development. In the long view, everyone will benefit. The sometimes absurd profits are explained — and defended — by Schumpeter (1942) as follows: Spectacular profits, much bigger than needed to motivate the action, are gained by a very small group of winners. The possibility, the dream, of these enormous gains stimulates normal business more efficiently than if everybody got only a very modest profit. They overestimate the possibility for the big win in the same way as a poker player or a lottery-ticket buyer. In Schumpeter's view, one should agree to these giant profits, and the thereby associated differences in income and wealth, because it is the best way to stimulate development. This is the essence also of others' defence for capitalism, even if modern proponents mostly express themselves a bit more politically correctly than Schumpeter did. Let us study this argument a bit more in depth and how well it passes the test of reality.

The right to extract wealth

Capitalism doesn't require limited companies, but it is the institution that most clearly epitomizes capitalism. Take the example of a small limited company.[170] It has existed for 10 years and its total revenue over 10 years is in the range of US\$ 30 million. The start-up capital invested in stocks was US\$ 60,000. The shareholders have got small dividends over the year, just corresponding to a typical interest rate; they wanted to keep the profits in the company to allow for rapid growth. Through accumulated profits, the company is now worth some US\$ 500,000; that is, the value of the 'investment' has increased eightfold in 10 years. This is nothing exceptional but all in all it is a reflection of a moderately successful company. During the 10 years of existence, some 10 people have been working in the company. They have had a good salary; they even have been part of a profit-sharing programme. One now wonders why the ones who invested the original capital 10 years earlier should be the owners of the whole company. Isn't it the fruits of the 10 people who work that has made it what it is today? The original capital, representing less than 1 person-year of work in value is now worth more than the 100 person-years of work dedicated to the company over the years. Strangely, this is accepted as the most natural thing.

The stated purpose of limited companies is to supply their owners with profits. This is motivated by the owners supplying the capital that is needed to start and invest in the operations. This is also the reality when many companies start. The bulk of the trade in stocks is not about company start-ups, however, but about stocks that are bought and sold purely speculatively. In the period 1900–1953, new stocks

---

[170]   The particular example is drawn from one of my own companies.

contributed less than 5% of the needed capital for business in the United States. At the end of the 1990s, only 1% of the value of the stock that was bought and sold reached the companies, the rest was only transfer between stockowners (Kelly 2001).

As the system is now, the stockowners have the power over companies and the exclusive rights to profits. Not the total right actually; the other power centre in society, the state, has ascertained its share of the profit through company taxes, in almost all societies. One can also see this as a payment for the services companies get from society, such as an educated work force, health system, legal system and protection. The employees, the clients, the local community or the environment have no right to a share in profits. Between 1987 and 1997, the stocks in the Dow Jones index went up 300%. In the same period, real wages in the United States dropped 7%. That can hardly be fair, reasonable or even efficient. As Schumpeter (1942) noted, when companies are not managed by entrepreneurs but by professional managers, and owned by investors, the whole idea of ownership and entrepreneurship is destroyed.

In a limited company, the owner has unlimited rights to the profits of the company, but only limited responsibility for the losses. This is the essence of limited companies; 200 years ago this didn't exist. It is an innovation and a privilege extracted by the rich in the same way as the nobles saw to it that their rights were enshrined in law. There were shareholding companies earlier, but the owners were personally responsible. The emergence of limited companies is the first example, and one of the clearer examples, of how the wealthy class tries to privatize gains and socialize losses, something that became very apparent during the financial crisis of 2008/2009 where a trillion of dollars or more were used by governments just to prime the financial markets because they were not willing to take any risks; the same markets that claims the legitimacy of their mere existence with that they are needed to provide businesses with risk capital!

It is an undisputed fact that capitalism is very good in making business out of innovation, but that doesn't mean that most ground-breaking innovations in society have been made by private companies. On the contrary, most big innovations have taken place in very different arenas. Automobiles, television, radio, nuclear power, antibiotics, wind power, sailing, electrification and the Internet are all examples of groundbreaking technologies that were driven primarily by governments or by curiosity and the 'wish to know' rather than by capitalist innovation (Hourihan and Atkinson 2011). When a big leap in technology has already been made, entrepreneurs find commercial uses for the new technology. The Internet is a very good example of this. Entrepreneurs are also very good at moving those technologies out to the people.

Technological development was rather slow before capitalism, and one of the reasons was that there were few incentives for people to improve efficiency. The rich could extract value added by serfs, tenants or slaves and didn't need mechanization and the poor had neither resources nor real possibilities to turn any innovation into business. Having said that, there is certainly no reason to believe that innovation will cease without the profit motive. To make the job easier or more entertaining is a driver for innovation as well. Human nature is curious and innovative and that will be the case also without profit motives. To do something new, something good, not only for oneself but for the whole community can be a forceful driver, and it is also something that could be highlighted and emphasized more in a way that it renders more status.

Property given by God?

Those who are in favour of property rights being almost unlimited and sacrosanct tend to be the same as those who want to limit the extent of the government. But they, conveniently, overlook a very fundamental fact. They have got their property from the government and it is only the government that can ensure their perpetuated right to the property. Before the existence of the institution of a government, of laws and of a state that could enforce these laws, there was no property. In most societies, most things 'belonged to' the community or the state. The root of private property is a privilege extracted from governments. This continues today where governments convert new things to goods and markets, for instance eco-system services. The market and control of such services is instituted by governments and through various processes allocated to private owners.

Over years, different groups, with the support of the state (or being part of the state) came to take over larger and larger parts of property, such as the warlords, later becoming the respected nobles or the financiers of the state's war, the emerging capitalists of high *finance*. It is estimated that some 1.5 million valuable items were stolen from the Old Summer Palace in China, when ransacked by the French and the British in 1860. Victor Hugo wrote:

> One day two bandits entered the Summer Palace. [. . .] All the treasures of all our cathedrals put together could not equal this formidable and splendid museum of the Orient. It contained not only

masterpieces of art, but masses of jewellery. What a great exploit, what a windfall! One of the two victors filled his pockets; when the other saw this he filled his coffers. And back they came to Europe, arm in arm, laughing away. (1985)

To stimulate the transcontinental railway building, the United States federal government gave a fourth of Minnesota and Washington and a fifth of Wisconsin, Iowa, Kansas, North Dakota and Montana to the railroad companies, an area bigger than France (Lindblom 2001). 'Land, natural resources, and government contracts or licenses are the predominant sources of the wealth of our billionaires, and all of these factors come from the government,' says Ragharam Rajan, an economist in Chicago (*International Herald Tribune* 2011), referring to the billionaires of India. And no doubt, enormous wealth and therefore property is generated by the American wars in Iraq and Afghanistan. How property in the collapsing Soviet Union was distributed was a crash course in how property is unequally distributed to those who are close to the power.

It is an entirely subjective and political decision whether some should be given unique rights to land or other resources or whether these should remain common. A government could as well declare a universal right to food as the private right to land. For those, like me, who are raised in a fully developed capitalist market society it is almost hard to conceptualize a world without private property. Still, in many parts of the world, forests and farmland are often owned by the state. The point, at this stage of discussion, is neither that private property by itself is good or bad nor that anyone with property is a thief (which would then include me as well). It is to clarify that it is a privilege based on state sanction, and that the unique rights of property have very little to do in a discussion about freedom, and even less that property rights should form a moral justification for limitation of other peoples' freedom. 'Imagine no possessions, I wonder if you can,' sings John Lennon in *Imagine*. This is probably the greatest challenge — to think outside the box.

Can capitalism be combined with sustainable development?

The financial crisis of 2008/2009 and the climax of attention to climate change gave new impetus to efforts to transform capitalism to also consider the environment. The New Green Deal was launched by the UNEP, and the renowned columnist Thomas L. Friedman wrote:

The old system, which has reached its financial and environmental limits, worked like this: We built more and more stores in America to sell more and more stuff, which was made in more and more Chinese factories powered by more and more coal that earned more and more dollars to buy more and more U.S. T-bills171 that got recycled back to America in the form of cheap credit to build more and more stores and more and more houses that gave rise to more and more Chinese factories. . . .This system was a powerful engine of wealth creation and lifted millions out of poverty, but it relied upon the risks to the Market and to Mother Nature being underpriced and to profits being privatized in good times and losses socialized in bad times. This capitalist engine doesn't need to be discarded; it needs some fixes. (International Herald Tribune 2009)

It appears to me that there is a stark contradiction between the harsh first sentence—that the system has reached its limits—and the last sentence's call for a mechanic who can fix the engine. Certainly, it is more than the 'engine' that needs to be fixed; the engine—rent-seeking and technological exploitation—is exactly what caused the problem and the fuel of the engine is fossil fuel. Capitalism is dependent on a thriving society and on the exploitation of nature's resources for its existence. At the same time, capitalism undermines both those fundaments of its own existence. This is the real crisis of capitalism.

Most of the discussions in this book have shown how the way human beings treat the globe, nature and each other undermines human survival. Many, but not all, of the problems are caused by, or at least closely linked to, the economic system and how we let it rule society. The question that begs an answer is whether the capitalist market economy that dominates and shapes the world can be improved or reshaped to become environmentally, socially, perhaps also economically sustainable. It is clear that capitalism can't be stationary and also that the system is on a bumpy road. Schumpeter speaks about the *creative destruction*:

Capitalism, then, is by nature a form or method of economic change and not only never is but never can be stationary. [. . .] The fundamental impulse that sets and keeps the capitalist engine in motion comes from the new consumers' goods, the new methods of production or transportation, the new markets, the new forms of industrial organization that capitalist enterprise creates. (1942: 82–83)

---

171   A short-term debt obligation backed by the United States government.

While it is clear that there will be constant motion and upheaval under capitalism that doesn't mean that it is not sustainable. There are those who are in total denial of the validity of the question of sustainability. They still think that there are no problems. Many voices simply state that capitalism is wrong and should be abolished. Other, more mainstream groups today severely criticize capitalism, but still think it can be reformed or reshaped to become sustainable. In the field of environment, these are represented by Jonathon Porritt, Paul Hawken, Amory Lovins and Hunter Lovins; in the field of society and economy, these are represented by George Soros, Amartya Sen and George Stiglitz. Porritt asks in *Capitalism as if the World Matters*:

> Is it possible to conceptualize and then operationalize an alternative model of capitalism — one that allows for the sustainable management of the different capital assets upon which we rely so that the yield from those different assets sustain us now, as well as in the future? (2007: 19)

He continues:

> Contemporary capitalism responds to the shortest of short terms, abominates the very notion of limits, celebrates excess, accepts that its 'invisible hand' will fashion as many losers as winners — and has no connectedness with the natural world other than as a dumping ground and a store of raw materials. (2007: 304)

> Hence this book has advocated the idea of capitalism as if the world matters; an evolved, intelligent and elegant form of capitalism that puts the Earth at its very centre and ensures that all people are its beneficiaries in recognition of our unavoidable interdependence. (2007: 324)

The discussion in *Natural Capitalism* by Hawken *et al.* (1999) is quite similar. Tim Jackson (2009), in *Prosperity without Growth: Economics for a Finite Planet*, speaks about the need for new macro-economic variables that, among others, reflect the resource dependency of the economy and the value of stocks of natural capital. The massive UNEP (2011) *Green Economy Report* is an extensive catalogue of things that are flawed and wrong in the current world. Still, UNEP is stuck in the growth paradigm and the paradigm of the market, and the UNEP proposes mainly market-based solutions. There is almost total unity among the concerned environmentalists that there should be fees on natural resource use, cap and trade on pollution, government regulations, abolition of sick subsidies (such as the ones to fisheries, fossil fuel and water use), ecological investments, ecological tax reform and revision of national accounts.

The end result will be a maze of regulations and fixes that will make the European Union's agriculture policy or the central planning of Soviet Union look liberal; even worse is that the end result will risk perpetuating the flaws of the system. It has proven to be difficult to internalize the cost for one little environmental aspect in the price of goods, on a national level, and even harder on the international level. The effect of a product and its production may also be quite different in different parts of the world. For instance, it is not so clear that global warming is bad for Scandinavia or Siberia; so why would people there want to include costs for global warming in the price of their goods? Depletion of phosphorus resources may not be of concern for Morocco that sits on large supplies. Depletion of water resources, understandably, is a bigger problem in arid zones than in the temperate North. Also, as the example of global warming so clearly shows, there is a very big time lag between doing something and seeing its large-scale effects.

Once measures such as environmental payments are installed, there are massive redistribution issues, between the rich and the poor, between consumers and producers, between past, present and future generations and their wealth. We discussed the 100,000 chemicals before. How would it be possible to calculate the external costs for these chemicals? In most cases, one has no idea of the costs. In many cases, if these are known, the more appropriate policy response is to ban the substances (e.g. ozone-depleting substances) or severely restrict them rather than to cap and trade.

To believe that prices could include all external social and environmental costs is just not realistic. Hawken *et al.* elaborate on this in *Natural Capitalism*:

> What might be called 'industrial capitalism' does not fully conform to its own accounting principles. It liquidates its capital and calls it income. It neglects to assign any value to the largest stock of capital it employs—the natural resources and living systems, as well as the social and cultural systems that are the basis of human capital. But this deficiency in business operation cannot be corrected simply by assigning monetary values to natural capital, for three reasons. First, many of the services we receive from living systems have no known substitutes at any price. [. . .]

> Second, valuing social capital is a difficult and imprecise exercise at best. [. . .]

> Additionally, just as technology cannot replace the planet's life-support systems, so, too, are machines unable to provide a substitute for human intelligence, knowledge, wisdom, organizational abilities, and culture. (1999: 5)

As there has to be economic growth for this system to work, it means that there will be more and more production of these externalities and that the market for them will be a new and growing market. None of the above makes clear *how* the incentive structure of capitalism can be changed from each one acting in his or her own interest to something else, or *what* that else will be. This is a great weakness as that is a core fundament of capitalism. Take that away and capitalism evaporates.

Those who believe that capitalism and the market economy can 'make' or 'fix' it have to go further and further. Today, farmers are paid to produce an appealing landscape, to save old breeds and, yes, even to keep alive some weeds threatened with extinction. Measures to reduce carbon emissions or to compensate for emissions already represented a market worth US$ 143 billion in 2009 (World Bank 2010b). Following this path, we see more and more ecosystem services being regulated by market mechanisms, which is a euphemism for privatization. The same apply to natural resources. The remaining natural resources are auctioned for exploitation. This kind of exploitation represents the only logical continuation of the capitalist paradigm—to apply capitalist principles to the management of these resources; to let entrepreneurs and speculators earn money on the damage caused by capitalism itself; to give them concessions and access to the resources, a 'property' they can later on sell to someone else.

This equals an expansion of the reach of capitalism, which in its effects on societies would be as important as the introduction of the labour market. The privatization of the last decades appears to be a small wave compared to this tsunami. Popular protests also reach the same magnitude, once people have seen through the scheme and realize that private companies own the air they breathe, the water they drink and the Peregrine falcon whose dive they try to follow. It is possible, but not plausible, that it could work, and it would represent an economic, social and ecological, high-risk project the failure of which would have devastating consequences. So why would one take those risks?

Looking more to the social side, Sen attacks those who believe that capitalism works only on the basis of greed and self-interest and states that 'capitalist economy is, in fact, dependent on powerful systems of values and norms. [. . .] Successful markets operate the way they do not just on basis of the exchanges being "allowed" but also on the solid foundation of institutions and behavioural ethics (which makes the negotiated contracts viable without the need for constant litigation to achieve compliance)' (1999: 262). The development of trust, social capital, is thus an important ingredient in the success of the system.

Sen, who certainly is no anti-capitalist, concludes that the capitalist ethic and system are very limited when it comes to inequality, environmental protection and forms of cooperation outside the market. His conclusion seems to be to let the market take care of what it is good at and to ensure that there are other mechanisms to check other problems, or to design correcting mechanisms within capitalism. Sen says that the problems with inequality and destruction of the environment will 'almost certainly call for institutions that *take us beyond the capitalist market economy*' (1999: 267). He refrains from expanding on what that means.

In *The Market System: What It Is, How It Works, and What to Make of It*, Lindblom (2001) states that while the 'market system' (the term Lindblom has chosen for the market economy) is the best mechanism yet devised for creating wealth and innovations, it is not very efficient at assigning non-economic values and distributing social or economic justice. It is efficient in offering a wealth of choices but, equally, it is inefficient because of all the choices it eliminates.

All these critics acknowledge that capitalism doesn't work for central parts of human life, but they hesitate in drawing the conclusion to its end. Perhaps they don't see any alternative; perhaps they want to remain in the main furrow and not appear to be silly Utopians. But the main furrow is completely awash with an uncritical and simple-minded homage to 'the market'.

I believe that the evidence I have presented in this book clearly shows that capitalism is working badly. That things are not worse is mainly because of all the corrective actions that individuals, civil society and governments are taking to mitigate the negative effects of capitalism. Those negative effects are sometimes completely rejected or neglected, which is one reason for why this book has so much data and scientific references. Sometimes they are acknowledged but called externalities. My view is that most of these 'externalities' are part of the business plan of industrial capitalism. Competition and profit-seeking, division of labour and technology, influence all economic activity under capitalism, and that of the public sector, workers' cooperatives and self-employed. To try to shift the burden to someone else, other people or future generations, to buy cheap and sell dearly (be it labour, raw materials or ready-made goods), to appropriate common resources for exploitation or as a dumping ground are just part of normal business.

Now, some readers will say that markets are most efficient. The question is efficient in relation to what? If global markets leave society with 1 billion hungry, 2 billion who have no electricity, the uncontrolled spread of 100,000 chemicals, climate change and dwindling natural capital, how can they be efficient? Ultimately, the flaw in the assumption that markets are always most efficient is that one extrapolates situations of simple distribution of goods to management of complex ecological and societal interactions. The flaws are most apparent in the food and agriculture sector, both because of its enormous social role and because farming doubles as a management system for the planet. And markets were never designed to be good management systems for planets.

A self-reproducing ideology and a self-reproducing system

Historically, it was the logic of war — which also was the fundament of the previous ruling class — that had the same kind of self-reproducing ideology. The logic of war works like this. One can't trust the other guys/tribes/nations; therefore, it is best to strike first and show aggression, which of course induces aggression by the others. And in human society, the logic implies that a strong leader with great powers will be chosen because that is useful in a warring culture. In this way, human beings reproduce the logic of war in social organizations. As long as war is the determining force in the environment, a society organized for war will be more successful than a society organized for peace, as the peaceful ones risk being conquered by militaristic neighbours. A strong and aggressive leader will also initiate a number of wars, because that is expected by his[172] people and by the neighbours, and ultimately it is also something that will strengthen his position. Economic activity, religion and other values are subjugated by the interest of war. The organization of society, its actual behaviour and the ideology and values embraced by people are in this way self-reinforcing and self-reproducing.

It is the same kind of logic that is expressed in economic 'laws', in the ideology of the 'invisible hand', in the organization of human societies and of the domination of economy over other aspects of society. This ideology, in the same way as the war ideology, penetrates other spheres so that governments, political parties, religious congregations and NGOs take over the capitalist logic and lingo. Abstraction and fetish money becomes the focus and the ruler against

---

172   It has almost uniquely been a 'he'.

which everything else is measured. Its message of the total inter-changeability between completely different things means that one can trade a permit to pollute the North Sea with a bucket of manure for a ticket to a concert of Brahms Concerto No. 4 by the London Philharmonic Orchestra or for two weeks of work by a cotton-grower. They all have the same value, the same 'meaning', or rather the total lack of meaning.

When a system is built on the concept that all individuals act with their narrow self-interest in mind, that kind of behaviour is rewarded. The system produces growing inequality and this inequality is at the same time a driver for more economic growth. The constant increased efficiency in production also drives growth, more inequality, more consumption, cementing the division of labour. The resulting scarcity of time leads to demands for more mechanization. We have discussed earlier how market imperatives such as competition together with fossil fuels and mechanization lead to constant need for economic growth, as standing still is no option. And this is reinforced by ever-increasing consumption, a consumption that motivates workers to agree to increased production rather than decreased work hours, which would be the more 'natural' reaction.

It was possible to leave the warring societies of the past and it is possible to leave the capitalist society of today. But one will have to have the courage to challenge the underlying assumption.

Why isn't it worse?

By and large, my discussion is a rather damning view of capitalism and the market economy. Capitalism destroys society and nature and we are undermining the very fundaments of our existence. And there are many reasons to be critical, even to despair at times. Still, the bigger mystery is why it isn't worse. I mean, to a large extent, birds are singing, trees are growing and fish are swimming. People live longer than ever before; they also live healthier; there is an incredible beauty around and, by and large, people behave nicely. People are also freer than they have been since thousands of years. The right of a human being to a dignified life is agreed upon by everybody and the notion that everyone is born free and equal is, at least in theory, accepted. If capitalism is so bad, how can that be?

My response is that capitalism is most likely better than the systems prevalent before and that it has developed an unprecedented material wealth. On the other hand, the combustion of fossil fuel can by itself explain most of the accomplishments. Any system would be able to generate wealth by spending millions of years of accumulated solar energy in just a hundred years. For all its might, capitalism still hasn't conquered all spheres of the economy and even less of human life. When I went by bicycle through Turkey in 2008, I noted the many wells at the roadside, freely available to travellers. They were constructed by local strongmen who want to do something good for the people. And if one looks around one sees plenty of examples of how people act on the basis of drivers other than economic self-interest. In most parts of daily life, people don't behave as economic agents seeking to maximize profit. Even in work, individuals act mostly like human beings and only secondarily as labourers, constantly revolting and objecting to the effects of capitalism. It is humanity as such that makes the world a decent place to live in, despite capitalism and not thanks to capitalism.

The political system, also, takes a lot of action to mitigate negative effects of unfettered capitalism, for instance in the labour market. Capitalists themselves are also human beings. While the forces of competition work on them, they also care about the environment, about the well-being of their employees and about the future of their children. Many farmers, artisans and family businesses continue to work in an outdated or old-fashioned way, trying to shield themselves from the market imperatives as much as possible. And even if the state, historically, is an oppressive institution, politicians and civil servants try to do something good most of the time. In the same way as humanity shows a lot of resilience towards the harmful effects of capitalism, nature does too. The negative effects of capitalism are real and harmful, even potentially devastating, but life is a strong force and nature is a real wonder. Life and nature have a lot of redundancy that enables them to survive, even thrive under hard conditions.

'Just as markets are bad at valuing social harm, they're also bad at valuing social benefits,' says Patel (2009: 76). I believe this is a very important observation. But this also means that one doesn't pay sufficient attention to those actions and institutions that actually increase human well-being or the well-being of the environment. And this is likely to lead to those actions not being pursued vigorously. Humanity is strong and one's sense of belonging to other people is strong. I believe that the future lies in nurturing and cultivating all sides of humanity that contribute to a good life and suppressing those that do not. And this leads us to the last part of this book, where we look ahead.

PART IV
WHAT LIES AHEAD OF US?

One doesn't discover new lands without consenting to lose sight of the shore for a very long time.

(Andre Gide,
The Counterfeiters, 1927)

We have made a rather extensive study over humanity's travel through millennia, and, in particular, we have studied the form of society that dominates us today, the capitalist market economy, and how it manages—or doesn't manage—human relationship with the rest of nature and with each other. Modern industrial society faces multiple challenges; in my opinion, it faces a crisis in the real sense—a process of transformation where the old system can no longer be maintained. Most of the challenges are a direct result or reflection of the economic system. It is not known which challenge will be the determining one. Some say it is climate change, but climate regulation is only one of many ecosystem services that have been stressed too hard; many more are on the verge of collapse.

Others say it is a sudden collapse of the energy supply that will bring down this society. Energy is the driver not only for climate change but also for the exploitation of most other mineral resources as well as the degradation and destruction of ecosystems. And there is certainly no doubt about the importance of energy in the whole modern world. The spread of 100,000 chemicals might also prove to be disastrous—perhaps it will provide a neat solution to an overcrowded world by simply destroying our ability to reproduce.

Population growth compounds the other problems. World population seems to move towards stabilization, but it is probably already too high. The issue at hand is not to 'feed all'—that is still not an urgent issue as long as society can distribute food equitably (which it doesn't do)—but to meet the challenges of overstressed ecosystems and overloaded biophysical cycles caused by, among others, how we farm. Parallel to the ecological challenges, the social and human challenges are equally huge. Worst are the glaring inequalities in the world; no just or peaceful society can be built on such a fundament, as it will pervert all efforts to create the trust and social coherence needed for society itself, but also for taking care of the ecological challenges. In addition, the constant race to the top with regard to individual skills, human consumption and competition with others is exhausting. Growth is proposed as a solution to individual, ecological and social challenges, but clearly aggravates the problems.

As we have seen from the past, most changes are a result of not one but several factors supporting each other. Judging from history, it is perhaps most likely that we will be taken by surprise; that the tipping point is reached because of something that is never guessed or envisioned.

Some people seek the solution to all the problems of this world in the values of people and how they are reflected in behaviour. Certainly, ideas and values play a big role in development. They can also not be seen in isolation from how society and production are organized. They all influence each other, and to only reform the people and expect the system to correct itself is an illusion.[173] Obviously, the opposite is also true: if only the system is changed and peoples' minds remain the same, they will rapidly replicate the mechanisms of the old system and thus turn things back to what they were.

Capitalism has so far shown a remarkable ability to adapt itself to new conditions. But it is not at all likely that capitalism can adapt itself to a

---

[173]   Even the Dalai Lama says he is a 'Marxist monk' and has made efforts to develop Buddhist economy, etc., with the knowledge that the material world also plays a very big role in human development (*Indian Express* 2008).

society with limited natural resources under socially acceptable forms. It is also apparent that capitalism has served the first entrants well, but mostly not the latecomers. The neo-liberals want to explain the failure of capitalism in developing countries, for a majority of humanity, by the fact that they are not capitalist enough, but they fail to see that the pressure of competition, accumulation and exploitation by the more developed capitalist systems makes it virtually impossible to 'repeat' the success of the early adopters; to follow the capitalist route for those countries is more likely to bring with it negative effects rather than material benefits. As is clear from the presentation so far, my assessment is that capitalism has passed its best-before date; that it produces more 'bads' than goods and that this is not a result of mistakes or of business leaders not being educated well enough, but a result of inherent mechanisms of capitalism.

In this last part of my presentation, I move away from facts and figures. I try to sketch ideas for the future, directions of change, values and language. They are not truths that shouldn't be challenged, but rather the opposite; I want to encourage freethinking and criticism by the reader. I can't stake out the path to a determined goal. I don't even try to set such a specific goal. I am convinced that it is dangerous to have too fixed ideas of how a future society should work. The combination of being unable to make any good prediction and of the fundamentalism that the end justifies the means, which always seems to affect ideologies based on a distant idealized goal, is a dangerous one. Communism,[174] as shaped by Lenin, Stalin and Mao, is the best example of how wrong it can go. The goal and ideal—to abolish the classes and to allow the people to share in solidarity both labour and the fruits of labour—were and still are, in my opinion, good and justifiable. But the means to the end, which included a violent revolution, the dictatorship of the proletariat (which translated into the dictatorship of the Party) and a highly centralized planned economy, were a disaster. Having said that, there is certainly a need to point in a direction, to engage people by a vision of how things might be, and to develop a set of criteria to assess future developments. At this point of human development, I believe the strategic points of intervention are mainly the mechanisms in the economic systems and the endless chase for more things and experiences.

---

[174]   Notably, there are other, and older, 'communist' traditions than the Marxist–Leninist version, for example, Christian communism and anarchism, but here I refer to communism as a social form as it was shaped in particular in the Soviet Union and later in China.

– 28 –
The End of Colonization

There is no land left to settle, the last 'frontier' we have left to civilize is ourselves.

(Jewel Kilcher,
the album cover of *This Way*, 2001)

In nature, some species are pioneers or colonizers; they are the first to invade 'new' lands, such as land created by lava from a volcano, the land after a forest fire, the naked land after a mudslide or other kind of erosion or land rising from the sea. These species prepare the land for a richer life. Normally, they are specialists in living on limited nutrients, such as lichens and moss. Human beings are also a kind of colonizer, even if certainly not the first ones to take new land into possession. The Swedish psychiatrist Nils Uddenberg (1993), in his book *Ett djur bland alla andra?*, writes that 'the colonization stage of humanity is over. There is no new land to conquer; we have to learn how to live from the land we already have. Other paradigms have to replace the pioneer mentality.'

This train of thought is not at all shared by those who see how the restless energy of capitalism constantly expands our world. 'The conquest of the air may well be more important than the conquest of India was—we must not confuse geographical frontiers with economic ones,' says Schumpeter (1942: 117), and his words are repeated today by almost all defenders of the existing system. In a limited sense he was correct; the airline industry perhaps has the same gross output as the GDP of India, and if it doesn't, one can throw in the mobile networks, radio and television, all using the air for the service they provide. But Schumpeter was also wrong because he didn't see that the economy is a subsystem of nature and not the other way round. Although it is true that economic, geographic or biological limits should not be mixed up, this doesn't mean that the economy can move any of the other limits. It just means that when one has reached the limit, expansion will have to take place somewhere else, which is exactly what is happening all the time and which is why capitalism is based on colonization and not on living within limits.

Growth as a symptom and not a problem?

When there is constant growth and when material wealth is distributed, not fairly but so that everyone gets a piece of the cake, capitalism can still do quite well, in the eyes of most people. But when growth ceases, the inherent contradictions and conflicts become visible. This is one of the many reasons for the obsession with growth in capitalist society. Another is that growth is hard-wired into the economic system.

The discussion is complicated by different people meaning different things by 'growth'. One meaning is the growth of the physical resources at hand. This is simply not possible, but, of course, when we expand (colonize) into places not explored before, such as to the moon, deeper and deeper into the earth's crust or into the sea, deeper inside molecules, we obviously expand the resources that can be exploited for economic activity. We can also expand, or reduce, the value of such physical resources, and indeed we do. Things that were valuable, for example flint, become redundant because of the discovery of iron, and things that were abundant and therefore of no value gradually become scarce and therefore valuable; today, even sand for construction is a desirable commodity and water is increasingly getting more and more expensive. The more a resource is depleted the more valuable it becomes, until it collapses as an interesting resource. So, in that sense, resource depletion actually increases the value of the stock.

The last decades have also created new growth sectors, such as the cleaning-up sector, where the massive flow of capital towards the mitigation of greenhouse gas is a good example, among many more. The ultimate growth prospect is in the idea of selling that which will not be used as a resource, to preserve it. And, of course, unlimited growth can be created from doing nothing. What will you pay me for not colonizing Mars or what will you pay me for going back to the hunter and gatherer society (negating the last millennia of environmental destruction) or what will you pay me for not having a child so that your child can take over my ecological footprint? That may sound absurd, but already 'services' are operating along those lines of thought, for instance PopOffsets (2011). That kind of growth has no limits for sure. Sometimes, one discusses how 'fixing' the problem of greenhouse gases will 'cost' so and so many percent of the GDP, but the reality is that it will increase the GDP. The final, and in the long run interesting, discussion with regard to growth is about whether it will increase human well-being.

It is a paradox that those who are the most positive towards capitalist society and modern technology don't seem to believe in it. Isn't it the overall idea that we can work a bit less and that industry, technology and energy should relieve us from the toil of work and give us more leisure and more time to enjoy culture or whatever else we want to do? How come those who envisioned this are against a society without growth, a society where people work less and not more? In a similar way, politicians who are sceptical over the growth paradigm are lured into it and speak about 'green growth' and 'green economy', which shows how strong the growth paradigm still is.

Perhaps the interesting thing is not to discuss growth or zero-growth but to free the mind from using economic growth as a good indicator of development of any kind, whether good or bad. Growth is, from this perspective, not the problem and certainly not the solution, but just a symptom of other things. First, it is the result of an economic system that needs to expand all the time; its 'stable' mode of operation is growth, whereas no growth is unstable and described as a crisis (which interestingly enough is the opposite to the normal definition of crisis). It seems to me that the primary drivers of growth are competition, technical developments, interest rates and the constant expectations of profit. Regarding profit, it might not make such a difference if the profit comes as rent on capital, speculation or dividends from stocks. In a stationary economy, all forms of profit would lead either to inflation or to the transfer of money from the poor to the rich. And, well, money is transferred from the poor to the rich constantly, and it is only by growth that the poor don't end up starving (which they still do in many parts of the world) or starting a revolution, which then is another reason for why growth is so necessary for the stability of human society.

Mirroring this, human beings also aim to increase material wealth, which is largely a function of one's strive to attain a status—a status that in today's society is acquired by a high income and wealth. The desire for status is rather hard-wired in humanity. Most societies have had mechanisms to keep the hunt for a status under control. So, even in societies where physical force gave a high status, there were limitations on how one could express that physical force. Many traditional societies have strong limitations on the possibility for the individual to amass property; the expectation is that one will share with all one's siblings, the clan or the community. And even after money became important for society, the handling of money was not seen as a status job; rather, on the contrary. The Spaniards, for instance, didn't take good care of all the silver they stole from America; their nobles didn't want to deal with money, so it was the Dutch and the English who transformed their money into luxury goods (that they did want) from China, and of course made large profits from it. The Spaniards' own money was managed by the Genovese. In Christian Europe, Jews were heavily involved in finance and money matters because it was not being considered an honourable activity. In the Ottoman empire, the infidel Greeks were the merchants and bankers as the noble Turks didn't want to soil their hands with such menial work. (In the light of the near bankruptcy of the Greek state in recent times, it seems ironical.) Capitalism did away with all these prejudices and turned money-making into one of the noblest endeavours and the state of being rich into a potent aphrodisiac.

Finally, the introduction of fossil fuel made all this possible, to cut the chains of in-built biological limitations, limitations that, in any case, made growth impossible. It is therefore more interesting to discuss profit, the status of materials wealth, the degradation of nature and the use of energy than to discuss growth per se.

Increase happiness productivity

An increase in the absolute income by a certain sum does a lot more good — results in more well-being — for a poor person than for a rich person. One could discuss the 'happiness productivity' of a certain resource, that is, how to use a resource to deliver as much happiness or satisfaction or well-being as possible. Compare a litre of water in the swimming pool of a rich person with a litre used for drinking or for cooking in the house of a poor person.[175] Isn't this perspective in itself enough to make one argue in favour of global redistribution of resources?

It is clear from what has been discussed that increased material wealth doesn't lead to more well-being; on the contrary, it appears that the human quest for more things is threatening not only our space on earth but, ultimately, also our own well-being. There is no reason to moralize over this; considering that scarcity has been the norm for millennia, no built-in barriers exist against over-consumption of food

---

[175]   To make things worse, the cost of a litre of water that one has to carry by hand is often higher in the slums in developing countries than a litre conveniently poured from the tap by the rich, and the quality of the water is also mostly better for the rich. So the poor are discriminated against thrice over.

or things. But now the damage is evident, for the physical environment, for society as a whole and for individual human beings. Both the values that hail consumption and the economic system that is driven by this consumption and that, at the same time, amplifies consumption have to be changed. And these two are strongly linked, one feeds on the other, therefore they need to be tackled simultaneously. Inequality adds to the equation by leading to people being more frustrated than they would be in a more just society. To compensate for this frustration they consume. Not only that, inequality itself drives comparison and competition, which had growth as its main expression.

There is another scenario as well, a scenario where, in a third step (the first step was farming and the second was industrialism), a new phase of expansion is developed, with many more people. But nobody has been able to show what such a third step would look like and how it would fit into planetary boundaries. A real physical expansion has to be based on a more efficient and direct use of solar energy, where foodstuff that is put together in some solar-driven food factories is eaten. I don't think this is physically possible, and, despite human ingenuity, rarely is something made that is more efficient than what nature has already done; the only thing we can do is to take more of the biomass for ourselves.[176]

The options seem to be expansion to new planets, or the introduction of a totally managed planet, with abundant energy sources (e.g. solar or nuclear), and the production of synthetic food on a large scale. Failing any of these, we have to make the best out of what is known and available. Part of that is keeping the population at reasonable levels.

Another, perhaps more realistic and less damaging, possibility is that the expansion is only virtual, where growth is in that virtual world; instead of increasing the physical population, people gradually migrate to a virtual world, where they have a family and then progeny. It is not difficult to imagine that they would be able to reproduce them-

---

[176]   In that sense, humankind's possibility to be 'more people' would be enhanced if we exterminated most animals and redirected biomass to feed humankind instead. It also entails ploughing the steppe and savannah and cutting down more forests. There are, however, severe limits to this. Human beings are not able to take over the role of many of the animals, particularly not of the grass-eating species, and a farming system based solely on crops is not likely to be sustainable.

selves on the Internet, like some kind of virus. In this way, all human beings become God—creators of images of oneself. Status can be gained in a largely dematerialized world and in this way the chase for status will be no more significant than success at the poker table. Of course, the physical world will still have to be in balance with nature and socially sustainable, and the majority of people would have to chose or be forced to live only on the Internet, while a minority will live in real life—or perhaps they are the ones who will have to be coerced into doing that. Ultimately, I assume, or hope, that living in a virtual world will be the choice of an individual. Me? Well, I probably will choose to plant potatoes.

– 29 –
The Path and the Goal

You never change things by fighting the existing reality. To change something, build a new model that makes the existing model obsolete.

(Richard Buckminster Fuller)

Some claim that one has to know now what an alternative way of organizing society will look like, and that even to be entitled to criticize the existing society I need to have an alternative to put in its place. But there is no reason to demand that how a future society would look be known in order to criticize the current one. Humanity didn't choose to become farmers instead of collectors; it was a slow process. Humanity didn't choose capitalism; it grew, stealthily, under feudalism, and once it had matured and the conditions were right it took over society, or at least tried to dominate it. At no particular point in time, in most countries, did 'the people' make a clear choice to introduce capitalism. When the masses finally got limited power through parliamentary democracy, capitalism was already an established fact.

In the same way as the empires that emerged with iron-making represented an aggressive impulse that forced others to become similar, capitalism forced others to adopt the same system, either by direct conquest (colonization) or by competition. That military dictatorship or capitalism spread and conquered other cultures is not a proof that they are, or were, either superior in any way or sustainable, in fact as little as the cancer cell is better than the normal cell, even if it sometimes, sadly, wins.

Neither when humanity went from a hunter and gatherer society to an agrarian society nor when it went from an agrarian society to an industrial society was it known where humanity was going. Maybe human beings wouldn't even have gone there if the consequences were known. Perhaps, humankind is more enlightened now; perhaps, people have understood what shapes the physical world, the economy, societies, relationships and individuals, and how individuals, in turn, shape them. However, it is more likely that we fail to see the full picture. Or as Lebanese-born essayist Nassim Nicholas Taleb says, 'Our forecast errors have traditionally been enormous, and there may be no reason for us to believe that we are suddenly in a more privileged position to see into the future than our blind predecessors' (2007: 162).

One change induces another and the first steps can be seen, but there are unpredictable influences for each step ahead. The deck might be reshuffled by some unexpected event such as the plague, the 'discovery of America' or the development of the World Wide Web. The only thing that seems to be absolutely clear is that things never develop as one thinks they will. It is not even known whether we are facing a big leap ahead or a very big dip in human development, a 'perfect recession', where not only the economy shrinks, but human population and the use of natural resources dwindle; a giant correction, such as the Black Death in Europe in the fourteenth century which wiped out perhaps 50% of the population. In my view, we still have the means to influence this, and even if we can't be sure or we know that we will never know for sure, it is our obligation to do our best to ensure that future generations will have a decent life.

I am convinced that humankind is in the midst of a transformation into a new era. Like in the two other main shifts in human development, human beings have reached biological and energy limits and the human population has to stabilize from a point of rapid growth (and it is stabilizing). Perhaps the change will be slow and stretch over generations, but I am inclined to believe that it will be rapid. The transformation from an agrarian society to an industrial society occurred over six to seven generations in countries that industrialized first, but occurred over only one or two generations among the latecomers. And today, everything seems to change at a breakneck speed. The distinction between evolution and revolution is not so clear at times.

Not to care much about the final goals, to find compromises between conflicting aspirations and tendencies, to muddle through with piecemeal engineering is what normal politics does. And it is good—in ordinary times probably the best—but not enough in current times. A vision is needed to inspire and to act as touchstones for all these daily choices. Even the current situation is much stronger ruled by such visions and paradigms than most people want to see because we are so ingrained in them: the idea of constantly improving material wealth driven by economic growth, the idea that the markets are the best tool for not only distribution but also production, the idea that one will be happier by consuming more and have ever more choices. In the same way as our parents accepted the dictatorship of the machine, the mortgage and the queues of commuting traffic or the packing of people like sardines in subway trains as 'part of the deal' for the television, the gramophone, the automobile, the holiday home or the nylon stockings, we and our children need to feel that we will be rewarded by a new behaviour, that we want to swap some of this for *a better life* based on a more balanced and harmonious development, where we give ourselves and nature more time and care.

Where, when and how can we change?

In *Leverage Points: Places to Intervene in a System*, Donella Meadows (1999) describes where one should intervene if one wants to change a system (ecological, economic or social) on different levels. She lists twelve different ways ranging from constants, parameters and numbers to transcending paradigms. The level where daily politics are performed, subsidies levels, tax rates, etc. rarely change the system. One has to go further in her list to find leverage points that are of fundamental importance. There one also finds feedback flows in the system. They can be negative, that is, things that correct or regulate inherent tendencies. For instance, because of the tendency that the richer get richer, societies have introduced property taxes or inheritance taxes to redistribute wealth. Positive feedback loops or self-reinforcing processes are dangerous for society or any other system. According to Meadows,

> A system with an unchecked positive loop ultimately will destroy itself. That's why there are so few of them. [. . .] (1999: 11)

> . . . if you keep raising the capital growth rate in the world model, eventually you get to a point where one tiny increase more will shift the economy from exponential growth to oscillation. Another nudge upward gives the oscillation a double beat. And just the tiniest further nudge sends it into chaos.' (1999: 12)

Examples of such positive loops are speculative bubbles in housing or finance, eutrophication of a lake or methane emission from thawing arctic tundra.

The rules of the game and the power over those rules are apparently very important. The whole economic system rests on a legal protection of property and a violent state to enforce it, the so-called rule of law. Other such cornerstones are democracy and liberties. It is no coincidence that constitutions of countries are a classical battleground for ideologies and that they are not changed so easily. The question today is not about what is the best thing to do within the rules as they are, but about how we can get away from the rules that have been operating since the birth of industrial capitalism.

To challenge the goals of a system, or even simply to ask what they are (!), can be very powerful—mostly they were never formulated in the first place. Most systems have evolved over a long time, in small steps, from a set of original goals, but mostly those goals are not present in everyday discourse; many people don't know them and, perhaps, those who do don't agree with them. Thus, to question them can be a seed of change. These kinds of questions most likely played a rather great role in the fall of the Soviet Union; when the difference between the actual situation (little freedom, low quality of life and drudgery) and the stated goals (solidarity and a non-exploitative society) became too apparent and too big, the credibility of the system collapsed. 'The emperor is naked' noted the child in Hans Christian Andersen's *The Emperor's New Clothes*, and we need more such children to ask such questions about present society.

Today, one can ask whether the purpose of the capitalist market economy is to create immense wealth for a few through profits or whether it is to create wealth or happiness or freedom for everyone. Is it the sanctity of private property, the freedom of the individual, the difference in income, the commercialization of all aspects of one's life or the levelling of differences in cultures and values across the globe? Can anyone still remember? Did you *choose* to live in this society? What exactly is it in the capitalist project that we so desperately hang on to? It is time to ask these questions, to critically ask ourselves what it was that we wanted to accomplish. If it was a high standard of living, we have it now; if it was freedom of the individual, we have that as well (well, at least in many countries). Do we need the same system for maintaining these as the one that created them? For clearing new land, forests are cut down, swamps drained, vegetation burnt, and heavy machinery is used to make the land suitable for farming, but once this is done, much simpler techniques can be used every year. Is it perhaps like that with capitalism? It has done its job, so is it time to retire?

How we measure

Human society is obsessed with indexes and rankings. One of the most pervasive measures has been the GDP, supposedly a measure of economic wealth. No one claims that it is a perfect index and many think it is not even a good index for economic things. For example, while GDP is supposed to measure the value of the output of goods and services, in one key sector—government—there is typically no way of doing it; so often the output is simply measured by the input. If the government spends more—even if inefficiently—output goes up.[177] It is also known that even directly harmful things like a car accident can increase the GDP. Or if I choose not to cook my dinner but to go out and buy food in a restaurant, suddenly our 'wealth' increases. We have discussed how the costs for curbing greenhouse gas emissions also add to the growth in the GDP, and how the exploitation of limited resources is reflected as an increase in the GDP.

The GDP is sometimes used as a statement of the 'standard of living', but there is no such direct correlation. GDP also doesn't reflect inequalities, so some people can be dead poor even in a country with a high GDP. The GDP also understates the benefits sometimes, because it is a measure of monetary values, the price of goods or services. A lot of consumer items, such as electronics and food, have long-term falling prices, which means that more and more 'stuff' can be got for the same money; so even with a stagnant GDP, material wealth can improve considerably (e.g. the size of your TV screen or your can of Coke can double every 10 years—for the same price). Simon Kuznets, the main architect of GDP, said that 'the welfare of a nation can scarcely be inferred from a measurement of national income' (in Talberth *et al.* 2006: 1) and later on he said that 'Distinctions must be kept in mind between quantity and quality of growth, between costs and returns, and between the short and long run. Goals for more growth should specify more growth of what and for what' (Kuznets 1962). One would wish that politicians and economists would follow that advice more often.

Clearly, GDP is not an appropriate measure of the progress of human societies. A number of alternatives have been promoted such as:

Human Development Index (HDI), promoted by the UNDP, uses GDP as a part of its calculation and then factors in indicators of life expectancy and education levels. Notably, it doesn't include anything on ecological sustainability. Scandinavia 'scores' well in the HDI (UNDP 2005).

---

[177]   In the last 60 years, the share of government output in GDP has increased from 21.4% to 38.6% in the United States, from 27.6% to 52.7% in France, from 34.2% to 47.6% in the United Kingdom and from 30.4% to 44.0% in Germany (Stiglitz 2009).

Gross national happiness is an indicator of the 'quality of life'. Bhutan is working on a complex set of subjective and objective indicators to measure 'national happiness' in various domains, such as living standards, health, education, ecosystem diversity and resilience, cultural vitality and diversity, time use and balance, good governance, community vitality and psychological well-being.

Genuine Progress Indicator (GPI) and Index of Sustainable Economic Welfare (ISEW) attempt to address many of the above criticisms by taking the same raw information supplied for GDP and then adjusting it for income distribution, adding for the value of household and volunteer work and subtracting for crime, pollution and depletion of national resources. For example, loss of farmland, erosion and compaction of farmland are joined together to be one of the 29 indicators. The GDP of Australia grew by 3.9% annually between 1950 and 2000, whereas the GPI only grew by 1.47%. The disconnect between GDP and GPI growth has increased, as can be seen in Figure 29.1 which shows the development of GPI and GDP per capita in the United States (Talberth *et al.* 2007).

Figure 29.1 GPI and GDP for the United States.

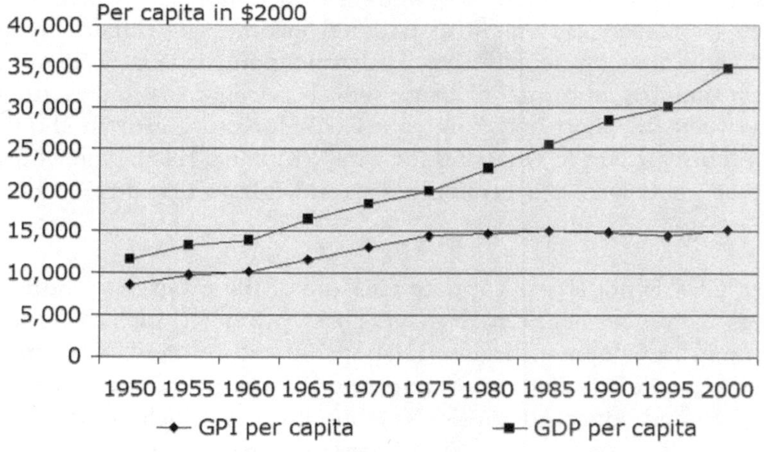

Source: Talberth *et al.* (2007).

The Ecological Wealth of Nations compares a nation's 'bio-capacity' with its 'ecological footprint'. This measure can be combined with other measures, for example, the HDI (Global Footprint Network 2010).

The Happy Planet Index (HPI) is an index of human well-being and environmental impact, introduced by the New Economics Foundation in 2006. It measures the environmental efficiency with which human well-being is achieved within a given country or group. Human well-being is defined in terms of subjective life satisfaction and life expectancy whereas environmental impact is defined by the ecological footprint. Latin America and the Caribbean score very well in the HPI (Abdallah *et al.* 2009).

Another way of looking at human development is through the five capitals framework: financial capital, nature capital, human capital, manufactured capital and social capital. Some also talk about intellectual capital or ecological capital. To use the five capitals model doesn't mean that one has to make balance sheets with the five capitals, but it provides an alternative perspective in the same way as other indexes (besides GDP). The perspective can be used to allocate values to all sorts of capitals and express changes in the capital stock in monetary terms, and in that way perhaps integrate them in the economic system. That one has to express the value of nature, of society and of the human being herself with the term capital shows the force of the economic paradigm, even among those like me who don't share it.

There are reasons to question measurements that come up with a single figure. For instance, it is not really meaningful to combine current well-being and sustainability into a single indicator. That amounts to mixing up the profit and loss statement with the balance sheet or to combining the speedometer and gas meter in a car. Even when speaking about sustainability, it is dangerous to treat natural and social capital interchangeably. Once a certain threshold of erosion of a natural resource has been crossed, the loss of that resource can't be balanced by any other resource. All in all, the various indexes have their strengths and weaknesses. It is not my task here to sort out which one is the best. Most people seem to be very impressed by these rankings and it is certainly a good way to make people more aware of the complexities in this world to adopt some other measures besides the GDP. Introducing other indexes and measurements has the benefit of making us think outside of the box.

While it would be good to find other measures, the effects of doing so should not be exaggerated. The numbers of hungry people in the world have been measured for many decades and the numbers are still appalling. Climate change too has been measured, but it hasn't impressed politicians or citizens enough to take radical action. GDP as an indicator wasn't developed until the 1930s, and clearly economies grew rapidly also without having any GDP targets. In reality, politicians are unable to influence growth rates as much as they pretend to.

Efforts to analyse the economy with alternative measures need to be kept separate from the terms and possibilities of *managing* the economy in that way. Companies don't try to increase the GDP but try to increase their profit or simply survive the competition. Even when they speak about 'triple bottom line' and other niceties, increasing the profit will always be the overarching driver. And this will remain the same even if societies trash GDP as a measure. I have explained earlier how the market and technology drive (GDP) growth. Similarly, consumers don't buy more stuff to contribute to the GDP but because it gives status or satisfaction or simply because they have money to spend, 'money burning a hole in my pocket' as the saying goes. The effect of these measurements is on the political discourse mainly, while a change in measurement will hardly affect the behaviour of economic agents.

How we talk

The measures we use are as important as the words we use. Inherent values are expressed in language. I will never forget Phuntso from Bhutan, who, at the final evaluation of a one-month training organized by my company, said that he had found the training very useful, but that he took exception to some animals being referred to as meat. We called them 'beef cattle', which is what cattle destined for the slaughter-house are called in Sweden. Our ideas and the language we use for technology, ecology and economy, even for people (who are workers or capitalists or consumers) are part of our cultural narrative, similar to the religious categories (sin, guilt, hell, heaven) of previous cultures.

We often smile over myths and fetishes of other cultures, but fail to see that the holy car is as comic as the sacred cow, the statues of the Easter Islands or the temples of Maya. Even in this book, I have used expressions such as 'social capital', 'ecosystem services' and 'externalities'. They all take the norm from the capitalist worldview. So-called economic externalities can also be called 'destruction' or even 'theft', because that is what it amounts to. Speaking about biomass points at the use of life as a raw material for further processing. Not only that, the use of the word 'biomass' makes algae, a rose, a cow and a forest interchangeable and sees them primarily as raw materials for mills.

The use of money as the sole ruler of value and the submission of human society to economics make us believe that market transactions are voluntary, that one is enabled by the market instead of being dependent on it, and that those transactions are equal and fair. How efficiency or productivity or technology is discussed has strong biases, clearly visible in agriculture, where the systems that waste most, pollute most and use much external energy are those that are modern, efficient and productive. The function of technology[178] to put other peoples' resources in the service of the already wealthy and to constantly increase the gap, is obscured by the myths about progress. Governments speak about their citizens as clients or customers, NGOs use market logic to achieve their goals (they 'sell their message', their magazines are closed down because they are not profitable) and churches compete for souls in the marketplace with simplistic television spots much in the same way as political parties. To change the world we need to change the language we use. We need to build new narratives, new tales, new heroes (or un-heroes) and new myths.

How we value

Smith (1776) stated that labour is the real standard by which the value of commodities can, at all times and places, be estimated and compared, and money is only a nominal price. How far we have come in these 240 years! Today, the notion that labour represents the real value is seen as quaint or perhaps communist, even if Marx was the one who realized that, through capitalism, labour is no longer a standard value but a commodity to be bought and sold for profit. According to Hornborg (2010), money is like an alphabet with only one sign. That is the reason why one can't communicate any meaning with money, anything more nuanced than a call to consume, that is, to use up resources. In the market society, the only way to assign a value to anything is to sell it, which means that things that are outside the market circulation by definition have no value. Well, reality is a bit more complex than that; for instance, personal relationships are not traded and certainly enjoyed, and most people still enjoy many free pleasures, but policy makers are increasingly discussing access to those freebies, such as a forest, in terms of 'willingness to pay' for them.

In economics, attempts are made to add more 'real' values to eco-system services and assign costs to the depletion of natural and social capital as well as to pollution. However, to have an impact on policy makers all this has to be expressed in dollars, that is, in terms of money. It's like the system of bride prices or dowry, where material

---

[178]   This is obviously not the only function, but one that is largely overlooked.

values are assigned to the union of two people. But as little as a dowry is an indicator of love so are the values assigned to ecosystem services an expression of their real values. With money as the determinant of value, nature is made a subsystem of the economy. That is one reason why the efforts of valuing environmental services are not sufficient to redirect the economy from the path of self-destruction—and why it even can be negative in the long run. In the short term, as measures to slow down or even reverse environmental destruction, assigning values to environmental services, charging fees for their use or assigning values to pollution makes a lot of sense, exactly because it works within the capitalist economic paradigm. In the longer term though, other ways need to be found because ecologists can't do a better job of managing capitalism than economists.

A similar problem arises in the valuation of work. By definition, market prices are always 'right'. This serves as a justification for a business leader earning 100 times as much as his workers, or for a worker in a rich country being worth 20 times as much as a worker in a poor country or 50 times as much as a smallholder farmer. In this way, money obscures the reality, and justifies the logic of exploitation and inequality.

The economic system can also be discussed in ecological or biophysical terms. This makes a lot of sense, at least for analytical purposes, because the economy is a subsystem of the physical system we live in. Instead of expressing the economic system in monetary values, the idea is to express it in 'real' values, such as land use, labour, tons of minerals extracted and so on, or composed values such as 'ecological footprints'. What comes out of that is certainly very interesting as it clearly exposes destruction of nature, increase in entropy and social disparities. From here, there is of course a big step to be taken to implement a system whereby the current currency is attached to one or more such 'real' values.

The impulses of change

New ideas can have tremendous power. For instance, new ideas or attitudes towards how people manage their bodies and diets spread rapidly. We have also seen how population growth has ceased at the same time in many countries, despite environmental factors being quite different, and therefore the need for stabilization of the population is different. This must be because of a strong idea that is spreading, such as the notion of the freedom of women to pursue an independent life rather than being primarily agents of reproduction.

The speed with which this is spreading has obviously increased with the Internet and other mass communication media, but it would be a mistake to believe that this is a new phenomenon.. For instance, in the 1960s the student revolution spread to countries where conditions (and certainly the conditions of students) were quite different. In the period just after the First World War workers rose against oppression in many countries inspired by the Russian revolution, and the same happened in Europe in 1848. The idea of nationalism spread, long before radio or television or even the telephone, in a rather short period, to a lot of people, some of who had never had their own 'nation' at any point in history, and to countries where the objective factors for nation-building appeared to be almost absent. When the time is ripe for new ideas, they have wings; when it isn't, they have prisoners' balls chained to their ankles.

Not only the brain but also technical–economic complexes have considerable inertia or 'path dependence'. People have invested so much in the existing systems that they are not ready to discard the infrastructure and the knowledge that was the basis for what they are and have become. This is, of course, particularly valid for those who are in control of or who profit from the infrastructure or the use of it. For example, carmakers have long had a very strong influence on society. To some extent, average citizens are the support troupes because society has evolved in a way where people actually need the car, and many of them have houses that would fall in value if cars were abolished. Cars themselves represent both 'hard' and 'soft' values, that is, a lot of capital is locked up in peoples' cars and in the car industry and associated infrastructure, but the car also symbolizes an individualistic society and a society without limitations where one can go when one likes as one like (of course, this assume that society has built roads before). Even within the prevailing car complex, the inertia is big. New types of cars, in particular electric cars, were developed by outsiders and neglected or even ridiculed by mainstream industry. The same holds true for wind and solar energy or organic farming. Most pioneer organic farmers were not even from a farming background, because a farming background makes one believe that organic farming is impossible.

One could assume that society, the norms, laws and conventions one follows, and the structure of organizations that constitute society should be easier to change, because they are 'soft', non-material and non-monetary, but nothing could be less true. Societal structures are self-preserving; the ruling groups profit from the status quo and the average citizen is comfortable most of the time or just simply afraid of change. Therefore, any radical societal change is mostly initiated by groups outside the mainstream.

We have seen how ideas, society, technology and economy are intertwined. The technical development complexes, societal structures and ideas (values and knowledge) are in a constant dance—they influence each other. Sometimes an impulse can shake the whole system and create new development chains. It can be an idea, such as Islam or Darwin's theory of evolution or a social innovation such as parliamentary democracy, money or the Declaration of Human Rights. It can be, and often is, a technological leap such as fire, iron or the steam engine, or the introduction of a new form of land use such as farming, or harnessing new energy such as the wind for ocean sailing or oil for combustion engines. We need an all-encompassing perspective on humanity and human evolution that includes the physical environment, organization in society, accumulated experience, wisdom-culture and, most fascinating of all, the human mind. With this view, we transcend the artificial divisions made between body and mind, between heritage and environment as drivers of evolution, between one's health and that of the environment, between culture and nature.

– 30 –
The Changes

*All revolutions are the sheerest fantasy until they happen;
then they become historical inevitabilities.*

(Sonmi~451,
in David Mitchell's Cloud Atlas, 2004)

In this chapter, we look at the changes that have started or are about to
happen in the years to come. Most predictions are wrong, and certainly
some of these will be too. Still, it is valuable to try to discern trends and
patterns of what is going on; it is the best that can be done. All societies
have their periods of growth, maturity and decline. Often, decline
starts long before there is definite proof. When a civilization builds the
biggest temples for its glory, mould, termites, rodents and rot are
already busy tearing it down. Today, we are there already. Shall we go
on building temples like the Maya? The good news is that a new
society is already emerging as we speak. It is not clearly seen yet, and
there is no master plan or blueprint or commander, but there are signs.
We should work in favour of this new society because it is already
coming into existence and because the current system is not socially or
environmentally or economically sustainable. An equally good reason
is that current society is based on an unjust system of privilege and
violence.

New times—new thoughts

Some people believe that two events in 1989 were the harbingers of a
totally new society: the fall of the Berlin Wall and the construction of
the World Wide Web. The first event marked the end of a bipolar
world and the second event meant the emergence of the information
revolution. Others say that the pictures of a lonely little blue and green
planet taken from the moon in 1969 by the Apollo 11 crew were images
that made everyone realize how vulnerable and dependent people are
on that little planet. Even more so as the remarkable achievement of
putting a man on the moon showed the extent of human capability, but
also showed that the dreams of human colonization of space remain—

for many centuries at least—just a dream. We simply have to make do with that little speck in the universe and take care of it. Without doubt, there is already a change in paradigm, in patterns of thought. This is also totally essential, a precondition, for any larger scale of change.

There is no doubt that our wants and desires are never fully met, which means that we are never fully satisfied. This restless quest for something else, something different, has been a major force in the development of humanity; it made human beings start wandering out of their home in Africa; it made human beings understand how to use fire and other parts of nature for our benefit. Because this is already a strong driver within us we don't need a society and culture that encourage it even more, and certainly not today.

One can satisfy one's wants in many different ways. Present-day society is built on a vision of unlimited material needs that can only be satisfied through more production. But a look at most, if not all, traditional hunter and gatherer societies shows that these societies satisfy the needs of their members well, and they do that with a very limited input of work and very limited use of natural resources.

However, we can't go back to that kind of life. One reason is that most people simply wouldn't want to; another reason is that, today, the human race is at least 50 times too many people to live in such an ecological niche. Finally, as Bateson says, 'Such a return would involve loss of the wisdom that prompted the return and would only start the whole process over' (2000: 503). But one can still look at these societies and the satisfiers they nurtured. Quality, care, feeling for nature and materials, satisfaction in one's work and pride in one's deeds and in life are all important for self-satisfaction. To live in a secure social context is another important factor. Interest could be renewed in myths and tales as carriers of new patterns of thought, like they always were. Human societies should aim at increasing these kinds of satisfiers and at giving them prestige and status over material wealth.

A new materialism

Many, including myself, say that this world is too materialistic, that other values are needed. However, one should be careful not to throw out the baby with the bath water. What one today calls materialism in daily talk is consumerism, where billions of things are churned from soulless production units by bored workers, used for a few days and then stowed away to finally end up on the rubbish heap. What one calls materialism is capitalism, where everything is given a market value. Materials and nature—the mother of all materials—are seen as something just to exploit.

Instead of rejecting materialism, we should free the materials and ourselves from consumerism. True materialism sees the beauty in the things, in their inherent history — from raw material to current use — as well as in their future destiny. A real materialist worships the processes that create the river, the rock or the butterfly. A real materialist chooses wood over concrete and plastic, not because concrete and plastic are unnatural, but because they are dull materials with low information content. The history of a board of wood, and the traces it leaves in the material, that is, the growth rings (annual rings), is simply so much more interesting.

A real materialist chooses the musty earthen taste of real food instead of the soulless industrial versions that taste the same from any of the 100 factories making identical products. Real materialism values things for their inherent quality and not for their market value. Real materialism recognizes the pricelessness of the world, rather than trying to allocate costs or values to everything. Real materialism makes us revel in the wonder of a little bug, a dandelion cracking the asphalt and a little grain of sand stuck to my feet when I walk in from the beach. Real materialism makes us see the value of nature and, ultimately, of ourselves.

The unlimited offer of goods and the enormous possibilities for choice spur a counter-reaction where non-consumption and non-choice become real alternatives. The curse of things is a heavy burden. Waste recycling makes one more aware that all one hauls home will also have to be taken care of, which makes one aware of the resources used and therefore less inclined to bring more stuff home. Already in high-income countries, people are increasingly seeking entertainment and experiences rather than things, even if there is a lot of hysteric consumption of these services as well.

Throughout history, there are many examples of people seeking *voluntary simplicity*, a simpler lifestyle. In most cases, these have been people motivated by a strong idea, such as a religious or political conviction. In the eighteenth and nineteenth centuries, many utopian projects were based on such ideas. Another wave of this kind of thinking came in the 1960s and 1970s, in the form of a 'back to the land movement'. It was somewhat absent for some decades — when the 'invest in yourself, be rich' paradigm ruled — but seems to pick up a renewed interest now. According to a recent survey in Britain, up to a quarter of the respondents opted for a lower income and for more time to pursue non-work activities (Jackson 2009).

From youth cult to experience

Those who think that human society will continue along the same path forget that the Western European countries and Japan have just undergone a 'once-in-a-millennium' demographic transition, which will not happen again, and that the United States is still in this transition and China has reached there as well. Throughout the history of humankind, a relatively stable population has been normal; it is the rapid population growth of the last 200 years that is abnormal. The end to population growth is perhaps a bigger challenge for the capitalist project rather than population explosion; after all, population explosion has gone hand in hand with capitalist expansion. A country like Japan has completed its transition and there has been virtually no economic growth for 20 years and thus a constant 'crisis'.

The changing population pyramid will change society. Many manufacturers, service providers and retailers are adapting their offers to appeal more to rapidly growing groups of elderly, who used to be active consumers (earlier generations of old people were not raised as 'consumers'). For governments, care for the elderly was earlier mainly a question of relief for the working population, and it was their perspective that dominated; the elderly were 'objects' rather than subjects. Today, it is the elderly themselves who shape how they are cared for, with a combination of political action and their own wealth. Less observed, but equally important, is the shift in values that arise.

The big cohorts of youth that characterized the systems of capitalism and industrialism are a demographic reflection of the expansion and pioneership of these systems. As with so many other things, one can also turn this on its head and say that it was this demographic structure that drove capitalism (and industrialism) in the early stages. It is not only economic growth that is propelled by large groups of youth but also the values typical to the young, such as rejection of traditions, rebellion, glorification of strength, risk-taking and body culture. Current society's glorification of the youth is in stark contrast to the attitudes of earlier cultures.

Society, including many of the elderly, has a kind of youth obsession. The elderly are expected to behave as 'old youth' in a similar way as children were treated as 'small adults' 100 years ago. There are examples of 80 year olds who engage in mountaineering, speed dating, wind surfing or parachuting:

Former President George H.W. Bush marked his 85th birthday on Friday the same way he did his 75th and 80th birthdays: He leaped

> from a plane and zoomed downward at more than 100 mph in free fall
> before parachuting safely to a spot near his ocean front home. [. . .]
>
> He told reporters that he jumped Friday for two reasons: to experience
> the exhilaration of free-falling and to show that seniors can remain
> active and do fun things.
>
> 'Just because you're an old guy, you don't have to sit around drooling
> in the corner,' Bush said. (Associated Press 2009)

This is about to change. Other properties, other characteristics of
being human, of the elderly will be held in higher esteem, properties
such as maturity and reflection and even slowness (e.g. the Slow
Food movement, as opposed to 'fast food'), in addition to values such
as care, spirituality and safety, as opposed to risk-taking. Care for
things dear to one and interest in the beauty of and satisfaction by the
small things in life are also typical characteristics of the elderly, in
contrast to the youth's taste for exploration, expansion and novelties.
The values of the elderly are better adapted to the values needed in a
sustainable society. The combined effect of the shift in values and the
effect on growth, as a result of an ageing population, will be a strong
determining factor in the coming century.

Global justice

Shortage of resources and the Internet together are making global
injustices more visible and also more unacceptable. Environmental
space or various footprint concepts are increasingly accepted categories of discussion. The whole debate about the Kyoto Protocol on
climate change is anchored around these concepts, even if horse-
trading will distort it in the end. Efforts to regulate whaling and
fisheries are similar in nature. One knows one has responsibility
towards others and that one's acts actually do influence, positively or
negatively, other's opportunities. Certainly, many people still resist
the ultimate consequences of this insight; those who already consume
a lot more than 'their share' — which includes myself — have all sorts
of excuses for why they will not change their lifestyle.

Global solidarity has been on the increase for a long time, at least
since the abolition of slavery, and with modern communication and
technology bringing people closer, it is bound to be an even stronger
factor. National borders are a result of chance, big historical migrations, wars and power. It is simply not tenable in the long run to
justify some peoples' privilege to them being born in a particular
country.

Society as an extension of the individual

The industrial and consumerist society tends to overemphasize the individual, its capabilities, its freedom and its shortcomings. The myth that the fortune of each person is a result of the individual's efforts and action has been debunked in the earlier chapters. Unfettered, unregulated capitalism, which is an expression of this view of the individual as separate from society, doesn't work, but on the contrary requires constant corrections and compensation from a society which it at the same time breaks down. Society is the carrier of all accumulated knowledge and experiences, and foolishnesses, of human beings from time immemorial, knowledge and experience without which we would still be apes. Instead of seeing society as a burden and limitation of the individual, one should see it as an extension of the power of the individual; we can do a lot more through society than without it. What would entrepreneurs, scientists, innovators or explorers have been without society? Thus, it is through society that one can realize oneself. Meanwhile, society must give the individual sufficient freedom to develop, to form another opinion or to live in the shadows, free from the meddling of others. From this perspective, we can break the false contradiction between the individual and society.

The right to free movement

Why are people not allowed to move freely, when capital, products and services can do it? Free, or at least considerably liberalized, movement of people will result in substantial global welfare gains and will reduce global differences. If global difference in income and standard of living is one of the biggest problems on the planet, it is immoral not to implement freer migration as one of the strategies to tackle it. It should be considered a human right to live where one wants (with the important limitation that one can't force others to move away to make space for oneself) and not be bound to the territory where one happens to be born.

Like any other freedom or right, the right to move where one wants cannot be unlimited or unconditional. But first and foremost, it is essential to state the right. It is important here to understand the difference between the right to move (i.e. mobility) and the action (i.e. migration). The right to move is a fundamental freedom, yet the reason why people migrate, in many cases, has nothing to do with freedom, because they are forced to move as a result of appalling conditions. To address those conditions is (or should be) an apparent priority.

There are obvious problems with increased migration. One is the problem of 'brain drain', that is, where well-educated professionals from low-income countries are recruited to better paid jobs in high-income countries. The phenomenon is not limited to so-called white-collar jobs, but extends to blue-collar jobs, to soccer players as well as to musicians. In today's world, however, people can move where they want as long as they are in demand, as most countries allow immigration by sought-after professionals. Free migration will not make this worse. There are also social challenges in mass migration, both for the people emigrating from a country and for the people in the countries they emigrate to. This may, at least in the short term, threaten social cohesion and trust, that is, erode the social capital. Mass migration of Africans to Europe and Latin Americans to the United States *will* press down salaries in those countries, but not as dramatically as many envision. Other challenges are posed by ecological disturbances caused by people moving to biomes they are not used to. This has been discussed in Chapter 10.

But are these inconveniences not worth paying for? Is it not an elitist notion that one should keep out other people to safeguard one's 'own' privileged position? And even if it hurts, is there a better proposal? The alternative is an increasing number of people being shot or electrocuted at borders, suffocated in containers, freezing to death on the bogie of a trailer or drowning when corroded ships sink. Most people want to stay in their country of origin. If dictatorships are abolished and global inequalities are reduced—things that in any case should be high on the agenda of humanity—flows of migration will not become tsunamis. It is not possible to lock out 'the others' or lock them up.

Work and leisure

Earlier societies had no strict divisions between production, reproduction, maintenance of the social fabric and values, consumption, work and leisure, taking care of the family and the local communities. All were part of the web of life. Still today, many people live like that and their roles are strongly interwoven; this is the case particularly in agrarian societies and generally in self-employed family operations. Industrialism and wage labour cut many of these bands and separated the activities, but the separation is essentially an artificial one, a separation for which human beings are not at all adapted, neither culturally nor genetically. This leads to living parallel, atomized lives.

The increased separate demands on human beings as producers, consumers, life companions (spouse, partner), parents and social beings lead to stress and a feeling of failure, of not 'being enough'. People try to fix it with more of the same, more consumption, more work, more relations, but individuals can't 'be more'. Human beings will have to recreate, but in new forms, the connections between these different parts of being a human. People will live on their hobbies, build on their house and grow their own food in lifestyle farms. Most farms in many high-income countries today are actually lifestyle farms already. Hunting went the same way long ago, from being essential production to being leisure for the rich. Distance work moves production back into homes, where it was during most stages of history in the first place.

What does new technology mean?

Humanity is in the middle of what is called the information society. Many speak about it as a real revolution that will change humankind forever. Like for any other process, when one is standing in the middle of it, it is hard to see clearly what it means or where it leads. The only thing that seems certain is that it generates new things all the time, needs that one didn't know one had. Who thought about Facebook 10 years ago? There are many questions to ask. Certainly, modern information and communication technology (ICT) has changed and will, in turn, change many businesses. Global outsourcing is on the rise and will continue, now not only of simple jobs but increasingly of more complex ones.

Improved access to information has tilted knowledge in favour of the buyer and makes profit-making in trading much more difficult. Trading itself seems to be commoditized to a very large extent. For instance, look at the book business, at how publishers and bookshops are squeezed between producers (authors) and consumers (readers). Three-dimensional 'printing' (i.e. producing a product directly from drawings) can perhaps become an important way of producing a product, or at least making spare parts and other things in short series; this is as close to teleportation one will come in the lifetime of those born now.

Many see the Internet as a freedom project; 'after two centuries when mechanisation was an alienating burden for many people, it now becomes possible to imagine machines as instruments of human freedom, allowing us, separately and together, to make ourselves at home in the world to an extent that once seemed gone for ever,' writes

Hart (2000: 48). It is hard to see this liberation. ICT can also bring alienation to new heights: it has no 'feel', doesn't know any concepts and can only handle exact expressions, which means that it fuels an ever more far-reaching standardization and the continued erosion of craftsmanship. To transform the process of making a good cheese or wine or the process of running a farm or planting a garden into a digital one is to make it uninteresting. There is a difference, of course, when a computer is used as a tool to help in the process, store data, look for information or monitor progress.

Under the hegemonic control in workplaces, ICT is increasingly used in the same way for mental work or artisan work as the assembly lines were used in industry, to control workers. The development of internal markets as management tools within institutions is also largely facilitated by ICT. Now departments are 'billed' for any little resource they use; overhead function costs can be assigned to them and they have to 'pay' for their use of not only space and machinery but every little resource, the pencil, the print jobs and, ultimately, every little bit of information. By this, market forces are deployed also within firms and organizations and don't only work from the outside.

Information technology and technology shifts are dramatic for individuals and businesses, but for our discussion here the main question is whether the digital revolution also will mean a radical shift in how society is organized, in the power relations between those owning and controlling the means of production and the rest, and in our relationship to nature and among each other. Is this technology shift as dramatic as the shift from the hunter and gatherer society to the agrarian society or the shift from the agrarian society to the industrial society? Or is it just a very strong change within the industrial society, perhaps on par with the mass introduction of electricity?

Wasn't the introduction of the telephone, steam liners, the railroad and the telegraph as remarkable as modern ICTs? Already then, more than 100 years ago, Indian villagers were directly affected by a crash at the London or New York stock exchange; or the price of pork in Germany was affected by the price of corn in Chicago. Already 100 years ago, global artists toured the world. Mass distribution of newspapers meant that one knew about an uprising on the other side of the world a day or two later. Global mass movements, such as the one for abolition of slavery or the Red Cross, existed more than 150 years ago. 'Distance and time have been so changed in our imaginations, that the globe has been practically reduced in magnitude, and there can be no doubt that our conceptions of its dimension is entirely different to that held by our forefathers' (Gleick 2011: 148). This quote is not from the 1990s, but was said by a telegraph engineer 150 years ago.

The American businessman and writer James Bennett (1999) says that the Internet means 'the end of capitalism and the triumph of the market economy'. He claims that the fall in price of digital power implies that the means of production (i.e. the computing power) can now easily be owned and controlled by the workers themselves, as opposed to machinery for industrial production. He also claims that the Internet is taking control out of the hands of the large corporations and that small players will begin to hold advantages over big ones. And finally he points to open-source software as representative of a totally new paradigm. He speculates that also industrial products, such as a jet airliner, can be based on an open-source design (Bennett 1999). If open source and the Internet indeed meant the end of patenting and intellectual property rights, well then he might be right. But it is hard to see that this is really happening on a large scale, even if there are some encouraging steps by the 'maker'[179] movement, 'open source ecology' and 'open source hardware'[180].

As most new technologies, the Internet and communications technology too have been incubators for small companies. But is the dominance of Microsoft, Google and Apple in their respective markets any sign that mega companies are losers in the new environment? On the contrary, I would say. They have a larger share in their respective global markets than almost any other companies have had in history. Another thing is that their life is precarious and that they can be easily beaten by another player. Mobility and the speed with which giant companies come and go certainly have increased.

But to a large extent the winners are the same; the winners are the investors whose money rapidly flows from one venture, from one brand, to the next. Management and ownership has since long been separated and, with the Internet, investors can be even more disloyal and just dump whoever is not performing well. Some claim that geography is less important with the Internet; that people all over the globe can compete. The software and Internet industry, however, is following the same clustering patterns as other industries and services,

---

[179]   The maker movement is a technology-based extension of Do It Yourself culture.

[180]   Open source hardware is hardware whose design is made publicly available so that anyone can study, modify, distribute, make, and sell the design or hardware based on that design.

much like the barbers around the same tree in a countryside town in Pakistan; the companies are all in a few clusters such as the Silicon Valley and Bangalore.

In agrarian societies, human beings were bound together by a strong web of religion, culture and economic interdependence, inside the family and in the village in the first place. With the advent of industrialism, the economic interdependence between one another and the people close to oneself, friends and family, became much weaker. Still, mass communication and mass media, such as radio, television and newspapers stimulated a kind of national culture and sense of belonging.[181] With the Internet, common points of references are rapidly lost. It is true that most people are hooked to the Internet, but one is very selective in what one does there. Most seeks people with similar interests and views to a very large extent, but they are rarely the girl or boy next door. So one will have less and less in common with the people one works with, with one's own family, even with the children one goes to school with. It is of course nice that gay Muslim players of World of Warcraft[182] or female carpenters interested in hip-hop can interact globally. Certainly, it means moving one more step away from identification with the nation-state and other social forms, also weakening the workplace as a 'second home', affecting the loyalty to these institutions.

An important result of information technology is the increase in 'free' services available for everybody. Access to a lot of information as well as to communication services is free on the Internet, admittedly in many cases subject to forced reception of commercial messages. The development of resources such as Wikipedia is almost like the communist agenda 'From each according to his ability, to each according to his needs', that is, those who are interested in using Wikipedia use it as much as they like for free, and those who want to contribute — by working or by making donations — do so clearly separate from any direct benefit they would have from it.

Many talk about the Internet as a groundbreaking tool for democracy. Certainly, it has increased spread of information and, in

---

[181]   Some point out that the first forms of mass communication, radio, television and movies, were much used by fascism and Nazism.

[182]   A very popular (at the time of writing in 2010) Internet-based game, probably forgotten 10 years ahead.

particular, the theoretical width of information, that is, there are many more voices; still, most of this is somewhat theoretical. Also on the Internet, main websites have millions of readers, while many blogs are just read by people who were going to meet anyway, if at all by anyone; one can as easily become a lone voice on the Internet as in the desert. Not only protesters use the Internet, it is increasingly used by oppressive governments and corporate interests to pervert information, and surveillance of people has never been easier than when one leaves a digital track all the time and when just one's pattern of browsing and friends on Facebook are enough to classify one as a white supremacist, Jihadist or revolutionary leftwing or simply the average Joneses.

The jury is still out. In my view, the biggest potential effect of the Internet is not at the level of production and the economic power relationships, but more at the level of politics, the mentality of people and how cooperation between people is organized in a direct way. The Internet is a rather unique platform for non-market interchange; it is a crossroad of markets, states, civil societies, communities and the private sphere, and a breeding ground for new commons. Open source plays an important role here, at least psychologically.

The nature of technology

My discussion might possibly seem hostile towards technological development. My basic assumption is that technology has great potential, but that technology needs to reorient itself to solve the great problems of society and not the short-term interests of investors. And this is actually not a new or very radical thought. Most technological development is a result of public investment and direction. For instance, the advances that allow you to fly in a jet plane for your holiday is, to a very large extent, based on technologies paid for by rich militaristic governments. The development of the Internet, too, was paid for by a government, also for military purposes. The expansion of communication technologies, shipping, railroads, etc. was always promoted by the state.

Most science that is the basis for technological advances today is a result of public funding. So there is nothing strange or abnormal to call for societal direction of scientific and technological development. Better tools are required for assessment of technologies, not only for their direct effects, which basically are the reasons for their development, or their indirect technical effects, but also for their effects on society at large and on power relations in particular. Very few technologies are 'neutral'.

The complex of technology, energy and division of labour is strongly linked to the capitalist market society, but we saw with the Soviet Union that the same complex can fit into other systems as well. The question is how much they have to change. I am convinced that there will have to be some radical changes also for this complex. Perhaps, these changes will emerge naturally once human beings free themselves from both capitalism and market imperatives.

Clearly, a lot can be done to adapt technologies to another way of interacting with nature. There are innovative approaches to design, such as bio-mimicry and cradle to cradle.[183] It must have its starting point in understanding the eco-socio-systems as the basis for human welfare. In the same way as in nature, one needs to work more with multi-functional systems. The inherent values of technologies also need to be better considered. Bookchin (1980) suggests that the main reason to promote alternative technologies is not because they are efficient and renewable, but because of their capacity and ability to reconnect man with nature, with the earth, with plants and animals, and thereby make us more sensitive to the biosphere. In addition, he emphasizes that the individual should be able to comprehend how technologies work so that one is self-confident and feels somewhat in control of what one does (Bookchin 1980). These are strong arguments for selection of simple technologies, as well as for a human scale of organization of production.

If you are a factory manager and you can choose between two new machines or tools, one that makes the work easier, more pleasant or more interesting and another that increases productivity of the worker, you will almost inevitably choose the one that increases productivity of the worker, because that is the one that gives you improved competitiveness. This is an effect of capitalism and market imperatives and not of technology itself. Without capitalism and market imperatives, the workers would most likely choose technologies that made work simpler or more pleasant. I believe in technology as something positive, if it is released from the demands of competition, profit-making, power usurpation and exploitation, and something used for one's joy, pleasure and improving the quality of life. But technology will only be used or applied in that way if society is governed in that way.

---

[183]   They are called innovations. However, they seem to do what indigenous and peasant technologies have done all the time, but on an industrial level.

Energy

Energy supply, as it has throughout history, will be in the centre of human efforts and attention. We have earlier discussed how critical energy is for development and how new forms and sources of energy are closely linked to technological leaps and to societal change. People in high-income countries and in some of the rapidly industrializing middle-income countries simply use far too much energy, regardless of how it is produced. Around a third of the world's population would need to use more energy to ever be able to climb out of poverty. But also for poor people, there are options for saving, in particular for cooking.

Many poor households cook over an open fire, which is highly inefficient and also causes considerable pollution and respiratory diseases. The poor of the world need more and better forms of energy, for example, electricity, at least for light. The International Energy Agency (admittedly a rather biased organization) says that US\$ 1 invested in energy efficiency in low- and middle-income countries gives back US\$ 3. Twenty-five million households already use biogas for cooking and light, biogas made in simple digesters with raw materials from farms, households and surrounding nature. Many other small wind energy units, hydro energy units or biomass energy units also provide a lot of energy (UNEP 2009). No single source of energy combines volume, ease and versatility of use, low cost of production and high returns in the same way as fossil fuels, so one should not expect energy to be cheap and abundant in the future. "The overwhelming likelihood is that, by 2100, global society will have less energy available for economic purposes, not more" (Heinberg 2012).

For principal, environmental and even economic reasons, future energy supply should be based on renewable energy. However, a panicky dismantling of dependency on fossil fuel may cause serious problems to society. It is nevertheless important, exactly for this reason, that the transformation starts as soon and as vigorously as possible. Fortunately, some impulses for transformation came in the 1970s with the first oil crisis and more came with the oil price hikes in the 2000s as well as with increasing awareness of the impact of climate change.

There is no point in determining how much of the future energy complex has to come from solar, wind or hydroelectric power or from one or the other forms of biomass; it is also not needed or possible to determine this. The drawbacks of large dams for hydroelectric power are known quite well, and the use of water for generating power might also compete with the use for irrigation, which sets quite some limitations. Too many drawbacks of wind turbines have not been figured out, even if there are objections regarding the impact on bird life, aesthetics, etc. The main issue around biomass is the competition for land between farming and wild 'nature'. Increased use of biomass for energy and industry is possible, but globally not at all on the level of replacing current petroleum consumption, even less the projected increase.

For sure the use of solar energy in all its three forms, and possibly some other hitherto unknown forms, needs to be expanded. In the longer term, it is likely that solar energy can provide most of the energy needed. Perhaps future photovoltaic cells can reach an efficiency of around 25%, which is double the current rate, and the cost of self-production of electricity[184] could come down to between 1 and 2 cents per kilowatt-hour (IIIEE 2007), 10% of current price.

Nuclear power is not an interesting technology from the perspective of energy efficiency or safety. It is a prime example of how something can be profitable because so many costs are externalized. It builds on the exploitation of nature, other people and future generations and on massive subsidies from society (in the form of infrastructure, research, insurance, etc.). In addition, the complexity in technology and society for operation of nuclear power is hardly desirable and not possible to guarantee over time.

Historically, it was natural for human beings to have a predisposition for sweet food; energy was always in short supply and our predecessors were rarely in the risk zone for diabetes. In modern society, where sugar is cheap and abundant, however, the craving for sweets needs to be kept in check. The consequences of not doing so are fatal. It is the same for society at large when it comes to energy. Fossil fuel is like an enormous bag of candy that someone left, and one just eats and eats. Too much cheap energy screws up the metabolism and we have to voluntarily restrict consumption. According to Janken Myrdal (2008), and I agree with him, with the human tendency for exaggeration we most certainly will destroy the basis for survival with an unlimited supply of energy. If we *first* adapt ourselves to a non-expansionist way of living, then cheap and easily available energy can be a boon.

---

[184]   The cost for solar electricity that has been fed into the grid and then distributed will obviously be higher.

More nature than nature

More and more ecosystem services are 'produced'. There are, roughly, two different tracks for production of ecosystem services. One is to set aside nature so that it can produce what people need, or believe they need, without human intervention. This is the philosophy behind national parks and protected areas. We need to continue to develop natural reserves, and, in particular, they need to be expanded in the ocean, in waterways and coastal zones. Many of the landscapes perceived as being wild are human creations, and in many other cases human beings play a considerable role; for example, in the African savannah, human beings and their herds were for a long time part of the ecosystem as much as the lion or the gazelle. So in most cases natural reserves are not, and should not be, about shutting out human beings altogether. One of the many reasons to still include substantial areas in reserves is that the effects of human actions are not fully understood, and no matter what one does, one is likely to lose some biodiversity. A wild nature represents buffers and refuges, or 'redundancies', which are of great importance for the stability of a system.

On land and sea that are not natural reserves, we need to more consciously produce the ecosystem services required; that is, we simply need to *produce more nature than nature,* or at least more nature that we can use, which is the essence of farming in the first place. Rather than focusing only on the reduction of negative human impacts on ecosystems, one should also try to increase *positive human impacts*. Instead of the monoculture mindset that has shaped land use since industrialism, one needs to act as a planet keeper, as a gardener, as the Gardener of Earth. Intensification doesn't have to mean more resource waste or less biodiversity, but it will mean another diversity.

In large parts of the world, especially in the tropics, farming systems can be very intensive and also very diverse. The forest-gardens and diverse smallholder systems are very good models. I am here talking about a bio-intensification rather than the intensification built on external resources. In the same way as capitalism is based on a rapid flow of money—so that profits can be generated again and again for each cycle of money use—natural resources such as land, water and nutrients need to be used efficiently, again and again. In each cycle, what is needed can be skimmed off, not eroding the capital but just taking away the surplus so to say.

In the same way as one creates space or room in a garden we need to do that on the planet at large. Cities, normally, should be dense and intensive, and the wild should be exposed to limited human intervention. Also in the cities, one needs to increase production of ecosystem services, grow more food, do more natural recycling, filter water through living roofs and the like. In between those two extremes, that is, in the majority of biomes, there are varying degrees of the intensity of use. There are semi-wild pastoralist landscapes or landscapes where wildlife is 'managed' to optimize the yield of meat and other useful resources as well as ecosystem services. There are farms similar to the farms of today, but there are, in particular, systems that are more intensive, more like gardening, many under protective structures and with more people engaged. These need to work in tandem with the cities to recycle city nutrients in the best way.

Long-distance trade in food is limited. The limitations mainly ensure that soils are not mined of their nutrients in exporting countries and that importing countries do not get overloaded with nutrients, but also to avoid a situation with unlimited competition and downward pressure on food prices. I doubt that there is any use, or any need for use, of chemical fertilizers or pesticides, and if they are used they will rather be the exception. Aquaculture of fish, shellfish, aquatic plants and algae is increasing substantially and systems have to be developed to better integrate aquaculture and farming,[185] inspired by existing systems primarily in Asia.

Biomass is produced for energy and for chemical industries from agriculture and forestland. The oceans are being foraged further down the trophic level.[186] Intensive large-scale animal-keeping doesn't exist, but small-scale poultry, dairy and pig production is well integrated with intensive gardens and aquaculture. Animals like cattle, sheep and goats are turning more to natural grazing but also to rather intensive rotational grazing where that is possible (mainly in moist climate). New forms for semi-wild farming have developed. Undoubtedly, less meat is being consumed in high-income countries, but hopefully more animal protein is being consumed in the countries that are poor today.

---

[185]   Nutrients that are lost from the soil and carried to the sea by sewage can through aquaculture be recycled into the food system, for example, by cultivation of mussels. Algae can be eaten, but also used as a fertilizer or as feed for animals. Unfortunately, much aquaculture today is based on predatory fish, such as salmon, which is fed on both fish and fodder from agriculture, a somewhat absurd production system.

[186]   A level in the food chain; herring is on a lower trophic level than cod as herring eats plankton and cod eats herring.

Many speak about the need to ruralize society; that is, the streams of population should be steered back to rural areas and people should create self-sufficient units in those areas. There are certainly some environmental and social advantages of more people being in the countryside; it should contribute to a sustainable social environment for those who take care of natural resources; it should allow small-scale distributed power generation and local cycles of nutrients and materials. However, for populations engaged in industry and commerce, a denser habitat can preserve resources and reduce demands on infrastructure and heating. Perhaps a network of small towns connected by rail traffic would be a good solution.

Ultimately, conditions vary in the world and the carrying capacity varies a lot. Therefore, solutions for places where there are hundreds, perhaps thousands of human beings per square kilometre will be different from those for places where there are just a few dozen people. This can be easily seen in the difference in historical population patterns as well. Population density and water and land resources are important factors that determine the optimal mix. In the end, people should have the right to live where they like and human settlement patterns will and should change by 'natural processes' rather than by Pol Pot-inspired localization politics.

Changing incentives and parameters will influence population patterns. Cheap fossil fuel leads to the creation of the automobile society and suburbia, to depopulation of the countryside and to increased global competition. Increased energy price is the parameter that most likely will change that. Authoritarian or unequal societies seem to drive urbanization, as everybody wants to be close to the power and the action. A more equal society will reduce this tendency.

More people, less nature

One doesn't have to be an economist to realize that people assign more value to something that is scarce than to something that is abundant. For a long time, nature was abundant, something just there to exploit. Human beings were few, and the deer, the fish and the trees were many. But now one can see that these resources are not unlimited and that one needs to take good care of them. I am not saying that one should compare trees and people, but use the example here as a drastic expression: as there are more and more people and less and less nature, one needs to assign higher values to trees and lower values to people than done earlier; that is, it is now more important to save nature than to save labour. It really makes little sense that only a few percent of the population is involved in natural resource management of all sorts, including agriculture, forestry, fisheries and nature conservation, while so many are engaged in commerce. It is ironic that, in an overpopulated world, the countryside is depopulated to the extent that sustainability is threatened.

Winners and losers

There will be winners and losers among countries, regions, organizations and individuals. People who are privileged today, like myself, will see their privileges diminish. But to lose one's privileges mustn't be a disaster; it can be liberating in the same way as the abolition of apartheid in South Africa didn't only liberate the blacks but also the whites from an undignified relationship. If I might venture a guess, the winners will be found in the middle-income countries. This may sound surprising because they are the ones that are well known for breakneck growth and environmental degradation. But they are not yet stuck in the high-energy society, and, in any case, energy doesn't offer them the same prospects as it did the high-income countries, as returns, both economic and energetic, on the use of more energy are diminishing. They have, hopefully, learned something from the mistakes of high-income countries and they can challenge the high-income countries politically and economically. There is a middle class that can be engaged in a project of political change.

The middle-income countries also comprise a very big part of the world's population, so their choice is to some extent the choice of humanity. In Europe and the United States, for decades environmentalists have used a rhetorical question: 'What will happen if the Chinese adopt western lifestyles and consumption patterns?' Now, this question is no longer a rhetorical question. It is reality; it will not be answered by western environmentalists but by the Chinese themselves, because no matter what the global impact, the impact in China will be even greater.

In 2006, there were 16 million electrical bicycles in China; in 2010 their numbers were probably 120 million (*New York Times* 2010b). Electrical bicycles are for sure no ideal and not unproblematic from an environmental perspective; they spread lead from batteries and need coal-generated electricity (ADB 2009). Nevertheless, they are clearly favourable compared to cars, and they represent just one of many examples of how middle-income countries can avoid the mistakes of high-income countries. China is also the number one producer of solar technology. It is also, perhaps surprising to many, a world leader in schemes paying for environmental services (Bennett 2009).

Countries, such as the United States, where all of society is stuck in and fed by high-energy consumption and a world order where society profits from the weaknesses of others will have difficulties to find new ways. The poorest countries, on the other hand, will be stuck in wishful thinking about being rich; they lack social, manufactured and human capital to make a comprehensive and fundamental change. Hopefully, a new world order will offer them a less raw deal and a pathway out of destitution.

The future is already here

Economic growth in most mature economies, such as Western Europe and Japan, has slowed down considerably. This should not be seen as a problem but as an opportunity. In most of them, and of course contributing to it, populations are no longer growing; in fact, population is decreasing in many countries or is stable or grows just because of immigration. This also means that these countries will be more positive towards, or will be coerced to accept, immigration.

Important technological developments, compatible with a sustainable society, are already in the pipeline and will continue. Organic farming is already practised widely in Europe, up to almost 20% in Austria and Sweden. Even poor countries like Moldova, Bhutan and Rwanda are keenly adopting organic agriculture and consumers are responding by buying more and more. Wind energy is rapidly expanding and solar energy is finally close to a massive breakthrough. Some countries are increasing their use of biomass, without threatening nature; for example, Sweden uses close to 40% renewable energy, of which biomass is a substantial part. Ecological houses or villages, passive houses (i.e. those without any active heating or cooling) and chimney-free or effluent-free factories are spreading.

Another interesting, and perhaps surprising, experience is the return of (some) wildlife to urban environments. There seem to be more deer in suburbia than in the wild. The Peregrine falcon breeds in London. Plants crack the asphalt and birds adapt their song to the noise of the city. New ecological niches are developed. Rabbits, raccoons, muskrats, skunks, and other small mammals, red-tailed hawks, ospreys, kestrels and other birds of prey are often spotted in densely packed neighbourhoods in New York. In 2007, a beaver was caught building a dam on a river in the heart of the Bronx, marking the animals' very first return to the area since the end of the fur trade (Greenwire 2010).

The Internet and globalization have brought people closer; freedom and human rights have become more or less global values. The respect for the environment is growing all the time and so is the awareness of our dependency both on the environment and on each other. The realization that some problems need global solutions is widespread.

Most people in high-income countries have already understood that a constant chase for more will not make one happier; most know that relationships and society are more important for the well-being of humankind than increased consumption. Human actions still follow mostly old patterns, but soon there will be a dip in Christmas shopping, not caused by economic crisis and/or guilt, but simply by lack of interest in more shopping. Gradually, people seeks to reduce risk, consumption and expansion and value safety, stability and proximity more.

The ever-increasing size and concentration in corporate businesses is already now counterbalanced by small-scale solutions, most visible in the food sector. On the one hand, the giant companies get bigger in terms of production, whole sales, processing and retail. On the other hand, as a counter reaction partly from consumers and partly from producers that are left behind in the process, new markets are created for speciality products. It is hard to know whether these local products will continue to be niche products. The strength in the 'local' is that it supports local economic development and that there is clear identity and responsibility for the production. It is rather the scale and the relationship that matter and not the distance as such; tightly knitted social networking on the Internet, in that sense, can also give the same feeling as shopping from the local farmer. It is more about connectedness than about physical proximity.

Movements such as the Transition Movement, initiated in the early 2000s in the town of Totnes in the United Kingdom, primarily focus on the process of 'energy descent', the transition that is seen as inevitable after reaching peak oil and to counter climate change. The initiative spread quickly, and, as of May 2010, over 300 communities are recognized as official Transition Towns in the United Kingdom, Ireland, Canada, Australia, New Zealand, the United States, Italy and Chile. Transition US has the vision 'that every community in the United States will have engaged its collective creativity to unleash an extraordinary and historic transition to a future beyond fossil fuels; a future that is more vibrant, abundant and resilient; one that is ultimately preferable to the present' (Transition US 2012).

Local currencies are also spreading. The idea is to build strength and interdependence in the local economy by keeping money circulating in the community and building new relationships. Apart from being an economic project local currencies also have a symbolic power, taking power from both nation-states and global markets into the hands of local communities. The association Jord Arbete Kapital[187] (JAK) in Sweden operates an interest-free bank since 1965 with 37,000 members. It expanded rapidly in the early 1990s when Sweden underwent a serious financial crisis — as a result of a real estate crash — with interests soaring at 500%. 'If my sister wants to borrow money, she can borrow it without interest,' says JAK in a film.

I believe that all these initiatives are laudable by themselves, but they are not enough. Ultimately, one has to change the logic of the economic system, and here I speak about what influences the daily choices of individuals as consumers, labourers or companies. Future development is largely shaped by those choices. To educate consumers and companies in sustainable or fair consumption or production is good. To change the system that determines the incentives is even better.

---

187   www.jak.se (in Swedish only).

Reorganizing the Economy

In nature's economy the currency is not money, it is life.
(Vandana Shiva,
Earth Democracy: Justice,
Sustainability, and Peace, 2005)

As we discussed earlier, it appears to be difficult to make a difference between capitalism and the market economy. Of course, one can envision market exchange without capitalism, after all market exchange is much older than capitalism. And markets in one form or the other are likely to be building blocks also of a future society. But it is not at all certain that they shall be the primary organizing and distributional principle, which is how I would define a market economy.

The peasant strategy

Let us consider how distribution of goods and services could take place outside of the market, and actually, it should not be too hard to do so. There are plenty of examples of economic activities that are based on neither capitalism nor monetized markets. To start with, the family, in whatever historical or modern constellations, was not and is still not based on either markets or capitalism. People provide services for each other and distribute goods among each other on the basis of kinship or good feelings.

Next to the family, most people have friends with whom they also exchange goods, services and gifts, on the basis of principles that are thousands of years old. Especially in groups a bit left behind by modern society, say the underworld in the cities or the rural outback, friendship relations are important for the social fabric and economic exchange in society; gifts and bartering is more common than cash sales.

After family and friends are communities and associations, where people participate in all kinds of activities and exchanges organized by civil society, often including production of common goods and services, but occasionally also of products that are sold commercially in bazaars or flea markets, firewood cut down in communal forests or fish from the common pond. The organizations of human society, the state and the local governments, are also not yet organized by market principles. They produce services that are most essential to humankind, which is why they are entrusted to common institutions. Ironically, they also produce the fundaments of the market economy and capitalism.

Several alternative systems run parallel to the prevailing system. An alternative that is often not noticed is the peasant or the small farmer. While indeed almost all farmers in the world are part of the global markets, or at least heavily influenced by them, peasants or small farmers all over the world[188] struggle to limit the influence of capitalism and market economy on their daily lives, for a multitude of reasons, but in particular to keep their autonomy. Traditional farmers not only produce but also re-produce the means of production. While the 'normal' economy is extractive and exploitative and assigns no value to natural capital, for farmers who are tied to their turf for life and for generations to come, reproduction of the natural capital is as natural as reproduction of their own species.[189]

The first economic strategy is with pluriactivity, that is, to generate income from outside of the farm. This is, and has been all along, a very important livelihood strategy for farm households in most parts of the world. The other, more important, strategy is to rely on the market for sales but to limit the market dependency on the input, that is, on the reproduction of the system. A farm that is based on family labour, has no debt, uses local natural resources, its own seeds, limited machinery and soil regenerating technologies such as crop rotations, instead of depending on the market for its production and re-production, can sell in the market without being so totally in the hands of the market. It is also resilient and can withstand social and natural shocks.

---

[188]   Normally this is mainly discussed as a phenomenon characteristic of developing countries, but many farmers in industrial countries use rather similar strategies (see Ploeg 2009).

[189]   Certainly, there are also examples of peasants who have destroyed their environment. Mostly that is a result of shocks to their system, such as migration, land expropriation and, lately, the introduction of commercialized farming. Forced resettlement of communities for reasons ranging from land redistribution, soil conservation, delivery of social services to political control have also played a big role.

Non-capitalist economic activity

There are really no difficulties to find economic activity organized in ways that are non-capitalist in their way of functioning, although still many of them operate in markets. There are many forms of cooperative operations in the world. Their roots go back to the guilds and village communities, and they became a popular tool to compete with capitalism when capitalist society emerged. They exist in all spheres. Many production cooperatives, factories and services are managed by workers and employees; services cooperatives, such as nursing or private schools, are managed by the employees or the users. Cooperatives exist for individuals in the role of consumers (e.g. shops) and people organize credit or insurance together. Housing is also often organized by cooperatives. And then there is the public sector.

In the 20 biggest cities of the United States, almost 40% of the 200 biggest operations are not for profit, such as most hospitals and universities. Forty million Americans get their electricity from municipal utilities. There are 48,000 cooperative shops with a total of 120 million members and 30% of primary agriculture production is sold through cooperatives (Wilkinson and Picket 2009). Land O'Lakes, Inc. (2010) in the United States is a national, farmer-owned food and agricultural cooperative with approximately 9000 employees, 3200 direct producer-members and 1000 member-cooperatives serving more than 300,000 agricultural producers. It handles 12 billion pounds of milk annually and produces a plethora of dairy food products. Land O'Lakes, Inc. does business in all 50 states and more than 50 countries.

In Great Britain, there are more than 700 cooperative credit institutions, 70 mutual insurance companies and 25 illness and retirement organizations. There are 170,000 charities with an annual income of some US$ 40 billion (Wilkinson and Picket 2009). Presently, there are over 2000 housing cooperatives with over 2 million apartments and over 3 million members in Germany. In India, apartments in housing cooperatives are very common in big cities like Mumbai. Owners have a share of the cooperative and not the actual real estate itself. Owners can sell the 'share' in the open market, but they have to get the approval of the housing cooperative to complete the transaction (Wikipedia 2010b). Car pools are a more recent way of sharing an expensive product; some of them are organized as cooperatives, others as capitalist businesses.

The Mondragón Corporation is a federation of worker cooperatives based in the Basque region of Spain. Currently, it is the seventh largest Spanish company in terms of turnover and the leading business group in the Basque Country (providing 6.6% of the GDP of the Basque economy). At the end of 2009, it was providing employment for 85,000 people working in 256 companies in four areas of activity: finance, industry, retail and knowledge. The cooperatives are owned by their worker-members and power is based on the principle of 'one person, one vote'. When asked whether their system is an alternative to capitalism, they say:

> 'We have no pretensions in this area. We simply believe that we have developed a way of making companies more human and participatory. It is an approach that, furthermore, fits in well with the latest and most advanced management models, which tend to place more value on workers themselves as the principal asset and source of competitive advantage of modern companies' (Mondragón 2010).

John Lewis Partnership in the United Kingdom has 70,000 permanent staff who own 31 John Lewis shops across the United Kingdom, 235 Waitrose supermarkets, an online and catalogue business (johnlewis.com), a production unit and a farm with a total turnover of nearly £7.4 billion as of 2009. Partners share in the benefits and profit of a business puts them first: 'The Partnership's ultimate purpose is the happiness of all its members, through their worthwhile and satisfying employment in a successful business. Because the Partnership is owned in trust for its members, they share the responsibilities of ownership as well as its rewards — profit, knowledge and power' (John Lewis Partnership 2010).

When the Argentinean economy collapsed in 2001, half of the population was thrown into poverty and many factories closed down; the owners just emptied the coffers and locked the door. This situation, like all crises, triggered responses of various kinds. Companies ranging from the ceramics factory Zanon, with almost 500 worker-members, to the four-star Hotel Bauen, to the suit factory Brukman, to the printing press Chilavert, and many others were taken over by cooperatives. As of 2007, the phenomenon involved roughly 185 mostly small- and medium-sized enterprises estimated to include between 9000 and 10,000 workers (Vieta and Ruggeri 2007). Some businesses have now been legally purchased by the workers for nominal fees; others remain 'occupied' by workers who have no legal standing with the state (and in some cases reject negotiation with the state on the grounds that working productively is its own justification). The cooperatives also help each other and many of them are funded by a cooperative micro-credit institution, La Base (Grus & Guld 2010).

In Friesland in The Netherlands, a cooperative of some 900 farmers and rural dwellers, the Noardlike Fryske Walden, manages an area of 50,000 hectares with new forms of self-regulation. It integrates the management of the farms with nature and society. To some extent it was a response to zealous regulation of the farm sector by the government in a piecemeal way by imposing a lot of limitations on farming, threatening the farms as well as many of the objectives of the regulations. The cooperation is based on join management of the landscape, which is a 'commons' even if each piece might have private ownership (Ploeg 2009).

By and large, capitalism still rules

Despite all the examples of economic activity carried out outside of the capitalist system, in most cases they are not radically different from capitalism. The capitalist paradigm and ideology have the upper hand in most places in the world; its logic of right to profit from private property and perpetual growth is not really challenged much. Further, there is a lot of interaction between those non-capitalist economic units and the capitalist society, and because of the efficiency of the capitalist system, but also because of its aggressiveness, it is hard to resist the capitalist logic. The production cooperatives, for goods or services, compete with capitalism in the market and the same holds true for consumer cooperative shops. Originally, they were very different from capitalism, but the Scandinavian cooperative shops today are much the same as their competitors; they stock a bit more organic and fair trade products. New cooperatives are often more different, but gradually they adapt themselves to capitalism. They start, say, with equal pay for workers, which then makes the workplace very attractive for those who would normally have a low pay, and those who are talented managers are recruited by capitalist operations.

The not-so-orthodox Marxist Meiksins Wood gives a plausible argument for why cooperatives fail to work in a truly non-capitalist way: 'even workers who own the means of production, individually or collectively, will be forced to respond to the market's imperatives — to compete and accumulate, to exploit themselves, and to let so-called "uncompetitive" enterprises and their workers go under' (1999). This is underlined by the fact, noted earlier, that in today's society the self-employed are often far more exploited than workers. As I explained in the Foreword, this is what I experienced as a farmer trying to live in collective simplicity. Still, even with these limitations, the real and symbolic importance of these alternatives should not be neglected.

It also begs the question of whether it is what Meiksins Wood (1999) calls the market imperatives — the imperatives of competition, accumulation, profit-maximization, and increasing labour productivity

—or whether it is capitalism governing the markets that is the bigger problem. Or perhaps this is the separating line between a market economy and a capitalist market economy. Can we have a market economy without endless competition?

Seeking the leverage points

If profit-seeking and private wealth accumulation are the main drivers, then those are also the targets for change. If, however, Meiksins Wood (1999) is right about market dependency being at the core of development, then the strategy is not so much about workers taking over production but rather about withdrawing as big spheres as possible from market imperatives, to de-commodify and democratize as many spheres of life as possible. I would argue that both are valid perspectives and that regardless of which one is first or which one is driving, they are mutually reinforcing.

Three key institutions in the capitalist market economy are markets, property and money. They are also intricately linked to each other. For example, property is sold in the market for money, and the profit in money is converted into more property, etc.

To decide to move some things out of the market may sound like a very difficult thing to do. In the last decades, the move has been in the opposite direction. But those are short-term swings. For instance, the market for slaves worked with some limited regulations, but ultimately humanity decided that it wasn't a good thing. Most countries have limitations for markets in human organs or for buying political votes or decisions. All in all, we have choices for what we want to subject to the markets. There is simply a difference between how much the market *can* coordinate and how much it *should* coordinate.

Some readers are probably by now upset and think that I am arguing for the total abolishment of markets. That is not the intention. The right to exchange goods with each other can also be seen as a basic human right. Sen says, 'The contribution of the market mechanism to economic growth is, of course, important, but this comes only after the direct significance of the freedom to interchange—words, goods, gifts—has been acknowledged' (1999: 6). I find this an appealing perspective. I do argue against a market society; I argue against using markets as a means to regulate human use of nature; I argue against using markets as a way to determine what people do (labour markets); I argue against the tendency to let more and more commons and public goods be managed by markets. I believe markets are working fairly well for the distribution of goods, and that they will work even better if one takes away profit-making and competition as the two main drivers in markets.

I believe we can apply a similar perspective to property. It is commonly agreed that one can't own everything (e.g. the air one breathes, other people or the government), and also that one can own some things that can't be sold (e.g. one's body). But by and large, as human society is organized most things one owns are subject to the market. To own something doesn't mean that one has unlimited rights to it. Most countries and cultures have some limitations, which means that there comes some responsibility with ownership; for example, one can't make a old house that is almost crumbling accessible to the public, because one jeopardizes other peoples' safety; or if one has a well in an open space, one should put a lid on it so that children will not fall into it; or if one dumps waste, it should not pollute the neighbours' water or land. The German Constitution says, 'Property entails obligations. Its use shall also serve the public good' (in Helfrich 2009: 3).

I believe that private ownership (in the sense of control) of one's clothes, house and basic means of production (which can mean farmland for a farmer) makes sense. The ownership of other peoples' homes, factories where other people work or minerals in the ground is a very different thing, however. If you own my house, I can't own it; your ownership clearly limits my ability to own it. Some property should be redistributed to commons (from which most of them emerged in the first place) and others redistributed to the users; for example, the workers of the factories or the local communities should own the factories rather than a few investors, which would ensure that they are developed for the good of the local community rather than for distant owners. Admittedly, the devil is in the detail, and there may be many impediments for this transformation. Nevertheless, I believe that it is very important to stress the principle.

Another institution that has to be carefully considered is money. I don't claim to know the truth when it comes to money. Hopefully, more knowledgeable and intelligent readers can contribute to the demystification of money and the design of new means of exchange. Hart says, 'The consequences of examining what money really is are so shocking . . . that the world prefers, for the most part, not to think about it' (2000: 245). Hornborg adds, 'the fundamental problem is that our global social system has adopted an institution, based on the idea that everything is interchangeable, that continues to encourage an accelerating dissipation of resources' (2009: 257; emphasis in original).

It seems to me that in order to make a real change one has to abandon money as the determinant, the measurement, for all human activity. This can be accomplished with a combination of several complementary strategies. First, substantial parts of human activities need to be pulled out from the market system, essentially land and labour and a lot of work related to daily life, housing, cooking, etc. Second, different kinds of currencies need to be developed for different circulations, which will mean that there will be several loops of exchange, with quite a few barriers between them. This is essential for reducing both the enormous power of competition that works on all aspects of life today and the exploitation of people in distant countries, and it is important to underline that not all things are interchangeable. Third, other measures of value need to be developed to train one to think outside the money box. Fourth, the functions of using money for exchange and as financial capital need to be separated. The ability to earn money from money should be limited by various means. Instead of constantly growing through interest rates and rents among other ways of making profit, money should perhaps be devalued over time, in the same way as manufactured capital loses its value. In this way, money could again become a tool primarily for exchange and not for profit-making.

Seeking autonomy

One challenge is that the reach of today's capitalist market economy and its associated technologies is both wide (i.e. all over the globe) and deep (i.e. affects most aspects of daily life). This means that there are few 'free zones' where alternatives can emerge. Part of the strategy for a new world is therefore to establish more such free zones — by refraining from letting the prevailing paradigm control us or by means of political reform creating such free zones. But in doing so we will also challenge strong interests in keeping things as they are, society at large, the ruling technical complexes as well as ordinary citizens who are worried about change, earn their living from some doomed trade or have their wealth tied up in the prevailing system.

The examples that are really different are different because they have excluded themselves to a very large extent from interaction with the capitalist economy, for example, self-sufficient consumption and production cooperatives in the countryside, or because they are held together by a very strong ideological or religious creed. However, most of them have failed to appeal to any larger number of people and some of them end up as rather insulated autarchic places. The appeal of autonomy is strong for some people, but autonomous societies end interaction with the rest of the world and therefore they influence little.

All in all, I believe that the peasant strategy of partial autonomy from the market combined with a communitarian political system is promising. The peasants don't represent an outlier or a curiosity. They still represent a third of the world population. While not always being conscious of its meaning and its character of civil obedience or sabotage, the peasant strategy is also behind the strive for mechanisms that put us outside the reach of the monetary system, such as bartering and/or swapping services as well as local exchange trading system (LETS) currencies. The informal economy and urban agriculture are also partly based on the 'peasant principle'.

Delinking from the global capitalist economy, or simply ignoring it, is one of the most important strategies for individuals, communities and countries that want to enter a new path of development. By creating several spheres of exchange, something that was normal before the breakthrough of capitalism and something that still exists, for example, in the exchange within the private sphere, civil society, the market and the state, some of the effects of the unfettered global capitalism can be mitigated. In this way, one market where all goods and services are equally exchangeable can be replaced with several markets operating with different currencies.[190]

---

[190]	Already by having a national currency that is controlled, governments can do this to some extent, but through increased globalization the possibility for countries to keep their own currency regime and capital regulations has been undermined, and most countries have deliberately chosen to deregulate their capital markets.

– 32 –
How We Organize Society

Freedom grows by living free.

(Vandana Shiva,
Earth Democracy: Justice,
Sustainability, and Peace, 2005)

Enormous challenges lie ahead for humanity, challenges for which one certainly needs societal organizations; human beings need and feel well with a society. The appreciation from others and a feeling of a common vision and interest are very important for human well-being, and we engage with others to fulfil our dreams. More and more aspects of life have been pulled away from common responsibility and solidarity to the logic of private capital and profit. Human beings need to strengthen forms of social cooperation and ensure that they lead to an increase of the freedom and reach of the individual and not the opposite. In this chapter, I discuss the institutions of human society and which institutions or spheres can take care of which kinds of activities and functions.

People's trust in their governments is fairly low. In most places, politicians, elected to manage society, have low credibility and are often seen as plainly incompetent or crooks, and often both. This leads some people to distance themselves from society; this is aggravated by the frequent confusion between the state (the government) and society (how one organizes oneself for cooperation). Historically, most states have been undemocratic and essentially tools for oppression and exploitation. However, the state has also protected its citizens against external aggression and internal violations. Without that, subordinates would most likely revolt. The state is, from this perspective, a remnant of the warlords and kings who established a mafia style of protection. Once a state institution, that is, some form of government, emerged, power over that institution became the subject of struggle between individuals or clans or social groups.

These multiple roles of the state still exist today, regardless of whether one speaks about Western parliamentary democracies or China. They also colour how one looks at the state today. Some see the state as a protector for the weak; others see it as a protector of the rich. On the one hand, it puts in place the institutions that enable the rich and wealthy to exploit others as well as nature and society at large. On the other hand, it takes measures to support the poor, to mitigate injustice and limit exploitation. Obviously, the political conclusion is quite different in the two scenarios, and the state normally fulfils both roles, as well as other roles, simultaneously.

The nation-state is losing ground

Humanity has, since the agrarian revolution, organized itself territorially. This way of organization has its origin in the need for safety and security; besides, a ruler's reach also needs demarcation against other rulers and other territories. With increasing globalization, with the insight that human beings are all together on this rather small planet, with global climate crisis, with global economic crises and with global terrorism (just to mention the most obvious manifestations), the nation-state is failing to meet all challenges and sometimes it is seen as an obstacle. Without doubt, the nation-state has had to devolve a lot of power in the latest century (see Figure 32.1).

Figure 32.1 Devolution of power from the nation-state.

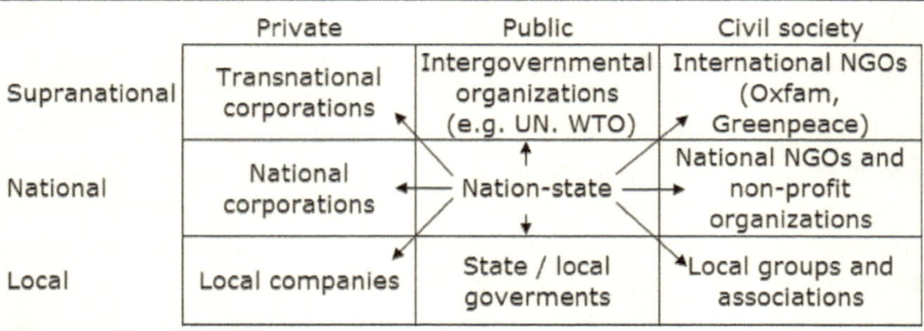

Source: Nye (2008).

Transnational corporations have certainly increased their power over the state, but one should not be fooled that they had no such powers earlier. 'What is good for General Motors is good for America' is just one of many expressions of how tightly knit the state and the big business has been for a long while. That General Motors was taken over by the United States government during the financial crisis in

2008/2009 just underscores this. Even the real pioneers of modern capitalism, the Dutch and British East India Companies, were semi-governmental transnational companies, and their proportional powers certainly exceeded those of any of the current transnational companies. Perhaps the biggest difference now is that a lot of the transnational companies don't pay allegiances to *any* particular state. Many of them have a rather anonymous ownership from many parts of the world, which means that their interest is in the welfare of the companies and nothing else.

Many things are determined outside the nation-state and inter-governmental organizations. I had the opportunity to serve as President of the International Federation of Organic Agriculture Move-ments (IFOAM) from 2000 to 2005. It gave me an insight into how people, just based on common interest, can work together globally, not only with activism and propaganda but also to build functioning institutions. Since 1972 IFOAM has been setting standards for organic agriculture, and since 1992 there is an international accreditation system to approve those who control the production, all in a voluntary organization with 800 members in over 100 countries and where 'one member, one vote' is the principle, not 'one country, one vote'. Almost all modern standardizations, one of the pillars of the modern world, such as the GSM standard, Internet protocol and e-mail addresses, just to mention a few, are determined outside the governmental sphere.

A private company, Microsoft, has determined the rules, the operating system, for what today is the most common tool after the telephone and the hoe, the computer. NGOs or celebrities such as Bono or Princess Diana have launched campaigns that have made even the United States change its policies. An international NGO, the Red Cross, and its sister organizations have succeeded in providing medical services in the most dreadful humanitarian and security conditions. In 2009, the President of the United States travelled to the meeting of the International Olympic Committee, a non-governmental organization, to lobby for his country getting to host the next Olympics — and lost. All air traffic, rules for air tickets, etc. are coordinated by the Inter-national Air Transport Association (IATA), a cooperation not between countries but between airlines. The Internet provides many examples of new forms of how communities manage their affairs. Finally, as a result of increased individualism, people simply have much less confi-dence in the state and challenge its authority in all sorts of fields: Who are they to tell us what to do? Who believes that the nation-state is still in the driving seat?

Global and local governance

The nation-state has also let go of power 'down' to local governments. This is perhaps both a function of local governments demanding more influence and, at least in some countries, central governments having 'passed the buck', shifting the responsibility for provision of essential, but costly, services (such as health care, schools and roads) to municipalities, districts or counties. The end result is in any case more power to the local government. Because of the many global challenges, nation-states have also let go of power to intergovernmental organizations such as the United Nations and regional organizations such as the European Union.

How international governance should be organized is a difficult question. Today, there are a large number of international organizations, many of which have overlapping mandates. Some have muscle; others are more 'talk shops'. Only for global environmental management, between 1992 and 2007 there were 540 meetings, which resulted in more than 5000 resolutions or decisions for 18 multilateral environmental agreements (UNEP 2010a). The urgency, real or imagined, of environmental challenges, in particular climate change, seems to favour strong global governance. It is not obvious, however, that all global challenges are best regulated globally. 'All politics is local,' the saying goes, and although there is certainly a need for truly global treaties and agreements, most issues still should be tackled at the local level.

Although almost all subscribe to the principle of democracy, not all agree upon what it means and even fewer agree upon how it ought to look at a global level. There are three main lines of thought about global democracy.[191] The first is the *human rights approach*, which suggests that 'rule by the people' is achieved when everyone everywhere on the planet enjoys certain fundamental enabling circumstances. The second is the *civil society approach*, which emphasizes transnational associations (often called 'civil society') such as NGOs and social movements that increasingly demand representation and power in global governance, a kind of activist and stakeholder democracy. The third is the *federalist-representative approach*, where global democracy is achieved through a world government with an elected global parliament, building on the nation-states as the democratic units.

---

[191]   The discussion builds to a large extent on 'Globalizing "rule by the people": A deliberative view' (Erman 2010).

None of these models pass the test of real democracy. The first is strong on liberties and rights but doesn't include any real influence and participation of the people in determining the affairs of the planet. To expand and ensure human rights and liberties is a good thing and a precondition for democracy, but it isn't democracy. The NGO approach is based on the assumptions that these 'popular' or 'grassroots' movements are truly democratic themselves, and that they organize most people. But this is not at all the case. It is one thing to acknowledge their importance and to be positive about their involvement in global politics, and quite another to assign formal powers to them. In a way, they are best when they are constantly fighting for influence but are never fully successful. Many NGOs are not built as institutions for global governance, and if they were given formal representative powers, they would be prone to corruption and manipulations of all sorts. In my view, civil society organizations should be given the authority to manage their own business, which they already do successfully. I believe they can also find new life in managing old and new commons. It is a very different story to allocate authority to them to mind other peoples' business.

The existing representative approach to global governance is different from having a global parliament. A global parliament, however, may or may not strengthen democracy. Because people tend to be closer to their national government than to global institutions, it is not at all clear that assigning more power to a global parliament will mean more or better democracy. There are obvious fears that a global government could become a monster and the distance between the average citizen and this institution would just be enormous.

Because none of the frameworks are ideal or even close to ideal, it is probable that we will muddle along with what already exists and use several tools and institutions instead of a unified global governance structure. Most likely global governance and global democracy will be a mixture of a strengthened framework for human rights, expanded also to socioeconomic rights, such as the right to food, increased role for global civil society organizations, further development of truly global institutions based on common interests and the continued intergovernmental cooperation of nation-states in UN-style organizations.

Parallel to building a new global architecture of governance there is a need to strengthen local communities or create new local communities. Ultimately, I believe that local governments can be a step towards closing one of the many gaps created, the one between society and government. Note, however, that local entities for social interaction will only be strong if they have a meaningful economic and institutional basis. Interdependency is a very strong maker of ties, sympathies and feeling of common cause. A village in mediaeval times was a strong unit because people needed each other, for organizing work on commons, helping each other in times of trouble and protecting the village from threats. But today's local communities lack much of this. Without such a role, a village association or a neighbourhood council will be nothing more than cultural clubs. A pleasant scenario perhaps, but without any real meaning. Is the development of local currencies a strategy to strengthen local communities? How can the direct interaction between people in neighbourhoods be increased?

How can the local communities and the global level be linked? Is that at all possible? The nation-state lacks resources and knowledge and it is also too slow and bureaucratic to really take care of all the needs and challenges. Nonetheless, the nation-state provides identity and a sense of security; it is also a democratic unit that has some trust, credibility and legitimacy among citizens. Even with corrupt, dictatorial or simply useless regimes, people tend to feel for their country or nation. This means that the nation-state most likely will continue to be an important player for decades, if not centuries, to come. With that in mind, most likely devolution of power will continue, in particular to new civil society organizations and institutions. A weaker state, less threatening for citizens, is a step in the right direction, if, but only if, power relationships in society are more equal, that is, if economic democracy is also installed. Failing that, a weak state is likely to benefit more the rich.

Democracy—muddling along?

Representative democracy rooted in political parties doesn't seem to provide all the answers. And why should it? In my view, it is an institution with roots in different social groups trying to control the state for their own interest. The partiers are competing in the voters' market, eloquently expressed by a local politician, Tom Chemisito, in Uganda: 'When people present themselves to run for political office, they become like commodities in the market where buyers are at liberty to buy an item of their choice' (The Vision 2012). Other forms of democracy are promoted as alternatives, in particular direct democracy, or more recently 'participatory democracy', in particular in Latin America. Porto Alegre in Brazil started a process of participatory budgeting in 1989. This has spread to hundreds of municipalities in Brazil as well as to other Latin American countries. *Consejos Autogestivos* (self-management councils) in Mexico are managing protected areas in a participatory way (Dagnino *et al.* 2008).

In Bolivia,[192] the new constitution defines three modes of democracy that are used in parallel in the process of deepening democracy: representative democracy (according to the western liberal democratic tradition), participatory democracy (incorporating the demands of radical democracy and the experiences of council democracy) and communal democracy (according to the traditions of self-governance of the indigenous peoples) (Lander 2010). This seems to me to be a promising avenue: participatory democracy, or an older term, council democracy, for workplaces and institutions where people work together and communal democracy for local matters. How to stitch it all together at the national and global level, limiting conflict between different forms of democracy, remains to be figured out.

Regardless of which systems ultimately rule, they will only be successful if people feel that they can contribute to development, that they are part of development and that they can influence as citizens. To a large extent, it is about developing social capital at the cost of financial capital, private property and consumerism. This social capital has to be built with the individual as a subject who voluntarily seeks association with other individuals, not as a collectivism that sees the main role for the individual as a cog in the societal wheel. It is also about creating public spaces that can serve as the platform for citizens' participation. Here, the role of the Internet is potentially very important, even if the physical space is likely to remain very crucial, as was demonstrated by the Arab Spring in 2011. Social and environmental activist Vandana Shiva says that we should 'reinvent the state in a way that is not centralized, bureaucratic, and controlling, in a way grounded in the community and responsible to community' (2005: 89). That sounds good, but is that still 'a state'? Being 'centralized, bureaucratic and controlling' seems to be defining characteristics of a state.

---

[192]   This interesting fact doesn't mean that I have studied modern Bolivia in any detail or that I recommend its policies, as little as this footnote should be construed as denouncing them.

It is not only about power

The eternal discussion about the market and the state is sometimes a bit confused by states doing very different things. They regulate our obligations and rights, as well as their own, through laws; they impose taxes and use that money to pursue things that have been decided, and they are responsible for stewardship of common resources. There is a traditional divide between those who think that a function shall be taken care of by the market and those who believe it should be taken care of by the state. Somewhat simplified, the former are liberals (or the Right) and the latter socialists (or the Left); the terminology depends on which country one is in.

One can look upon both markets and the state as social institutions to regulate relationships between people and distribution of property and products. Which institution one likes the most depends perhaps on one's own situation, how the institution is organized and which context it is in. The same person who is very sceptical to market solutions is perhaps perfectly comfortable with her relationship to the local organic farmer and finds it deplorable that the government determined that organic chicken were not allowed to go out during the bird-flu scare in 2006, or thinks that the government is over-regulating small-scale food processing. And the person who is against government control still applauds government takeover of bankrupt banks or car-makers during the financial crisis of 2008/2009.

Citizens have very little access to the state; one's role is to pay taxes to finance it and regularly vote who shall rule it and therefore rule all citizens. Sometimes one interacts with the state through civil action, direct contact with a politician, participation in a demonstration or by writing a letter to the editor in a newspaper or a complaint (or more rarely praise) on a website. People are constantly participating in the markets and, in theory, everybody in the same market is equal, as long as one has money to pay. The market can be seen as a giant social decentralized network.

A problem with the market is that it favours those who are already rich and have better information and these differences tend to increase rather than decrease over time. The other problem with markets in their current shape is that they pitch people against each other through competition, something that in turn is bad for society. Despite all the hype about the market as a democratic institution, it is a place of very imbalanced relationships.

State takeover of markets, in most cases, has been disastrous and market takeover of the state is almost as bad. Obviously, a certain balance is the best. But one needs to look *beyond* markets and the state.

By concentrating the debate around the market or the state, one tends to lose sight of other possibilities. There are many other possibilities as we discussed earlier. Voluntary organization of people to accomplish certain objectives is a very old process, much older than both markets and the state. Because of their communitarian nature, there are few champions for these solutions because nobody can profit from them. There is no indication that such arrangements would have been inefficient. And even if they were 'inefficient' this may well be a price worth paying for other benefits, such as more equality or better consideration of nature—after all the most valuable moments of one's life are inefficient. Enjoying culture, good food, socializing with family or friends or love-making are all 'inefficient'.

That it is hard to run a company with 50,000 employees with direct democracy in cooperative form can be used as an argument against direct democracy and cooperative forms, but it can as well be an argument against having companies with 50,000 employees. That smaller units thrive on social capital and at the same time nurture it is another argument in their favour. In the smaller scale, equity and efficiency go hand in hand (Serageldin and Grootaert 1999). The same is true for the relationship between scale and division of labour. Smaller-scale appropriate technology, less division of labour and more local power are therefore all part of another world. Globalization and large units have less operational self-correcting mechanisms, while activities embedded in the local situation respond to ecological, economic, social, cultural and political feedback. Tying in to the earlier discussion about democracy, more activities and functions being taken over by civil society organizations will also strengthen a more participatory democracy. A new alliance between civil society organizations and local governments and local communities seems to be a promising avenue.

Last but not least, individuals and their small units, such as the family represent an important level. Much that is important takes place at that level, for example love, reproduction, food preparation, housing maintenance, etc. The private sphere has let go of a lot of power and a lot of activity, and its main institution—the family—is weakened. Also the next level—the clan, the tribe or the local communities—has lost a lot of power. This means that human beings become individual atoms in an unlimited universe, colliding randomly with the other seven billion atoms. But we are not well adapted to this extreme individualism. It is a bit hard to see which units are the future ones for private organizations. A resurrection of the family seems less likely for a multitude of reasons; today, it is not even an institution for reproduction in many industrial societies. Its death is somewhat exaggerated, however, even if its form may be more varied than before.

Figure 32.2 The movement of activity between spheres.

Function and activity during capitalism

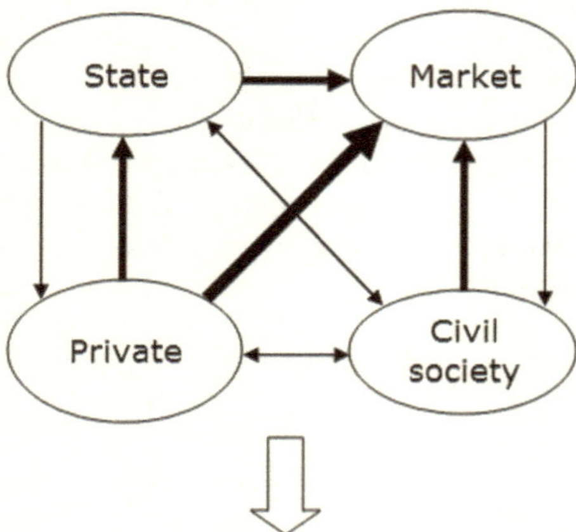

A future movement?

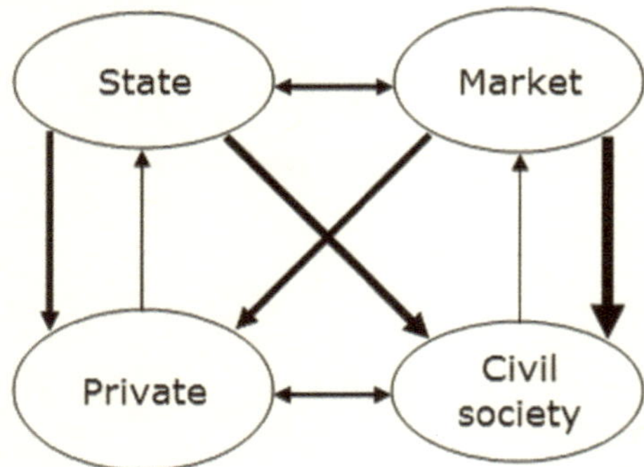

Source: Author.

Figure 32.2 tries to depict how various activities have moved from different spheres during the growth of the capitalist society (on the Top) and where they might go in an alternative development (in the Bottom). With the introduction of capitalism and market economy, many things were moved out from the family, local communities and civil society (the local communities, e.g. the villages, are here seen as part of civil society rather than as part of the state). The role of the state also increased initially: the state organized education, built infrastructure, health care, etc. It also regulated the family and civil society increasingly. In the later phase of capitalism, more and more has been moved to the market; some aspects of life, like religion, have been relegated to the individual, mainly because they are no longer seen as essential for social order. Civil society organizations have been further weakened, mainly because the market has penetrated most of its space and realm and because of the attitude and values of the market economy. Individualism and competition have not been conducive for civil society.

In the future, we should strive for more activity in the local communities and civil society and less activity in the market sphere as well as in the central government. Also the private sphere should regain some of its relevance, in particular regarding food production and preparation and other household services.

– 33 –
Garden Earth

You may say that I'm a dreamer
But I'm not the only one
I hope someday you'll join us
And the world will live as one

(John Lennon,
'Imagine,' 1971)

Hunters and gatherers lived in a wild nature, although they manipulated it more than most are aware. Their values, religion and society were shaped by their environment and ecological niche. A small fraction of the primary production of ecosystems was taken care of by human beings. With agriculture, it all changed. A part of nature, gradually growing, was occupied fully by the human species, cleared from undesired species and put directly, and uniquely, in our service. In this way, we maximized our share of the biological production, and favoured those species that produced what we needed. Because of more intensive use, buffers were smaller and agrarian society was thus vulnerable to disturbances and shocks, such as climatic events, disease of crops and animals as well as of ourselves. Soil erosion was an ever-present threat and a cause of societal collapses or simply slow decline. Our relationship changed not only with nature but also with each other and with the society that was established to manage this new system. Human society changed from a basically egalitarian society of small bands to agglomerations of authoritarian lordships, kingdoms and empires, and from a society with very limited division of labour to a stratified society where each had his or her well-defined role.

With the Industrial Revolution, based on fossil fuel, human control and reach expanded even more, to almost all nooks and corners of the globe. Humankind went further, climbed and flew higher, dived and dug deeper—all in the effort to find more resources to exploit. More and more parts of the living as well as the dead were put in direct service of humanity. Industrialism also brought new impetus to farming. The first stage brought tools and the power to extend the land under plough, but gradually also more and more machinery to increase productivity per labourer. This was stimulated by an increased use of fossil fuel so that farming from being a net producer of energy became a net consumer. Chemical fertilizers, extracted, manufactured and shipped from far away, were a step towards freeing agriculture from the limitations of the site, to manipulate life processes. These changes were also accompanied by increased use of pesticides and medicines. Now, when the rich supply of genes in nature doesn't suffice, or there are natural barriers for combining them as desired, genetic manipulation is used.

At the same time as we are dependent on more and more of the bio-geosphere, people live lives increasingly more distanced from it. The warmth doesn't come from wood that we cut in forests, but from distant power plants, district heating or oil and gas pumped from far away. We rarely touch the soil; the green plants most touch are retarded, tamed and flower-induced pot plants, if they are not plastic. We rarely feel the hot wind on our cheek, the rain in our hair or the bitter cold in our marrow. Still, we are more dependent on nature than ever before. Nature doesn't care for humankind because it will survive well without it, but humankind has to care about nature.

Modern society has been successful in delivering personal wealth and economic growth, for a few people, that is. It has also broken down old prejudice and expanded human rights, freed people from bonded labour and other oppressive institutions. It has come a long way in ending discrimination of women and other groups often victimized, and been extremely successful in producing more things for the health, comfort and joy of human beings. The system, however, has not been successful in replacing the old system with a new one that gives people a feeling of meaning. Personal wealth and growth have become ends in themselves to such an extent that they remind of cancer.

The economic system and society have reached a critical limit through population explosion, unscrupulous and reckless exploitation of natural resources and constantly increasing energy use. For some resources, the limits have been passed already, which puts increasing pressure on societies, the global community of human beings and the foundations for human existence. We hunt for more and more things and satisfaction through consumption; the enormous choices haven't made anyone happier. Unprecedented wealth is created, but it has not spread to most citizens of the world, despite everyone living in the same global economy. The gap between the rich and the poor is widening. Despite constant increased productivity leisure time doesn't increase; contrarily, one's leisure time and private sphere are invaded by the capitalist paradigm and behaviour patterns — to consume becomes a compulsion as strong as to work.

The capitalist system is a system of colonization, bent and suitable for geographical, physical and economic expansion. Society has now reached a stage of saturation; there is nothing left to colonize. The behaviours that were appropriate or at least acceptable for a species in rapid expansion are not the ones suitable for a life in balance. In some ways humanity seems to adapt to it — how else would one explain population ceasing to grow in developed countries? Organic farming, passive heating, electrical bicycles and solar energy represent technological developments better adapted to 'householding' societies with regenerative economics. The wild finds new ways of cohabiting with human beings and human beings find new ways of interacting with the wild. Cooperation within civil society has increased tremendously globally and is revived locally. There is a shift in values, where more and more people want to jump off the treadmill and seek that which gives 'real' well-being. This maturity is reinforced by the population ageing. The values of older people are adverse to risk and expansionism, and are more reflective and caring.

In the discussion about growth and limitations for human use of natural resources, it should not be forgotten that the poorest third of humanity needs to increase its use of natural resources, while the rest of humanity takes a break and enjoys life. This increased use is certainly not decoupled from direct exploitation of resources; it is the development of very real things such as electricity, houses, roads, water and sewage. Still, many old and new technologies can be deployed to make this growth less wasteful than previous modernization; these technologies include solar energy, distributed power, separation of water toilets from other water use, or buildings made from adobe, etc.

Most people have not yet realized the extent of the change required; humankind needs a completely new economy, a new society. Another society has to be built on values and conditions other than those of the capitalist society. It is a society oriented to closing gaps — gaps between people, between man and woman, between ruler and ruled, between 'we' and 'them', between one country and another and between man and nature. It is about unifying the many divisions that currently exist — division of labour, division between production and consumption, between work and leisure, between the individual and society, between economy and society and between beauty and efficiency.

As an alternative vision to the capitalist ideology, I offer Garden Earth.

Where capitalism sees self-interest as the major human driver, Garden Earth is built on human needs satisfied in many ways and with voluntary cooperation.

Where capitalism is based on private property, the foundation for Garden Earth is stewardship of common resources.

Where capitalism sees nature as commons free to exploit, to mine and to use as a dumping ground, Garden Earth sees nature as our home.

Where capitalism sees specialization and division of labour as something very good, Garden Earth sees it as a potential evil, a risk that needs to be limited and its effects on individuals and society mitigated.

Where human energy in capitalism is used for salaried labour that a few profit from, in Garden Earth people work for their satisfaction and for doing something for the community.

Where technology in capitalism is developed and chosen to maximize some peoples' profits, in Garden Earth it is selected to make life more enjoyable or transcend resource limitations.

We need each other and the existing form for our cooperation is what is called society or community. It is ruled by laws of nature, ecological conditions, human laws, norms inherent in our culture and values. The human being is part of society and vice versa. The well-being of humankind and society is intricately intertwined forever. Markets, the state, civil society and the private have roles to play in the future society. It is about expanding liberty and the notion of liberty to include also the capabilities of the individual and to resolve the false contradiction between the freedom of the individual and that of society.

We simply have to go back to the Garden of Eden. But we can no longer do so as the hunters and gatherers who were driven out. Humankind is no longer equal to the giraffe, the carrot or the sheep, even less to the stone. Man is part of nature; thus, the well-being of human beings and of other species is intertwined. Because of the numbers, intelligence and development of the human species, we have a special responsibility for the survival of other species and for the relative stability and productivity of the earth's ecosystems. Whether or not we like it, the survival of not only the human species but also many other species is in the hands of the human being.

References

Ababouch, Lahsen. 2009. 'Fish utilization and trade', in *Fisheries, Sustainability and Development: Fifty-Two Authors on Coexistence and Development of Fisheries and Aquaculture in Developing and Developed Countries*, edited by P. Wrammer, H. Ackefors and M. Cullberg. Stockholm, Sweden: Royal Swedish Academy of Agriculture and Forestry.

Abdallah, S., S. Thompson, J. Michaelson, N. Marks and N Steuer. 2009. *The Happy Planet Index 2.0*. London: New Economics Foundation; http://www.happyplanetindex.org/public-data/files/happy-planet-index-2-0.pdf

ADB. 2009. Electric Bikes in the People's Republic of China: Impact on the Environment and Prospects for Growth. People's Republic of China: Asian Development Bank.

Apoteket. 2005. *Läkemedel och miljö*. Stockholm, Sweden: Apoteket AB / Stockholms läns landsting / Stockholms universitet; http://www.apoteket.se/privatpersoner/om/Documents/Om%20Apoteket/Milj%C3%B6/L%C3%A4kemedelochmilj%C3%B6_2005.pdf

Arbets och Miljömedicin. 2009. 'Pressmedelande', 23 November; http://www.ammuppsala.se/upload/File/Press/Pressmedelande%20ML%206.5%20milj%20frn%20Formas.pdf

Arrighi, Giovanni. 2010. *The Long Twentieth Century, Money, Power and the Origins of Our Times*, new and updated edition. London: Verso.

Associated Press. 2009. 'Elder Bush completes birthday parachute jump', 12 June 2009; http://www.msnbc.msn.com/id/31301377/

Atkinson, Anthony B. and Andrew Leigh. 2010. 'The distribution of top incomes in five Anglo-Saxon countries over the twentieth century', *IZA DP No. 4937*; http://ftp.iza.org/dp4937.pdf

ATL. 2009. 'Nödvändigt minska matsvinnet', Jerry Simonson, 17 February 2009.

Austen, Ralph A. 1987. *African Economic History*. Oxford: James Currey Ltd.

Australian Government. 2008. *Restoring the Balance in the Murray-Darling Basin*. Australia: Department of Environment, Heritage and the Arts; http://www.environment.gov.au/water/publications/mdb/restoring-balance.html

Ayres, Robert U. 1989. *Technological Transformations and Long Waves*. Laxenburg: International Institute for Applied System Analysis.

Ayres, Robert U. (ed.). 1998. *Eco-Restructuring: Implications for Sustainable Development*. Tokyo, New York and Paris: United Nations University Press.

Baltzer, Kenneth, Henrik Hansen and Kim Martin Lind. 2008. *A Note on the Causes and Consequences of the Rapidly Increasing International Food Prices*. Denmark: Institute of Food and Resource Economics, University of Copenhagen.

Bartsch, Kristina J. 2009. 'Employment projections, 2008–18'. *Monthly Labor Review* (November): 3–10; http://www.bls.gov/opub/mlr/2009/11/art1full.pdf

Bateson, Gregory. 2000. *Steps to an Ecology of Mind*, with a new foreword by Mary Christine Bateson. Chicago: The University of Chicago Press.

Bath, Bernard Slicher van. 1976. *De agrarische geschiedenis van West-Europa (500–1850)*. Utrecht-Antwerp: Het Spectrum.

Bayliss-Smith, T. P. 1982. *The Ecology of Agricultural Systems*. Cambridge: Cambridge University Press.

BBC. 2010. 'UK water use "worsening global crisis" ', 19 April 2010; http://news.bbc.co.uk/2/hi/science/nature/8628832.stm

Bennett, James C. 1999. 'The end of capitalism and the triumph of the market economy'; http://www.pattern.com/bennettj-endcap.html (accessed 10 September 2010).

Bennett, Michael T. 2009. *Markets for Ecosystem Services in China: An Exploration of China's 'Eco-compensation' and Other Market-based Environmental Policies*. Washington, DC: Forest Trends; http://www.forest-trends.org/documents/files/doc_2317.pdf

Bhopal.net. 2010. 'The 1984 Union Carbide gas disaster in Bhopal'; http://bhopal.net/the-1984-gas-leak/ (see www.bhopal.net) (accessed 17 May 2010).

Bigsiteofamazingfacts.com 2009. 'How are bricks made'; http://www.bigsiteofamazingfacts.com/how-are-bricks-made (accessed 10 February 2009).

Bigsten, Arne and Dick Durevall. 2007. *The African Economy and Its Role in the World Economy*. Uppsala: Nordiska Afrikainstitutet.

Björklund, Johanna, Pär Holmgren and Susanne Johansson. 2008. *Mat & klimat*. Sweden: Medström.

Blanden, Jo, Paul Gregg and Stephen Machin. 2005. *Intergenerational Mobility in Europe and North America*. London: Centre for Economic Performance.

Bloomberg Businessweek. 2008. 'There will be water', 12 June 2008; http://www.businessweek.com/magazine/content/08_25/b4089040017753.htm (accessed 2 October 2010).

Bookchin, Murray. 1971. Post-Society Anarchism. San Francisco: Ramparts Press. [Originally published in Bookchin's newsletter *Comment* in 1964 and republished in the British monthly *Anarchy* in 1965.]

Bookchin, Murray. 1980. *Toward an Ecological Society*. Montreal: Black Rose Book.

Bond, Michael and Colette Braeckman. 2001. 'A moral minefield'. *New Scientist* 170(2285) (7 April): 46.

Borgström, Georg. 1976. *Gränser för vår tillvaro*, 5th edition. Stockholm, Sweden: LT.

Borlaug, Norman E. 2008. 'Stem rust never sleeps'. *New York Times*, 26 April 2008; http://www.nytimes.com/2008/04/26/opinion/26borlaug.html

Boserup, Ester. 2005. *The Conditions of Agriculture Growth: The Economics of Agrarian Change under Population Pressure*. New Brunswick, NJ: Transaction Publishers.

Braudel, Fernand. 1982. *The Wheels of Commerce*. New York: Harper & Row.

Braun, Joachim von. 2010. 'Food security and the futures of farms: 2020 and toward 2050', *Bertebos Prize Lecture in Proceedings of the Bertebos Conference 2010*, 29–31 August 2010, Falkenberg, Sweden, The Royal Swedish Academy of Agriculture and Forestry.

Brewtopia. 2010. 'About us: What is Brewtopia?'; http://www.brewtopia.com.au/about-us/about-brewtopia-custom-labels-beer-wine-labels-bottled-water

Brown, Lester. 2008. *Plan B 3.0: Mobilizing to Save Civilization*. New York and London: W. W. Norton & Company; www.earthpolicy.org/Books/PB3/index.htm (accessed 14 December 2009).

Bull, Jamie. 2010. 'EROEI of electricity generation', 19 May; http://oco-carbon.com/2010/05/19/eroei-of-electricity-generation/

Burenstam Linder, Staffan. 1970. *The Harried Leisure Class*. Columbia: Columbia University Press.

Carley, Michael and Philippe Spapens. 1998. *Sharing the World: Sustainable Living and Global Equity in the 21st Century*. London: Earthscan.

CAST. 1999. *Animal Agriculture and Global Food Supply*. Ames, IA: Council for Agricultural Science & Technology.

CBD. 2010. *Global Biodiversity Outlook 3*. Montréal: Secretariat of the Convention on Biological Diversity.

Chapman, J. L. and M. J. Reiss. 1999. *Ecology: Principles and Applications*, 2nd edition. Cambridge, UK: Cambridge University Press.

Chunliu, Zhan and Marlene R. Miller. 2003. 'Excess length of stay, charges, and mortality attributable to medical injuries during hospitalization'. *JAMA* 290(14): 1868–1874.

Chunliu, Zhan, Judith Sangl, Arlene S. Bierman, Marlene R. Miller, Bruce Friedman, Steve W. Wickizer and Gregg S. Meyer. 2001. 'Potentially inappropriate medication use in the community-dwelling elderly: Findings from the 1996 Medical Expenditure Panel Survey'. *JAMA* 286(22): 2823–2829.

CIHI. 2002. *Motor Vehicle Collisions Most Frequent Cause of Severe Injuries*. Ottawa, Ontario: Canadian Institute for Health Information; http://secure.cihi.ca/cihiweb

Clastres, Pierre. 1974. *La société contre l'état: Recherches d'anthropologie politique*. Paris: Minuit.

ClimSoil. 2008. *Review of Existing Information on the Interrelations between Soil and Climate Change*. Brussels: European Commission.

Clingingsmith, David and Jeffrey G. Williamson. 2005. *Mughal Decline, Climate Change, and Britain's Industrial Ascent: An Integrated Perspective on India's 18th and 19th Century Deindustrialization*. USA: National Bureau of Economic Research.

Cohen, David. 2007. 'Earth's natural wealth: an audit'. *New Scientist* 2605 (23 May): 34–41.

Colborn, Theo, Dianne Dumanoski and John Peterson Myers. 1997. *Our Stolen Future. Are We Threatening Our Fertility, Intelligence and Survival? – A Scientific Detective Story*. New York: Plume.

Collodi, Jason and Freida M'Cormack. 2009. *Population Growth, Environment and Food Security: What Does the Future Hold?* Sussex, UK: Horizon, Institute of Development Studies, University of Sussex; http://www.eldis.org/vfile/upload/1/document/0908/Horizon_Briefing_August 2.pdf

Cox, Caroline. 2003 [1996]. 'Nonyl phenol and related chemicals'. *Journal of Pesticide Reform* 16(1): 15–20 (corrected 4/2003).

Cunfer, Geoff. 2005. *On the Great Plains: Agriculture and Environment*. College Station, Texas: Texas A&M University Press.

Dagnino, Evelina, Alberto Olvera and Aldo Panfichi. 2008. 'Democratic innovation in Latin America: a first look at the Democratic Participatory Project', in *Democratic Innovation in the South: Participation and Representation in Asia, Africa and Latin America*, compiled by Ciska Raventós, 1st edition. Buenos Aires: Consejo Latinoamericano de Ciencias Sociales (CLACSO).

Danielsson, Albert. 1977. 'Faktorer som styr teknisk utveckling', in *Människan i Tekniksamhället*, Sweden: Kungl. Vitterhetsakademien.

Danish Ministry of Environment. 2005. *Nature and Environment 2004. Theme: Nature in Denmark*. Denmark: Danish Environmental Protection Agency, Ministry of Environment; http://www2.mst.dk/udgiv/publications/2005/87-7614-725-8/pdf/87-7614-726-6.pdf

Davies, James B., Susanna Sandström, Anthony Shorrocks and Edward N. Wolff. 2008. *The World Distribution of Household Wealth*. Helsinki, Finland: United Nations University, World Institute for Development Economics Research (UNU-WIDER).

Davis, Donald R., Melvin D. Epp and Hugh D. Riordan. 2004. 'Changes in USDA food composition data for 43 garden crops, 1950 to 1999'. *Journal of the American College of Nutrition* 23(6): 669–682.

Davis, Steven J. and Ken Caldeira. 2010. 'Consumption-based accounting of CO_2 emissions'. *Proceedings of the National Academy of Sciences* 107(12): 5687–5692 (Epub 8 March 2010; http://www.pnas.org/content/107/12/5687.full).

de Larosière, Jacques. 2007. 'Reflections on the history of globalisation'; http://www.canalacademie.com/ida1594-Reflections-on-the-history-of-globalisation.html (accessed 1 April 2009).

Diamond, Jared. 1997. *Guns, Germs, and Steel: The Fates of Human Societies*. New York: W. W. Norton & Company, Inc.

Diamond, Jared. 2005. *Collapse: How Societies Choose to Fail or Succeed*. New York: Viking Penguin.

Diener, Ed and Martin E. P. Seligman. 2005. 'Beyond money, Toward an economy of well-being'. *Psychological Science in the Public Interest* 5(1): 1–31.

Economist, The. 2009a. 'What's cooking: The evolutionary role of cookery' [the American Association for the Advancement of Science reporting on a study by Richard Wrangham], from the print edition of 21 February; http://www.economist.com/node/13139619

Economist, The. 2009b. 'The hourglass effect: Seven maids with seven mops might fail, but Singapore gets close' [a report on Singapore's sand shortage], from the print edition of 10 October; http://www.economist.com/node/14588255

Economist, The. 2009c. 'Fishy tales: Only radical actions can save Europe's fish – and its fishermen', from the print edition of 25 April; http://www.economist.com/node/13526726

Economist, The. 2010a. 'We all want to change the world', from the print edition of 3 April; http://www.economist.com/node/15814427

Economist, The. 2010b. 'Shuffle up and deal', A special report on gambling, from the print edition of 8 July 2010; http://www.economist.com/node/16507670 (accessed 10 July 2010).

Economist, The. 2011a. 'The drying of the West', from the print edition of 29 February; http://www.economist.com/node/18013810

Economist, The. 2011b. 'Internal affairs', from the print edition of 12 March; http://www.economist.com/node/18332880.

Economist, The. 2011c. 'Economies of scale made steel', from the print edition of 12 November; http://www.economist.com/node/21538156

Edberg, Rolf. 1976. *Dalens ande*. Stockholm, Sweden: Norstedts.

EENGO. 2008. EENGO Submission to the Irish National Sustainable Development Strategy. Final Report, 24 January 2008, produced by the Environmental (Ecological) NGO; http://michaelsmithie.files.wordpress.com/2008/03/eengo-submission-on-nsds.pdf

EERE. 2009. *Solar Energy Technologies Program*. Washington, DC: Energy Efficiency and Renewable Energy, US Department of Energy; http://www1.eere.energy.gov/solar/about.html (accessed 10 May 2009).

Ehrlich, Paul. 1968. *The Population Bomb*. Cutchogue, New York: Buccaneer Books.

EIA. 2010. *Electric Power Industry 2008: Year in Review*; Washington, DC: US Energy Information Administration; http://www.eia.doe.gov/cneaf/electricity/epa/epa_sum.html

EIA. 2011. *International Energy Statistics*. Washington, DC: US Energy Information Administration; http://tonto.eia.doe.gov (accessed 27 April 2011).

Eklund, Klas. 2000. *Jorden gick inte under*. Stockholm, Sweden: Sveriges industriförbund.

Ellis, Erle C. and Navin Ramankutty. 2008. 'Putting people in the map: Anthropogenic biomes of the world'. *Frontiers in Ecology and the Environment* 6(8): 439–447; doi 10.1890/070062.

Elmqvist, Thomas. 2007. *Tänkvärt om hållbarhet*. Sweden: Systemekologiska institutionen, Stockholms universitet.

EPA. 2010. 'Mitigation banking factsheet: Compensating for impacts to wetlands and streams'; http://www.epa.gov/wetlands/facts/fact16.html (accessed 20 July 2010).

Erb, Karl-Heinz, Helmut Haberl, Fridolin Krausmann, Christian Lauk, Christoph Plutzar, Julia K. Steinberger, Christoph Müller, Alberte Bondeau, Katharina Waha and Gudrun Pollack. 2009. *Eating the Planet: Feeding and Fuelling the World Sustainably, Fairly and Humanely – A Scoping Study*. Commissioned by Compassion in World Farming and Friends of the Earth UK. Vienna, Potsdam: Institute of Social Ecology and PIK Potsdam.

Erman, Eva. 2010. 'Globalizing "rule by the people": A deliberative view', Part of the Building Global Democracy (BGD) Programme's 'Conceptualising Global Democracy' project; http://www.buildingglobaldemocracy.org/content/globalizing-rule-people-deliberative-view; www.buildingglobaldemocracy.org (accessed 13 July 2010).

European Commission. 2003. 'Towards a thematic strategy on the sustainable use of natural resources', *Communication from the Commission to the Council and the European Parliament*, COM(2003)572. Brussels: European Commission.

European Commission. 2010a. *Energy: Statistical Pocketbook 2010*. Brussels: European Commission; http://ec.europa.eu/energy/publications/statistics/statistics_en.htm

European Commission. 2010b. 'Europe 2020: A European strategy for smart, sustainable and inclusive growth', COM(2010)2020; http://ec.europa.eu/eu2020/pdf/COMPLET%20EN%20BARROSO%20%20%20007%20-%20Europe%202020%20-%20EN%20version.pdf

EU–US Summit. 2009. 'EU–U.S. Summit declaration', 3 November 2009; http://eeas.europa.eu/us/sum11_09/docs/declaration_en.pdf

Fall, Alioune and Adama Faye. 1999. 'Minimum tillage for soil and water management with animal traction in the West-African region', in *Conservation Tillage with Animal Traction: A Resource Book of the Animal Traction Network for Eastern and Southern Africa (ATNESA)*, edited by P. G. Kaumbutho and T. E. Simalenga. Harare, Zimbabwe: ATNESA; http://www.atnesa.org/contil/contil-fall-westafrica.pdf

FAO. 2000. 'The energy and agriculture nexus'. *Environment and Natural Resources Working Paper No. 4*. Rome: Food and Agriculture Organization of the United States.

FAO. 2003. *World Agriculture: Towards 2015/2030. An FAO Perspective.*
London: Earthscan.

FAO. 2006. *Livestock's Long Shadow.* Rome: Food and Agriculture Organization of the
United Nations.

FAO. 2007. *The State of Food and Agriculture.* Rome: Food and Agriculture Organization of
the United Nations.

FAO. 2008. 'The world only needs 30 billion dollars a year to eradicate the scourge of
hunger', Press release 3 June 2008;
http://www.fao.org/newsroom/en/news/2008/1000853/index.html

FAO. 2009. *Review of Evidence on Dryland Pastoral Systems and Climate Change,* edited by C.
Neely, S. Bunning and A. Wilkes. Rome: Food and Agriculture Organization of the
United Nations.

FAOstat. 2011. http://faostat.fao.org/

Farb, Peter. 1968. *Man's Rise to Civilization: As Shown by the Indians of North America from
Primeval Times to the Coming of the Industrial State.*
New York: E. P. Dutton & Co., Inc.

Fast Company. 2010. 'Unpacking the global Human egg trade'. *FastCompany* 148
(September); http://www.fastcompany.com/magazine/148/eggs-for-sale.html

Filmer, Deon. 1995. 'Estimating the world at work: A background paper for *World
Development Report 1995'. Policy Research Working Paper 1488.* Washington, DC:
World Bank.

Fisher, William W. III. 1999. 'The growth of intellectual property: A history of the
ownership of ideas in the United States';
http://cyber.law.harvard.edu/people/tfisher/iphistory.pdf

Florida Panther Conservation Bank. 2011. 'Overview';
http://pantherconservation.com/index.html (accessed 28 April 2011).

Forbes.com. 2010. 'The Global 2000'; http://www.forbes.com/lists/2010/18/global-
2000-10_The-Global-2000_Sales.html (accessed 25 May 2010).

Friedman, Thomas L. 2005. *The World Is Flat: A Brief History of the Globalized World in the
Twenty-First Century.* London: Allen Lane.

FSC. 2009. 'Forest Stewardship Council'; www.fsc-sverige.org (accessed 10 May 2009).

Gadd, Carl-Johan. 2000. *Det svenska jordbrukets historia, bd 3: Den agrara revolutionen 1700–
1870.* Stockholm: Natur och Kultur.

Gallai, Nicola, Jean-Michel Sallesc, Josef Setteled and Bernard E Vaissièrea. 2009.
'Economic valuation of the vulnerability of world agriculture confronted with
pollinator decline'. *Ecological Economics* 68(3): 810–821.

Gambling Commission. 2006. *Basic Facts about the British Gambling Industry.* Birmingham:
Gambling Commission.

Gärdenfors, Peter. 2003. *How Homo Became Sapiens: On The Evolution of Thinking.* Oxford,
UK: Oxford University Press.

Gerbehaye, Claude Moreau de. 2002. 'Järnhanteringens utveckling i södra delen av
provinsen Luxemburg 1500–1800', in *Valloner-Järnets människor,* edited by Anders
Florén and Gunnar Ternhag. Sweden: Gidlunds.

Giampietro, Mario and David Pimentel. 1993. 'The tightening conflict: Population,
energy use, and the ecology of agriculture'. NPG Forum Series (October): 1–8.

Gibbons, Peter and Stefano Ponte. 2005. *Trading Down: Africa, Value Chains, and the Global
Economy.* Philadelphia, PA: Temple University Press.

Giljum, Stefan and Friederich Hinterberger. 2004. *Economic Growth and Material Use in the European Union and the World Economy*. Vienna, Austria: Sustainable Europe Research Institute (SERI);
http://www.euroecolecon.org/pdf/SERIMaterialIntensities.pdf

Gleick, James. 2011. *The Information: A History, a Theory, a Flood*. New York: HarperCollins.

Global Footprint Network. 2010. *The Ecological Wealth of Nations*. Oakland, California: Global Footprint Network;
http://www.footprintnetwork.org/en/index.php/GFN/page/ecological_wealth_of_nations_en; www.footprintnetwork.org

Global Property Guide. 2010. *Global Property Guide*;
http://www.globalpropertyguide.com (accessed 20 August 2010).

Goldstein, Joshua S. 2001. *War and Gender: How Gender Shapes the War System and Vice Versa*. New York: Cambridge University Press.

Gonzalez-Martinez, Ana Citlalic and Heinz Schandl. 2007. 'The biophysical perspective of a middle income economy: Material flows in Mexico'. *Socio-Economics and the Environment in Discussion (SEED) Working Paper 10, CSIRO Working Paper Series 2007–10*. Canberra: CSIRO Sustainable Ecosystems.

Grameen Bank. 2009. http://www.grameen-info.org/ (accessed 17 February 2009).

Greenpeace. 2010. *Make IT Green: Cloud Computing and Its Contribution to Climate Change*. Amsterdam, The Netherlands: Greenpeace International;
http://www.greenpeace.org/international/en/publications/reports/make-it-green-cloud-computing/ (accessed 20 July 2010).

Greenwire. 2010. 'WILDLIFE: An urban jungle grows wild as it greens'; 9 August 2010;
http://www.eenews.net/public/Greenwire/2010/08/09/2

Grus & Guld. 2010. 'Alternativa lån räddar jobb i Argentina', *GRUS* 4/2010: 4–7;
http://www.hoglundreportage.se/wp-content/uploads/2011/01/argentina.pdf

Guardian, The. 2009a. 'Shell: Market alone cannot deliver green energy', 24 November;
http://www.guardian.co.uk/business/2009/nov/24/shell-chief-carbon-tax?INTCMP=SRCH

Guardian, The. 2009b. 'Paying to keep oil in the ground', 7 August;
http://www.guardian.co.uk/commentisfree/cifamerica/2009/aug/07/ecuador-carbon-emissions

Günther Folke. 2011. 'The price for energy';
http://www.holon.se/folke/worries/oildepl/WORKFOReng.shtml (accessed 15 March 2011).

H. John Heinz III Center. 2008. *The State of the Nation's Ecosystems 2008: Measuring the Lands, Waters, and Living Resources of the United States.*
Washington, DC: Island Press.

Hall, Charles A.S. and Kent A. Klitgaard. 2006. 'The need for a new, biophysical-based paradigm in economics for the second half of the age of oil'. *International Journal of Transdisciplinary Research* 1(1): 4–22.

Hall, Charles, Pradeep Tharakan, John Hallock, Cutler Cleveland and Michael Jefferson. 2003. 'Hydrocarbons and the evolution of human culture'. *Nature* 426: 318–322.

Hall, Charles, Stephen Balogh and David J. R. Murphy. 2009. 'What is the minimum EROI that a sustainable society must have?' *Energies* 2: 25–47;
doi:10.3390/en20100025.

Hamilton, Henry Sackville. 2008. 'The pesticide paradox'. *Rice Today* 7(1) (January–March): 32–33.

Hård, Mikael and Andrew Jamison. 2005. *Hubris and Hybrids: A Cultural History of Technology and Science*. New York: Routledge.

Hardin, Garrett. 1968. 'Tragedy of the commons'. *Science* 162(3859): 1243–1248.

Hardin, Garrett. 2003. 'Extension of the tragedy of the commons'; http://www.garretthardinsociety.org/articles/art_extension_tragedy_commons.html

Hart, Keith. 2000. *The Memory Bank: Money in an Unequal World*. New York: Texere.

Hart, Keith. 2003. 'The political economy of food in an unequal world', in *The Politics of Food*, edited by Marianne Elisabeth Lien and Brigitte Nerlich. Oxford: Berg.

Hawken, Paul, Amory B. Lovins and Hunter L. Lovins. 1999. *Natural Capitalism: Creating the Next Industrial Revolution*. Boston: Little, Brown and Co.

Hayek, Friedrich von. 2001. *The Road to Serfdom*. London: Routledge Classics.

HealthGrades. 2004. 'In-hospital deaths from medical errors at 195,000 per year, HealthGrades Study finds', July 27 2004. Lakewood, Colorado: HealthGrades.

Heinberg, Richard (2012-07-09). *The End of Growth: Adapting to our new economic reality*. Booksource. Kindle Edition.

Helfrich, Silke. 2009. *Strengthen the Commons – Now!*, translated by Michelle Thorne, Silke Helfrich and David Bollier. Germany: Heinrich Böll Foundation; http://commonsblog.files.wordpress.com/2009/12/commonsmanifesto-engl.pdf

Helmfrid, Hillevi and Andrew Haden. 2006. *Efter oljetoppen: Hur bygger vi beredskap när framtidsbilderna går isär?* Sweden: Royal Swedish Academy of Agriculture and Forestry.

Helpman, Elhanan. 1993. 'Innovation, imitation, and intellectual property rights'. *Econometrica* 61(6): 1247–1280.

Hendrickson, John. 1994. *Energy Use in the U.S. Food System: A Summary of Existing Research and Analysis*. Madison, WI: Center for Integrated Agricultural Systems, UW-Madison.

Herrmann, Michael. 2009. 'Food security and agricultural development in times of high commodity prices', Discussion paper of the United Nations Conference on Trade and Development (UNCTAD), No. 196, November 2009, UNCTAD/OSG/DP/2009/4; http://www.unctad.org/en/docs/osgdp20094_en.pdf

Hirsch, Robert. 2008. 'Mitigation of maximum world oil production: Shortage scenarios'. *Energy Policy* 36(2): 881–889.

Hobsbawm, Eric J. 1991. *Nations and Nationalism since 1780: Programme, Myth, Reality*. Cambridge, UK: Cambridge University Press.

Hoffmann Ulrich. 1999. 'Requirements for environmentally sound and economically viable management of lead as important natural resource and hazardous waste in the wake of trade restrictions on secondary lead by decision III/1 of the Basel Convention: The case of used lead-acid batteries in the Philippines, Geneva', Draft study, July 1999; www.unctad.org/trade_env/test1/publications/battery1.pdf

Hoffmann, Ulrich. 2011. 'Assuring food security in developing countries under the challenges of climate change: key trade and development issues of a fundamental transformation of agriculture', Discussion paper of the United Nations Conference on Trade and Development (UNCTAD), No. 201, February 2011;

Hofstadter, Richard. 1973. 'White servitude', in *America at 1750: A Social Portrait*. New York: Random House; http://www.montgomerycollege.edu/Departments/hpolscrv/whiteser.html

Hourihan Matt and Robert D. Atkinson. 2011. *Inducing Innovation: What a Carbon Price Can and Can't Do*. Washington, DC: The Information Technology & Innovation Foundation; http://www.itif.org/files/2011-inducing-innovation.pdf

Honda, Gail. 1997. 'Differential structure, differential health: Industrialization in Japan, 1868–1940', in *Health and Welfare During Industrialization*, edited by Richard H. Steckel and Roderick Floud. Chicago: University of Chicago Press.

Hornborg, Alf. 2006. 'Footprints in the cotton fields: The Industrial Revolution as time-space appropriation and environmental load displacement', *Ecological Economics* 59(1): 74–81.

Hornborg, Alf. 2009. 'Zero-sum world challenges in conceptualizing environmental load displacement and ecologically unequal exchange in the world-system'. *International Journal of Comparative Sociology* 50(3–4): 237–262.

Hornborg, Alf. 2010. *Myten om maskinen. essäer om makt, modernitet och miljö*. Göteborg: Daidalos AB.

Hugo, Victor. 1985. 'The sack of the summer palace'. *UNESCO Courier*, November 1985; http://findarticles.com/p/articles/mi_m1310/is_1985_Nov/ai_4003606/ (accessed November 2010).

Huxley, Thomas H. 1891. 'The struggle for existence in human society', introductory essay in *Social Diseases and Worse Remedies. Letters to the 'Times' on Mr. Booth's Scheme*. London: Macmillan and Co.; digitized by the Internet Archive in 2007.

Hylland Eriksen, Thomas. 2008. *Jakten på lycka i överflödssamhället*, translated by Jim Jakobsson. Nora, Sweden: Nya Doxa.

IAASTD. 2009. *Agriculture at a Crossroads: Global Report*. Washington, DC: International Assessment of Agricultural Knowledge, Science and Technology for Development.

ICTSD. 2009. 'Who should pay for embedded carbon?' *Bridges* 13(1) (March); http://ictsd.org/i/news/bridges/44150/

IEA. 2006. *World Energy Outlook 2006*. Paris: OECD/IEA.

IEA. 2008. *Key World Energy Statistics 2008*. Paris: International Energy Agency.

IEA. 2009. *Key World Energy Statistics 2009*. Paris: International Energy Agency.

IFOAM. 2005. *First Legal Defeat of a Biopiracy Patent: The Neem Case*. Bonn, Germany: International Federation of Organic Agriculture Movements.

IFRTD. 2009. 'Animal traction'. London, UK: International Forum for Rural Transport and Development; http://www.ifrtd.org/new/issues/an_trac.php (accessed 13 December 2009).

International Herald Tribune. 2005. 'Class in America: The very rich leave the plain rich behind', 6 June 2005; http://www.iht.com/articles/2005/06/05/news/class.php (accessed 12 February 2009).

International Herald Tribune. 2009. 'The price is not right', 2 April 2009; http://www.nytimes.com/2009/04/01/opinion/01friedman.html

International Herald Tribune. 2011. 'The superrich pull ever farther away', 26 January 2011.

IIIEE. 2007. *Konkurrenskraft för nätansluten solel i Sverige – sett ur kraftföretagens och nätägarnas perspektiv*. Sweden: International Institute for Industrial Environmental Economics, Lund University; http://www.solelprogrammet.se/PageFiles/328/solel_071122_konkurrenskraft.pdf?epslanguage=sv

Illich, Ivan. 1973. *Tools for Conviviality*. New York: Harper & Row Publishers.

Illich, Ivan. 1978. *Toward a History of Needs*. New York: Pantheon Books.

ILO. 2010. *International Labour Migration: A Rights-based Approach*. Geneva: International Labour Office.

Indian Express. 2008. ' "I am a Marxist monk," says Dalai Lama', 19 January; http://www.indianexpress.com/news/i-am-a-marxist-monk-says-dalai-lama/263170/

Inglehart, Ronald, Robert Foa, Christopher Peterson and Christian Welzel. 2008. 'Development, freedom, and rising happiness: A global perspective (1981–2007)'. *Perspectives on Psychological Science* 3(4): 264–285.

ISDA. 2009. *ISDA Market Survey, 1987–Present*. New York: International Swaps and Derivatives Association, Inc.; http://www.isda.org/statistics/pdf/ISDA-Market-Survey-historical-data.pdf

IUCN. 2009. 'Easing the pressure'. *World Conservation: The Magazine of the International Union for Conservation of Nature* (April): 8–9; http://cmsdata.iucn.org/downloads/wc_issue1_2009_en.pdf

IUCN. 2010. 'TEEB, public goods and forests'. *Arborvitae: The IUCN Conservation Programme Newsletter* (41): 8–9; http://cmsdata.iucn.org/downloads/av41_english__3_.pdf

Jackson, Tim. 2009. *Prosperity without Growth: Economics for a Finite Planet*. London: Earthscan.

Johansson, Kersti, Karin Liljequist, Lars Ohlander and Kjell Aleklett. 2010. 'Agriculture as provider of both food and fuel'. *Ambio* 39(2): 91–99.

Johansson, Thomas B. and Måns Lönnroth. 1975. *Energianalyser: En introduktion.* Stockholm, Sweden: Sekretariatet för framtidsstudier.

John Lewis Partnership. 2010. www.johnlewispartnership.co.uk (accessed 2 November 2010).

Karsten, Per and Bo Knarrström. 2002. 'Jägarsamhällets brytningstid', *Forskning & Framsteg*, 1/2002; http://fof.se/tidning/2002/1/jagarsamhallets-brytningstid

Kauppi, Pekka E., Jesse H. Ausubel, Jingyun Fang, Alexander S. Mather, Roger A. Sedjo and Paul E. Waggoner. 2006. Returning forests analyzed with the forest identity'. *Proceedings of the National Academy of Sciences* 103(46): 17574–17579 (Epub 13 November 2006; http://www.pnas.org/content/103/46/17574.short).

Kay, R. N. B. 1970. 'Meat production from wild herbivores'. *Proceedings of the Nutrition Society* 29(2) (December): 271–278.

Kelly, Marjorie. 2001. *The Divine Right of Capital: Dethroning the Corporate Aristocracy*. San Francisco, CA: Berrett-Koehler.

Kijne, Jacob, Jennie Barron, Holger Hoff, Johan Rockström, Louise Karlberg, John Gowing, Suhas P. Wani and Dennis Wichelns. 2009. *Opportunities to Increase Water Productivity in Agriculture with Special Reference to Africa and South Asia [A report prepared by Stockholm Environment Institute for the Swedish Ministry of Environment].* Stockholm, Sweden: Stockholm Environment Institute; http://www.sei-international.org/mediamanager/documents/Publications/Water-sanitation/opportunitiestoincreasewaterprod2009.pdf

Kinney, Jay P. 1975. *A Continent Lost, a Civilization Won: Indian Land Tenure in America* (American Farmers and the Rise of Agribusiness), reprint edition. USA: Arno Press, Inc.

Krausmann, Fridolin, Simone Gingrich, Nina Eisenmenger, Karl-Heinz Erb. 2009. 'Growth in global materials use, GDP and population during the 20th century'. *Ecological Economics* 68(10): 2696–2705; doi:10.1016/j.ecolecon.2009.05.007.

Kropotkin, Peter. 1902. *Mutual Aid: A Factor of Evolution*. London: William Heinemann.

Kuznets, Simon. 1962. 'How to judge quality'. *The New Republic*, 20 October 1962; quoted in http://en.wikipedia.org/wiki/GDP

Land O'Lakes, Inc. 2010. 'Welcome to Land O'Lakes, Inc.'; http://www.landolakesinc.com/company/default.aspx (accessed 1 November 2010).

Lander, Edgardo. 2010. 'The decolonisation of global democracy', Part of the Building Global Democracy (BGD) Programme's 'Conceptualising Global Democracy' project; http://www.buildingglobaldemocracy.org/content/decolonisation-global-democracy; www.buildingglobaldemocracy.org (accessed 13 July 2010).

Landes, David S. 1998. *The Wealth and Poverty of Nations: Why Some Are So Rich and Some So Poor*. New York: W. W. Norton & Company, Inc.

Lazere, Donald. 2009. 'A rising tide lifts all boats: Has the right been misusing JFK's quote? http://hnn.us/articles/73227.html

Legget, Jeremy. 2005. *Half Gone: Oil, Gas, Hot Air and the Global Energy Crisis*. London: Portobello Books Ltd.

Lewis, John Spedan. 1957. 'Dear to my heart', Speech recorded on the BBC on 15 April 1957.

Lichtheim, Miriam. 1980. *Ancient Egyptian Literature. Volume III: The Late Period*. Berkeley and Los Angeles, CA: University of California Press.

Lindblom, Charles E. 2001. *The Market System: What It Is, How It Works, and What to Make of It*. New Haven, CT: Yale University Press.

Linnaeus, Carl. 1929 [1734]. *Iter Dalecarlium*. Undomsresor, Stockholm: P. A. Norstedt & Söner.

Lithander, Leif. 2003. 'Människan, naturen, kulturen och historien'. *Göteborgs Naturhistoriska Museum Årstryck* 2003: 67–80.

Livi-Bacci, Massimo. 1992. *A Concise History of World Population*, translated by Carl Ipsen. Cambridge, MA: Blackwell Publishers.

LO. 2009. *Maktelien – Mycket vill ha mer*. Sweden: Landsorganisationen.

Lönnroth, Erik. 1977. 'Faktorer som styr teknisk utveckling', in *Människan i Tekniksamhället*. Sweden: Kungl. Vitterhetsakademien.

Lucassen, Jan. 2005. 'Proletarianization in Western Europe and India: concept and methods', paper presented in *The Rise, Organization, and Institutional Framework of Factor Markets, Workshop on Global Economic History Network (GEHN)*, Utrecht, 23–25 June 2005, unpublished.

Maarse, Lucy. 2010. 'Livestock, a smart solution for food and farming'. *Farming Matters* 26(1) (March): 7–12; www.ileia.org

Maathai, Wangari Muta. 2006. *Unbowed: A Memoir*. New York: Alfred A. Knopf.

Maddison, Agnus. 2001. *The World Economy: A Millennial Perspective*. Paris: OECD; www.oecd.org

Maddison, Agnus. 2008. *Historical Statistics for the World Economy: 1–2006 AD*; http://www.ggdc.net/maddison/ (accessed January 2009).

Malmberg, Bo and Lena Sommestad. 2000. 'The hidden pulse of history: Age transition and economic change in Sweden, 1820–2000'. *Scandinavian Journal of History* 25(1): 131–146.

Mann, Charles C. 2005. *1491: New Revelations of the American Before Columbus*. New York: Alfred A. Knopf.

Marks, Robert. 2002. *The Origins of the Modern World: A Global and Ecological Narrative*. Oxford: Rowman & Littlefield; http://hem.passagen.se/tips1x2/historien/industrialisering.htm (accessed 15 February 2009).

Martin, Emily. 1991. 'The egg and the sperm: How science has constructed a romance based on stereotypical male-female roles'. *Journal of Women in Culture and Society* 16(31): 485–501.

Marx, Karl. 1844. 'Wages of labour', in *Economic and Philosophic Manuscripts of 1844*; http://www.marxists.org/archive/marx/works/1844/manuscripts/wages.htm

Marx, Karl. 1867. *Capital: A Critique of Political Economy, Volume 1*, translated by Samuel Moore and Edward Aveling; edited by Frederick Engels. Moscow: Progress Publishers.

Marx, Karl and Friedrich Engels. 1848. *Manifesto of the Communist Party*; digitized and freely available.

Maslow, Abraham. 1943. 'A theory of human motivation'. *Psychological Review* 50: 370–396.

Max-Neef, Manfred, Antonio Elizalde and Martin Hopenhayn. 1989. 'Human scale development: An option for the future'. *Development Dialogue* 1: 5–81; http://www.dhf.uu.se/pdffiler/89_1.pdf

Mazoyer, Marcel and Laurance Roudart. 2006. *A History of World Agriculture: From the Neolithic Age to Current Crisis*, translated by James H. Membrez. New York: Monthly Review Press.

McDonald's. 2012. *Vårt Miljöarbete*; http://www.mcdonalds.se/se/om_mcdonald_s/vart_miljoarbete.html (accessed 6 March 2012).

McKinsey & Company. 2009. *Pathways to a Low-Carbon Economy, Version 2 of the Global Greenhouse Gas Abatement Cost Curve*; http://www.worldwildlife.org/climate/WWFBinaryitem11334.pdf

MEA. 2005. *Millennium Ecosystem Assessment: Ecosystems and Human Well-Being – Synthesis*. Washington, DC: World Resource Institute.

Meadows, Donella. 1999. *Leverage Points: Places to Intervene in a System*. Hartland, VT: Sustainability Institute.

Meiksins Wood, Ellen. 1995. *Democracy Against Capitalism: Renewing Historical Materialism*. Cambridge, UK: Cambridge University Press.

Meiksins Wood, Ellen. 1998. 'The agrarian origins of capitalism'. *Monthly Review* 50(3); http://monthlyreview.org/1998/07/01/the-agrarian-origins-of-capitalism#top

Meiksins Wood, Ellen. 1999. 'The politics of capitalism'. *Monthly Review* 51(4); http://monthlyreview.org/1999/09/01/the-politics-of-capitalism

MercoPress. 2006. 'Maya civilization collapsed upon learning Kings weren't Gods', 27 August; http://en.mercopress.com/2006/08/27/maya-civilization-collapsed-upon-learning-kings-weren-t-gods

Meyer, W. S. 1997. 'Water for food: The continuing debate', Draft paper for *1997 Irrigation Conference*. New South Wales: CSIRO Land and Water, Griffith Laboratory.

Minagri. 2009. *Strategic Plan for the Transformation of Agriculture in Rwanda – Phase II (PSTA II). Final Report*. Republic of Rwanda: Ministry of Agriculture and Animal Resources; http://www.gafspfund.org/gafsp/sites/gafspfund.org/files/Documents/Rwanda_StrategicPlan.pdf

MistraFarma. 2010. 'World's highest drug levels entering India stream'; http://www.mistra.org/program/mistrapharma/home/pressandmedia/newsarchive/news/worldshighestdruglevelsenteringindiastream.5.589e653711f5b17101b800012468.html (accessed 17 May 2010).

Mondragón. 2010. 'Mondragon Corporation: Frequently asked questions'; http://www.mondragon-corporation.com/, http://www.mondragon-corporation.com/language/en-US/ENG/Frequently-asked-questions/Co-operativism.aspx (11 September 2010).

Montgomery, David R. 2007. *Dirt: The Erosion of Civilizations*. Berkeley and Los Angeles, CA: University of California Press.

Mother Jones Magazine. 1995. 'Dwayne's world', July–August 1995, Berkeley, California.

Myrdal, Janken. 1999. *Det svenska jordbrukets historia, bd 2: Jordbruket under feodalismen 1000–1700*. Stockholm, Sweden: Natur och Kultur.

Myrdal, Janken. 2008. *Framtiden – om femtio år: Global utveckling och ruralt-urbant i Norden*. Sweden: Institutet för framtidsstudier.

Naess, Arne. 1974. *Økologi, samfunn og livsstil: Utkast til en økosofi*. Oslo: Universitetsforlaget.

New York Times. 2010a. 'Rising threat of infections unfazed by antibiotics', 26 February 2010; http://www.nytimes.com/2010/02/27/business/27germ.html?scp=3&sq=&st=nyt

New York Times. 2010b. 'An electric boost for bicyclists', 31 January 2010; http://www.nytimes.com/2010/02/01/business/global/01ebike.html

Nielsenwire. 2010. 'U.S. ad spend falls nine percent in 2009, Nielsen says' 24, February; http://blog.nielsen.com/nielsenwire/consumer/u-s-ad-spend-falls-nine-percent-in-2009-nielsen-says/

Norberg, Johan. 2001. *In Defense of Global Capitalism*, translated by Roger Tanner and Julian Sanchez. Washington, DC: Cato Institute.

Norris, Pippa and Ronald Inglehart. 2004. *Sacred and Secular: Religion and Politics Worldwide*. Cambridge, UK: Cambridge University Press.

NRM. 2008. 'Naturhistoriska Riksmuseet'; http://www.nrm.se/sv/meny/faktaomnaturen/djur/faglar/utrotadefaglar.195.html, 30 December (accessed 14 December 2009).

NSERC. 2006. 'Hooked on fishing, and we're heading for the bottom, says scientist', *EurekAlert!*, 17 February 2006, Natural Sciences and Engineering Research Council; http://www.eurekalert.org/pub_releases/2006-02/nsae-hof021706.php

Ny Teknik. 2008. 'Trafikdöden i EU ska halveras', 30 April 2008; http://www.nyteknik.se/popular_teknik/kaianders/article240207.ece (accessed 14 December 2009).

Nye, Joseph S. 2008. *Understanding International Conflicts: An Introduction to Theory and History*. Harlow: Pearson Education.

OECD. 2001. *Environmental Indicators for Agriculture, Volume 3: Methods and Results*. Paris: Organisation for Economic Co-operation and Development.

OECD. 2002. *Hazard Assessment of Perfluorooctane Sulfonate (PFOS) and Its Salts*. ENV/JM/RD(2002)17/FINAL. Paris: Organisation for Economic Co-operation and Development. Available online: http://www.oecd.org/dataoecd/23/18/2382880.pdf

OECD. 2009. 'Producer and consumer support estimates', *OECD Database 1986–2003*. Available online: http://www.oecd.org/document/58/0,3746,en_2649_33773_32264698_1_1_1_1,00.html

Ostrom, Elinor. 2009. 'Whiteboard seminar with Elinor Ostrom: Going beyond the tragedy of the commons'; http://www.stockholmresilience.org (accessed 14 December 2009).

Our Stolen Future. 2010. Various data accessed on www.ourstolenfuture.org (17 May 2010).

Oxfam. 2009. *Changing Food Consumption in the UK to Benefit People and Planet.* Oxfam GB Briefing Paper. UK: Oxfam.

Patel, Raj. 2009. *The Value of Nothing.* London: Portobello Books Ltd.

Perry, David A. 1994. *Forest Ecosystems.* Baltimore, MD: Johns Hopkins University Press.

Peters, Glen P., Jan C. Minx, Christopher L. Weber and Ottmar Edenhofer. 2011. 'Growth in emission transfers via international trade from 1990 to 2008'. *Proceedings of the National Academy of Sciences* 108(21): 8903–8908 (Epub 25 April 2011; http://www.pnas.org/content/108/21/8903.full).

Pimentel, David, Maria Tort, Linda D'Anna, Anne Krawic, Joshua Berger, Jessica Rossman, Fridah Mugo, Nancy Doon, Michael Shriberg, Erica Howard, Susan Lee and Jonathan Talbot. 1998. 'Ecology of increasing disease'. *Bioscience* 48(10) (October): 817–827.

Plato. 360 BCE. *Critias,* translated by Benjamin Jowett; http://classics.mit.edu/Plato/critias.html (accessed 14 December 2009).

Ploeg, Jan Douwe van der. 2009. *The New Peasantries: Struggles for Autonomy and Sustainability in the Era of Empire and Globalization.* London: Earthscan.

Polanyi, Karl. 1944. *The Great Transformation: The Political and Economic Origins of Our Times.* New York and Toronto: Farrar & Rinehart.

PopOffsets. 2011. http://www.popoffsets.com/ (accessed 30 April 2011).

Porritt, Jonathon. 2007. *Capitalism as if the World Matters,* revised edition. London: Earthscan.

President's Cancer Panel. 2010. *Reducing Environmental Cancer Risk: What We Can Do Now. 2008–2009 Annual Report.* Washington, DC: US Department of Health and Human Services / National Institutes of Health / National Cancer Institute.

Pretty, J. and H. Waibel. 2005. 'Paying the price: The full cost of pesticides', in *The Pesticide Detox: Towards a More Sustainable Agriculture,* edited by J. Pretty. London: Earthscan.

Pretty, J., A. S. Ball, T. Lang and J. I. L. Morison. 2005. 'Farm costs and food miles: An assessment of the full cost of the UK weekly food basket'. *Food Policy* 30: 1–19.

Pretty, Jules. 1995. *Regenerating Agriculture: Policies and Practice for Sustainability and Self-Reliance.* London: Earthscan.

Pretty, Jules. 2007. *The Earth Only Endures: On Reconnecting with Nature and Our Place in It.* London: Earthscan.

QLIF. 2009. 'Effects of production methods', QualityLowInputFood (QLIF) subproject 2: Quantifying the effect of organic and low input production methods on food quality and safety and human health; http://www.qlif.org/Library/leaflets/folder_2_small.pdf (accessed 14 December 2009).

Radkau, Joachim. 2008. *Nature and Power: A Global History of Environment* (Publications of the German Historical Institute). New York: Cambridge University Press.

Rajaee, Farhang. 2000. *Globalization in Trial: The Human Condition and the Information Civilization.* Ottawa: International Development Research Centre.

Rand, Ayn. 1976. *The Ayn Rand Letter.* Irvine, CA: The Ayn Rand Institute.

Raymond, Eric S. 2000. 'The social context of open-source software', in *The Cathedral and the Bazaar,* version 3.0; http://www.catb.org/~esr/writings/cathedral-bazaar/cathedral-bazaar/

Redman, Mark and Melike Hemmami. 2008. *Developing a National Agri-environment Programme for Turkey*. London: Institute for European Environmental Policy.

Region Skåne. 2005. 'Faktablad om läkemedel och miljön reviderad 7 juni 2006'; www.sjukvardsradgivningen.se (accessed 13 December 2009).

Regmi, Anita and Mark Gehlhar. 2005a. 'Processed food trade pressured by evolving global supply chains'. *Amber Waves* 3(1): 12–19; http://www.ers.usda.gov/AmberWaves/February05/pdf/FeatureProcessedFoodFeb05.pdf

Regmi, Anita and Mark Gehlhar (eds). 2005b. *New Directions in Global Food Markets*, USDA Agriculture Information Bulletin Number 794. Washington, DC: Economic Research Service, United States Department of Agriculture; http://www.ers.usda.gov/publications/aib794/aib794.pdf

Regmi, Anita, Hiroyuki Takeshima and Laurian Unnevehr. 2008. *Convergence in Global Food Demand and Delivery*, Economic Research Report No. ERR-56. Washington, DC: Economic Research Service, United States Department of Agriculture; http://www.ers.usda.gov/publications/err56/err56.pdf

Renault, Daniel. 2002. *Value of Virtual Water in Food: Principles And Virtues*. Rome: FAO; http://www.fao.org/nr/water/docs/VirtualWater.pdf

Reuters. 2007. 'Global stock values top $50 trln: industry data', 21 March 2007; http://www.reuters.com/article/2007/03/21/us-markets-shares-values-idUSL2144839620070321

Robelius, F. 2007. *Giant Oil Fields – The Highway to Oil*. Acta Universitatis Upsaliensis. Digital Comprehensive Summaries of Uppsala Dissertations from the Faculty of Science and Technology.

Rockström, Johan, Will Steffen, Kevin Noone, Asa Persson, F. Stuart Chapin, Eric Lambin, Timothy M. Lenton, Marten Scheffer, Carl Folke, Hans Joachim Schellnhuber *et al*. 2009. 'Planetary boundaries: Exploring the safe operating space for humanity'. *Ecology and Society* 14(2): 32.

Rosenberg, Göran. 2002. 'Sjukvårdsindustrins ohållbara löften', *Dagens Nyheter* 13 September 2002; http://www.dn.se/ledare/kolumner/sjukvardsindustrins-ohallbara-loften

Rousseau, Jean-Jacques. 1964. *The First and Second Discourses*, translated by Roger D. Masters and Judith R. Masters. New York: St Martin's Press.

Ruddiman, William F. 2005. *Plows, Plagues and Petroleum: How Humans Took Control of Climate*. Princeton, NJ: Princeton University Press.

Satoyama. 2010. 'Biodiversity and livelihoods, the Satoyama initiative concepts in practice'; www.satoyama-initiative.org (accessed 30 December 2010).

SCB. 1959. *Historisk statistik för Sverige t.o.m. 1955*. Stockholm, Sweden: Statistiska Centralbyrån.

SCB. 2003. *Indikatorer för hållbar utveckling, rapport 2003:3*. Stockholm, Sweden: Statistiska Centralbyrån.

SCB. 2004. *Pressmeddelande nr 2004:172*. Stockholm, Sweden: Statistiska Centralbyrån.

SCB. 2008. *Markanvändningen i Sverige*. Stockholm, Sweden: Statistiska Centralbyrån.

Schiermeier, Quirin, Jeff Tollefson, Tony Scully, Alexandra Witze and Oliver Morton. 2008. 'Electricity without carbon'. *Nature* 454 (14 August): 816–823.

Schumacher, E. F. 1973. *Small Is Beautiful: Economics as if People Mattered*. London: Blond & Briggs Ltd.

Schumacher, E. F. 1979. *Good Work*. London: Jonathan Cape.

Schumpeter, Joseph A. 1942. *Capitalism, Socialism and Democracy*. London: Allen &
    Unwin.

Schwarz, Barry. 2005. 'On the paradox of choice', TED talk;
    http://www.ted.com/talks/barry_schwartz_on_the_paradox_of_choice.html

Science Commons. 2010. 'About Science Commons';
    http://sciencecommons.org/about/

Scientific American. 2008. 'Plastic (not) fantastic: Food containers leach a potentially
    harmful chemical', 19 February;
    http://www.scientificamerican.com/article.cfm?id=plastic-not-fantastic-with-
    bisphenol-a

Seavoy, Ronald E. 2000. *Subsistence and Economic Development*. Westport, CT: Praeger.

Sen, Amartya. 1999. *Development as Freedom*. Oxford: Oxford University Press.

Sennett, Richard. 2006. *The Culture of the New Capitalism*. New Haven, CT: Yale
    University Press.

Serageldin I. and C. Grootaert. 1999. 'Defining social capital: An integrating view', in
    Social Capital: A Multifaceted Perspective, edited by P. Dasgupta and I. Serageldin.
    Washington, DC: World Bank, pp. 40–58;
    http://www.exclusion.net/images/pdf/778_okili_serageldin_grootaert.pdf

SERI. 2009. *Overconsumption? Our Use of the World's Natural Resources*. Vienna, Austria:
    Friends of the Earth.

Siegel, Lee. 2003. 'Bad mileage: 98 tons of plants per gallon'. *EurekAlert!*, 26 October;
    http://www.eurekalert.org/pub_releases/2003-10/uou-bm9102603.php

Sjukvårdsrådgivningen. 2007 'Att ta medicin när man är äldre / Biverkningar';
    http://vard.vgregion.se (accessed 13 December 2009).

Shiva, Vandana. 2005. *Earth Democracy: Justice, Sustainability, and Peace*. Cambridge, MA:
    South End Press.

Skogforsk. 2010. Swedish Forestry Research Institute; www.skogforsk.se (accessed 10
    January 2010).

Sloan Wilson, David. 2003. *Darwin's Cathedral: Evolution, Religion, and the Nature of
    Society*. Chicago: University of Chicago Press.

Smith, Adam. 1776. *An Inquiry into the Nature and Causes of the Wealth of Nations*. London:
    W. Strahan and T. Cadell.

SNF. 2007. 'Miljögåta har fått sin lösning: Miljögiftet nonylfenol i vatten kommer från
    handdukar', 11 September 2007; www.snf.se (accessed 15 May 2009).

SNF. 2008. 'EU:s kemikalielagstiftning', 9 October 2008; www.snf.se (accessed 15 May
    2009).

Society For Conservation Biology. 2003. 'Just how many species are there, anyway?'
    ScienceDaily; http://www.sciencedaily.com/releases/2003/05/030526103731.htm
    (accessed 23 July 2010).

Solberg, Birger. Undated. 'Historical trends in forest products markets in Europe', A
    study implemented in the framework of the European Forest Sector Outlook Study
    (EFSOS), Geneva Timber and Forest Discussion Papers. ISSN 1020 7228, Geneva,
    Switzerland: UNECE/FAO Timber Section;
    http://www.unece.org/fileadmin/DAM/timber/docs/efsos/03-sept/dp-a.pdf

Soto, Hernando de. 2000. *The Mystery of Capital: Why Capitalism Triumphs in the West and
    Fails Everywhere Else*. London: Bantam.

Stanton, Tracy, Marta Echavarria, Katherine Hamilton and Caroline Ott. 2010. *State of Watershed Payments: An Emerging Marketplace*. Ecosystem Marketplace; http://moderncms.ecosystemmarketplace.com/repository/moderncms_document s/state_of_water_2010_exec.pdf (accessed 28 April 2011).

Steckel, Richard H. and Roderick Floud (eds). 1997. *Health and Welfare During Industrialization*. Chicago: University of Chicago Press.

Stiglitz, Joseph. 2002. *Globalization and Its Discontents*. London and New York: W. W. Norton & Company, Inc.

Stiglitz, Joseph. 2009. 'GDP fetishism'. *Project Syndicate*; http://www.project-syndicate.org/commentary/stiglitz116/English

Stolton, Sue. 2002. *Organic Agriculture and Biodiversity*. Bonn, Germany: International Federation of Agriculture Movement (IFOAM).

Stuart, Tristam. 2010. Food waste facts; http://www.tristramstuart.co.uk/FoodWasteFacts.html (accessed 1 June 2010).

Sullivan, Sian. 2010. 'The Environmentality of "Earth Incorporated": on contemporary primitive accumulation and the financialisation of environmental conservation'. Paper presented at the *Conference An Environmental History of Neoliberalism*, Lund, 6–8 May 2010.

SvD. 2007. 'Miljögiftet Nonylfenol finns i svenska vatten'. *Svenska Dagbladet*, 27 September; http://www.svd.se/nyheter/inrikes/miljogiftet-nonylfenol-finns-i-svenska-vatten_259607.svd

SvD. 2008a. 'Civilisationens vagga byggdes av trä'. *Svenska Dagbladet*, 12 May; http://www.svd.se/kultur/understrecket/civilisationens-vagga-byggdes-av-tra_1236809.svd

SvD. 2008b. 'Oväntat stora mängder kol binds i naturbetesmark'. *Svenska Dagbladet*, 28 August; http://www.svd.se/nyheter/inrikes/ovantat-stora-mangder-kol-binds-i-naturbetesmark_1632583.svd

SvD. 2009a. 'Trafiken större hot än malaria i fattiga länder'. *Svenska Dagbladet* 7 May; http://www.svd.se/nyheter/utrikes/trafiken-storre-hot-an-malaria-i-fattiga-lander_2849221.svd

SvD. 2009b. 'Stor brist på ny antibiotika', *Svenska Dagbladet* 17 September 2009; http://www.svd.se/nyheter/utrikes/stor-brist-pa-ny-antibiotika_3533189.svd

SvD. 2010. 'Solenergin billigare med ny nanoteknik', *Svenska Dagbladet*, 25 January; http://www.svd.se/naringsliv/solenergin-billigare-med-ny-nanoteknik_4147207.svd

Tainter, Joseph. 2009. 'Human resource use: timing and implications for sustainability', in *Proceedings of the 94th Annual Meeting of Ecological Society of America in New Mexico*, 9 September 2009; http://www.theoildrum.com/node/5745 (accessed 10 January 2010).

Talberth, John, Clifford Cobb and Noah Slattery. 2007. *The Genuine Progress Indicator 2006: A Tool for Sustainable Development*. Oakland, CA: Redefining Progress; http://www.environmental-expert.com/Files%5C24200%5Carticles%5C12128%5CGPI202006.pdf

Taleb, Nassim Nicholas. 2007. *The Black Swan: The Impact of the Highly Improbable*. New York: Random House.

Tansey, Geoff and Tony Worsley. 1995. *The Food System: A Guide*. London: Earthscan.

TEEB. 2010. *The Economics of Ecosystems and Biodiversity Report for Business – Executive Summary 2010*. Geneva: UNEP;
http://www.teebweb.org/Portals/25/Documents/TEEB%20for%20Business/TEEB%20for%20Bus%20Exec%20English.pdf

Tehelka. 2009. 'The Double Bind Principle', 20 February;
http://www.tehelka.com/story_main41.asp?filename=Ws280209Double_Bind.asp

Thurstan, Ruth H., Simon Brockington and Callum M. Roberts. 2010. 'The effects of 118 years of industrial fishing on UK bottom trawl fisheries'. *Nature Communications* 1; Article number 15; doi:10.1038/ncomms1013

Times. 2006. 'David Baddiel reviews full-body medical screening', April 22 2006;
http://www.timesonline.co.uk/tol/life_and_style/health/features/article707573.ece

Trainer, Ted. 2007. *Renewable Energy Cannot Sustain an Energy Intensive Society*. Dordrecht, The Netherlands: Springer.

Transition US. 2012. *Our Story*; http://transitionus.org/our-story (accessed 1 April 2012).

Uddenberg, Nils. 1993. *Ett djur bland alla andra? Biologin och människans uppfattning av sin plats i naturen*. Nora, Sweden: Nya Doxa.

Uhlin, Hans Erik. 1997. *Energiflöden i livsmedelskedjan*. Stockholm, Sweden: Naturvårdsverket.

UNDP. 2005. *Human Development Report 2005. International Cooperation at Crossroads: Aid, Trade and Security in an Unequal World*. New York: United Nations Development Programme.

UNDP. 2009. *Human Development Report 2009. Overcoming Barriers: Human Mobility and Development*. New York: Palgrave Macmillan.

UNEP. 2008. *Green Jobs: Towards Decent Work in a Sustainable, Low-Carbon World*. Nairobi, Kenya: United Nations Environment Programme;
http://www.unep.org/labour_environment/PDFs/Greenjobs/UNEP-Green-Jobs-Report.pdf

UNEP. 2009. *A Global Green New Deal, First Draft Report*, by Edward B. Barbier for the United Nations Environment Programme, February 2009.

UNEP. 2010a. *Metal Stocks in Society: Scientific Synthesis*. Nairobi, Kenya: United Nations Environment Programme.

UNEP. 2010b. *UNEP Year Book 2010: New Science and Developments in Our Changing Environments*. Nairobi, Kenya: United Nations Environment Programme.

UNEP. 2011. *Green Economy Report: Pathways to Sustainable Development and Poverty Eradication*. Advance copy, online release 23 February 2011;
http://www.unep.org/greeneconomy/GreenEconomyReport/tabid/29846/Default.aspx

UNEP-EEA. 2007. 'Sustainable consumption and production in South East Europe and Eastern Europe, Caucasus and Central Asia', *EEA Report No. 3/2007*. Copenhagen, Denmark: UNEP-EEA (joint publication).

UNEP-FI. 2010. *Universal Ownership: Why Environmental Externalities Matter to Institutional Investors*. Geneva, Switzerland: UNEP-Financial Initiative and Principles for Responsible Investment.

UNESCAP. 2008. *Statistical Yearbook for Asia and the Pacific 2007*. Statistics Division, United Nations Economic and Social Commission for Asia and the Pacific.

UNFPA. 2011. *State of World Population 2011 – People and Possibilities in a World of 7 Billion*. New York: United Nations Population Fund;
http://foweb.unfpa.org/SWP2011/reports/EN-SWOP2011-FINAL.pdf

UNIDO. 2008. *Energy, Development and Security: Energy Issues in the Current Macroeconomic Context*. Vienna, Austria: UN-Energy, United Nations Industrial Development Organization.

United Nations. 1987. *Our Common Future: The World Commission on Environment and Development*. Oxford: Oxford University Press.

UNT. 2008. 'Läkemedelscocktail i kranarna?', Ingrid Eckerman m.fl., *Upsala Nya Tidning*, 11 November 2008.

US CRS. 2004. *Energy Use in Agriculture: Background and Issues*, 19 November. Washington, DC: US Congressional Research Services.

USA Today. 2008. 'AP: Drugs found in drinking water'; http://www.usatoday.com/news/nation/2008-03-10-drugs-tap-water_N.htm (accessed 16 March 2009).

USDA. 2002. *GAIN Report #SA2022*. Washington, DC: Global Agricultural Information Network Online, United States Department of Agriculture; http://www.fas.usda.gov/gainfiles/200211/145784677.pdf

Vieta, Marcelo and Andrés Ruggeri. 2007. 'Worker-recovered enterprises as workers' cooperatives the conjunctures, challenges, and innovations of self-management in Argentina', in *Co-operatives in a Global Economy: The Challenges and Innovations of Co-operation Across Borders*, edited by Darryl Reed and J. J. McMurtry. Newcastle upon Tyne, UK: Cambridge Scholars Publishing, pp. 178–225.

Vision, The 2012. 'Father, son fight for Kween top seat', The Vision, 17 November 2012.

Vitousek, Peter M., Harold A. Mooney, Jane Lubchenco and Jerry M. Melillo. 1997. 'Human domination of earth's ecosystems'. *Science* 277: 494–99.

Vylder, Stefan de. 2002. *Utvecklingens drivkrafter: Om fattigdom, rikedom och rättvisa i världen*. Stockholm, Sweden: Forum Syd.

Ward, Peter. 2009. *The Medea Hypothesis: Is Life on Earth Ultimately Self-Destructive?* Princeton, NJ: Princeton University Press.

Weber, Max. 1986. 'Kapitalismens uppkomst', selected and translated from *Gesammelte Aufsätze zur Religionssoziologie* [Wirtschaftsgeschichte]. Stuttgart: UTB.

Welzer, Harald. 2011. *Mental Infrastructures: How Growth Entered the World and Our Souls*, Volume 14 of the Publication Series on Ecology, edited by Heinrich Böll Foundation; translated by John Hayduska. Berlin: Heinrich Böll Foundation; http://www.boell.de/downloads/Ecology-14-Mental-Infrastructures.pdf

WHO. 2006. *Fuel for Life: Household Energy and Health*. Geneva, Switzerland: World Health Organisation; http://www.who.int/indoorair/publications/fuelforlife.pdf

Wikipedia. 2009. 'Tragedy of the anticommons'; http://en.wikipedia.org/wiki/Tragedy_of_the_anticommons, 9 May 2009 (accessed 16 March 2009).

Wikipedia. 2010a. 'List of countries by traffic-related death rate'; http://en.wikipedia.org/wiki/List_of_countries_by_traffic-related_death_rate (and subsequent links to death rates in Japan and the United States) (accessed 8 August 2010).

Wikipedia. 2010b. 'Housing cooperative'; http://en.wikipedia.org/wiki/Housing_cooperative (accessed 11 September 2010).

Wikipedia. 2010c. 'Venetian Arsenal'; http://en.wikipedia.org/wiki/Venice_Arsenal (accessed 5 May 2010).

Wikipedia. 2011. 'Apache HTTP server'; http://en.wikipedia.org/wiki/Apache_HTTP_Server (accessed 1 November 2011).

Wikipedia. 2012. 'Wikipedia:About'; http://en.wikipedia.org/wiki/Wikipedia:About (accessed 27 December 2012).

Wilkinson, Richard and Kate Pickett. 2009. *The Spirit Level: Why More Equal Societies Almost Always Do Better*. London: Allen Lane.

Willer, Helga and Lukas Kilcher (eds). 2011. *The World of Organic Agriculture: Statistics and Emerging Trends 2011*. Bonn/Frick: IFOAM/FiBL.

Williamson, Samuel H. 2011. 'Seven ways to compute the relative value of a U.S. dollar amount, 1774 to present'. *MeasuringWorth* (March); www.measuringworth.com/uscompare/

Wired. 1997. 'The Doomslayer, Wired 5.02'; http://www.wired.com/wired/archive/5.02/ffsimon_pr.html (accessed 16 August 2010).

Woods, Jeremy, Adrian Williams, John K. Hughes, Mairi Black and Richard Murphy. 2010. 'Energy and the food system'. *Philosophical Transactions of the Royal Society B* 365: 2991–3006.

World Bank. 2007. *World Development Report 2007*. Washington, DC: World Bank.

World Bank. 2009. *Rural Roads: The Development Impact Evaluation Initiative*; www.worldbank.org

World Bank. 2010a. 'Doing business: Measuring business regulations, Starting a business'; http://www.doingbusiness.org/ExploreTopics/StartingBusiness/ (accessed 26 July 2010).

World Bank. 2010b. *State and Trends of the Carbon Market 2010*. Washington, DC: World Bank.

Worldwatch Institute. 2000. *State of the World 2000*. New York and London: W. W. Norton & Company, Inc.

Worldwatch Institute. 2004. *State of the World 2004. Special Focus: The Consumer Society*. New York and London: W. W. Norton & Company, Inc.

Worldwatch Institute. 2006. *State of the World 2006. Special Focus: China and India*. New York and London: W. W. Norton & Company, Inc.

WRI. 1997. *Resource Flows: The Material Basis for Industrial Economies*. Washington, DC: World Resource Institute.

WRI. 2008. 'High-value assets: Belize's coral reefs and mangroves', by Emily Cooper for the World Resource Institute, 16 March 2009; http://www.wri.org/stories/2008/11/high-value-assets-belizes-coral-reefs-and-mangroves

WWF. 2008. *The Living Planet Report 2008*. Gland, Switzerland: WWF—World Wide Fund for Nature (formerly World Wildlife Fund).

Zeller, Dirk, William Cheung, Chris Close and Daniel Pauly. 2009. 'Trends in global marine fisheries—a critical view', in *Fisheries, Sustainability and Development: Fifty-Two Authors on Coexistence and Development of Fisheries and Aquaculture in Developing and Developed Countries*, edited by P. Wrammer, H. Ackefors and M. Cullberg. Stockholm, Sweden: Royal Swedish Academy of Agriculture and Forestry.

Zetterberg, Hans. 1977. *Arbete, livsstil och motivation*. Sweden: SAF.

Index

About the Author

Gunnar Rundgren is a critical thinker with experiences from more than 100 countries in the world.

He has worked with most parts of the organic agriculture sector — from farming to policy — since 1977, starting on the pioneer organic farm, Torfolk. He is the founder and a senior consultant of Grolink AB (www.grolink.se), a consultancy company engaged in certification development, policy development, project development, marketing strategies and international training programmes — mainly targeting developing countries. He has been engaged as a consultant by NGOs, the government, private companies and intergovernmental organisations such as OECD, UNEP, UNCTAD, the World Bank and the FAO.

Rundgren is the initiator of several organisations for organic agriculture in Sweden, including its main eco-label KRAV (www.krav.se) where he was the director for the first eight years. He served as the first President of the Accreditation Programme Board of the International Federation of Organic Agriculture Movements (IFOAM) 1992-1997. He was an IFOAM World Board member in 1998 and the IFOAM President during the period 2000-2005. In 2002 he was a founding board member of the ISEAL Alliance.

He has published several books related to organic farming. In 2010, he published a book about the major social and environmental challenges of our world, *Trädgården Jorden*, which was translated into Japanese in 2012 and is now published in English. He also co-authored the book *Jorden vi Äter*, published in 2012, about the challenge of feeding the world's population in a sustainable way.

Rundgren was awarded an honorary doctorate in Science at the Uganda Martyrs University 2009. The same year, he was appointed a Member of the Royal Swedish Academy of Agriculture and Forestry.

Internet: gardenearth.blogspot.com/
Contact: info@gardenearth.info